Industry 4.0 - Transforming the Future Beyond Manufacturing - Volume 2, Human-Centred, Organizational and Societal Transformations

Edited by Miguel Delgado-Prieto and Luis Romeral Martinez

Published in London, United Kingdom

Industry 4.0 - Transforming the Future Beyond Manufacturing - Volume 2, Human-Centred, Organizational and Societal Transformations
http://dx.doi.org/10.5772/intechopen.1014267
Edited by Miguel Delgado-Prieto and Luis Romeral Martinez

Contributors
Ahmed Abi Sen Adnan, Anita Rani Chopra, Ashok Chopra, Athanasios Priftis, Beata Mrugalska, Brigita Gajšek, Catherine Equey, Daniela Schmidt, Dzhamilia Nugmanova, Elsa Negre, Günter Müller-Czygan, Hans-Dieter Wehle, Inna Kozlova, Inna Kozlova Mikurova, Irem Ünal, Jannis Pfister, Joseph Mukuni, Krisztián Attila Bakon, Lilia Dvořáková, Marek Segeč, Matthias Jenner, Mauro De Bari, Miroslav Svitek, Mitt Nowshade Kabir, Mohammed Alshammari, Mohammed Gronfula, Rainer Müller, Rogério Dionísio, Romain Stevens, Roman Kupriyanov, Sergej Kozhevnikov, Till Günther, Viktoriya Tarasyuk, Vladimir Marik

First published in London, United Kingdom, 2026 by IntechOpen
IntechOpen is the global imprint of INTECHOPEN LIMITED, registered in England and Wales, registration number: 11086078, 167-169 Great Portland Street, London, W1W 5PF, United Kingdom

For EU product safety concerns: IN TECH d.o.o., Prolaz Marije Krucifikse Kozulić 3, 51000 Rijeka, Croatia, info@intechopen.com or visit our website at intechopen.com.

British Library Cataloguing-in-Publication Data
A catalogue record for this book is available from the British Library

Industry 4.0 - Transforming the Future Beyond Manufacturing - Volume 2, Human-Centred, Organizational and Societal Transformations
Edited by Miguel Delgado-Prieto and Luis Romeral Martinez
p. cm.
Print ISBN 978-0-85014-730-8
Online ISBN 978-0-85014-729-2
eBook (PDF) ISBN 978-0-85014-731-5

Meet the editor

Dr. Miguel Delgado-Prieto received the M. Sc. (2007) and Ph.D. (2012) degrees in Electronics Engineering from the Universitat Politècnica de Catalunya (UPC), Barcelona, Spain, where he is an Associate Professor in the Department of Automatic Control. His work focuses on the digitalization of industrial and energy systems through predictive approaches that enhance reliability, efficiency, and data-driven decision-making. He is actively involved in research collaboration and technology transfer with industry, and has contributed to scientific dissemination as an editor and author of book volumes and publications.

Dr. Luis Romeral was born in Asturias, Spain, in 1960. He received his electrical engineering degree and his Ph.D. from the Technical University of Catalonia (UPC) in 1985 and 1995, respectively. In 1988, he joined the Electronic Engineering Department of the UPC, where he is currently an Associate Professor. His research interests include electric machines, power electronics converters and modulation strategies, variable-speed drive systems, and fault detection algorithms. He has authored more than 250 peer-reviewed scientific papers published in technical journals and conference proceedings. Over the last ten years, 10 Ph.D. dissertations have been completed under his supervision. He participates as a partner or as a responsible person in European Union-funded projects and in research projects funded by Spanish agencies. He has also participated as an engineer or project manager for research and development projects funded by local private companies, in the areas of control automation, industrial networking, supervision, and data acquisition.

Contents

Preface

If the first volume of *Industry 4.0 - Transforming the Future beyond Manufacturing* focused on the technological and operational backbone of industrial transformation, this second volume turns explicitly toward its human, organizational, territorial, and societal dimensions. Taken together, the chapters collected here show that Industry 4.0 cannot be understood solely as automation, connectivity, and data-driven efficiency. Its broader significance lies in how these technological developments reshape work, skills, institutions, public services, cultural systems, and the ethical frameworks that govern digital transformation. In this sense, Volume 2 reflects a shift from technology-centered implementation toward a more human-centered, resilient, and socially embedded understanding of industrial change.

The volume opens with a first section devoted to human-centred Industry 4.0, workforce transformation, and organizational resilience. *Perspective Chapter: Industry 4.0's Artificial Intelligence - A Blessing and a Nightmare for the Workforce* introduces the ambivalent role of AI in the workplace, highlighting both its productivity-enhancing potential and its implications for job insecurity, worker well-being, and pedagogical change. This perspective is expanded in *Cognitive Industry 4.0 Reimagined – Human-Centered Intelligence and Organizational Resilience in Post-Industrial Systems*, which proposes a socio-technical framework centered on co-adaptive intelligence and organizational resilience. The following two chapters, *Perspective Chapter: The Imperative of Proactive Workforce Management - A Methodology for a Human Capital Digital Twin* and *Perspective Chapter: The Role of Artificial Intelligence in Workforce Training and Talent Acquisition under Industry 4.0*, explore how organizations may anticipate skill needs, redesign workforce planning, and integrate AI into recruitment and professional development. The section closes with *Knowledge Management in Production: Human-Centered and AI-Supported Approaches for the Shop Floor* and *Filling Data Gaps with Operational and Experiential Knowledge from Employees – Opportunities and Challenges for AI-Based Solutions*, two contributions that emphasize the centrality of employee knowledge, tacit experience, and human–AI collaboration in sustaining robust socio-technical systems.

A second section turns to digitalisation in SMEs, education, and skills formation. Here, the emphasis shifts from organizational resilience to capability-building across firms and learning environments. *Perspective Chapter: Factors Influencing Digitalisation in Small and Medium-Sized Enterprises* examines the conditions under which SMEs can implement digitalization effectively, emphasizing the roles of information systems, human resources, and internal readiness. *Adaptation of Bachelor's and Master's Students in Engineering to a Changing Environment* then addresses how future engineers respond to changing educational and social conditions, underlining the importance of resilience, well-being, and sociocultural integration in preparing graduates for Industry 4.0. The educational dimension is further broadened by *Human-Centered Digitalization in Training Language Industry Specialists*, which explores how digital learning environments can remain dialogic and human-centered, and by *Industry 4.0 in Logistics Higher Education*, which analyzes competence gaps and curriculum needs in an area undergoing rapid transformation through digitalization, automation, and sustainability pressures. Together, these chapters show that the success of industrial transformation depends not only on technologies but also on the capacity of educational institutions and organizations to cultivate adaptable, reflective, and future-oriented learners and professionals.

The third section develops the "beyond manufacturing" perspective most explicitly by extending Industry 4.0 and 5.0 concepts into territories, infrastructures, and cultural ecosystems. *Perspective Chapter: Transforming Public Infrastructure through Industry 4.0 - Systems Integration and Governance in the MENA and EU Contexts* explores how digital infrastructures, interoperability, cloud systems, and governance frameworks are reshaping public services. *Establishing a Regional Industry 4.0 Electronics Hub: The GreenUpPCB Model* presents a regional innovation model that integrates circularity, repair, training, and technological sovereignty to support sustainable territorial development. *Applying Industry 4.0 Principles to Smart and Resilient Cities* then shows how cyber-physical and data-driven logics can be translated into urban systems, smart services, and resilient city infrastructures. The cultural and heritage dimension is addressed in *Perspective Chapter: From Industry 4.0 to 5.0 - Human-Centered Digitalization for Sustainable Heritage Tourism* and *Artificial Intelligence and Eco-Museums: Climatic Challenges and Ethical Perspectives within European Cultural Sustainability*, both of which reposition digital technologies as tools for cultural preservation, community agency, inclusion, and sustainability rather than as ends in themselves. This section demonstrates that the relevance of Industry 4.0 increasingly lies in its capacity to support more resilient, participatory, and place-based systems well beyond the factory floor.

The final section focuses on governance, ethics, sustainability, and data. *Security and Privacy in Human-Centered Modern Technologies* addresses the tensions between personalization, human-centered digitalization, and privacy and security protection in increasingly connected environments. *Perspective Chapter: AI-Enhanced Digital Product Passports: Enabling ESG Transparency in Industry 5.0* examines how AI-supported data architectures can contribute to environmental, social, and governance transparency across industrial value chains, particularly in response to new regulatory demands. Finally, *Perspective Chapter: Social Innovation, Ethical Digital Services, and Personal Data in an AI Era: A Case Study from Switzerland* presents a compelling example of citizen-centered digital service design that privileges trust, transparency, and data sovereignty over premature or opaque AI adoption. Taken together, these chapters underline that the future of Industry 4.0 and 5.0 will be determined not only by technical capability but also by the governance models, ethical safeguards, and institutional choices through which such capability is deployed.

Taken as a whole, the contributions in this volume show that digital transformation is no longer a narrowly technical project. It is also a matter of workforce adaptation, educational renewal, institutional redesign, territorial development, cultural sustainability, and responsible governance. For this reason, Volume 2 is intended not simply as a complement to the first volume, but as an essential extension of it:

a reminder that the meaning of Industry 4.0 ultimately depends on how technologies are aligned with people, organizations, communities, and social values.

We hope that the chapters that follow will not only serve as a useful reference but also inspire new collaborations and research questions, ultimately helping to shape more human-centered, resilient, and socially responsible pathways of digital transformation.

As editors, we would like to thank all authors for their contributions and commitment throughout the editorial process. We are also grateful to the reviewers and to the publishing team for their support in bringing this volume to completion.

Miguel Delgado Prieto
Automatic Control Department,
Polytechnic University of Catalonia,
Terrassa, Spain

Luis Romeral Martínez
Electronic Engineering Department,
Polytechnic University of Catalonia,
Terrassa, Spain

Section 1

Human-Centred Industry 4.0, Workforce and Organizational Resilience

Chapter 1

Perspective Chapter: Industry 4.0's Artificial Intelligence – A Blessing and a Nightmare for the Workforce

Joseph Mukuni

Abstract

The nature of work has been changing since the invention of machine tools during the First Industrial Revolutions in the mid-1700s. In recent years, advancements in technology have accelerated the rate of change in work, particularly with the onset of the Fourth Industrial Revolution, which is also referred to as Industry 4.0. One of the notable fruits of Industry 4.0 is artificial intelligence (AI). It has come as a blessing to industry because of its tremendous contribution to improved productivity. Industry is now able to produce more for less with more precision, thanks to advancements in technology that have brought about AI. For the workforce, AI can be seen as a blessing to the extent that it is a tool that makes work light. This has the potential to lower labor costs because work that needs to be done by many people can be done by fewer people. However, artificial intelligence is also a nightmare because of its potential to threaten job security, workplace safety, and worker well-being. This chapter discusses the advantages of AI, the challenges that it poses (including ethical issues, fears of the unknown, alienation, worker safety, and job insecurity), and the pedagogical implications of AI (including reskilling, adoption of strategies for life-long learning, experiential learning, holistic learning, and multidisciplinary learning). The chapter also recommends that workforce development institutions should adopt learner-centered pedagogical approaches which foster experiential learning, prepare learners for lifelong learning, and facilitate acquisition of soft skills (such as communication, problem-solving, creativity, and critical thinking).

Keywords: Industry 4.0, Industry 5.0, alienation, environment, human-centricity, artificial intelligence, job insecurity

1. Introduction

This chapter discusses the advantages of artificial intelligence (AI), the challenges that it poses, and the pedagogical implications of AI. The chapter also heralds Industry 5.0 as a cure for some of the side effects of Industry 4.0.

The history of man-made tools can be traced back to 2.5 million years ago with the invention of stone tool technology [1]. Since the adoption of stone tool technology, to overcome innate human limitations that make work laborious, man has made and used tools. Technological innovations of the First Industrial Revolution (also referred to as Industry 1.0), which happened around 1750–1,840 did not, therefore, mark the beginning of the use of technology in man's quest to use tools to perform tasks for which human hands alone were not sufficient. However, Industry 1.0, through the invention of the steam engine, made a huge impact on the rate of industrialization and the displacement of many workers [2] and laid a foundation for today's workplace technologies.

On the heels of Industry 1.0 in the late 1800s and early 1900s came Industry 2.0, which is remembered for the advent of electricity, electronic and mechanical devices, and cars [3]. Highlights of Industry 2.0 included a huge increase in mass production, a significant drop in costs, a rise in labor unions and social welfare programs, and workers' demand for better working conditions and better wages [4].

Industry 2.0 was followed by Industry 3.0 (starting in the1950s), which was associated with "the movement from mechanical and analogue electronic technology to digital electronics, such as green buildings, electric cars, and distributed manufacturing. It is based on energy transition and digital technologies, and the internet" [5]. Then, in the 1990s, Industry 3.0 seamlessly transitioned into Industry 4.0. Commenting on Industry 4.0, some scholars have said:

> *As industrial revolutions have moved from the mechanization of production in the first industrial revolution, to the mass production in the second, and then to the automation of production in third, the standards of living for most people around the world have greatly improved. Undoubtedly, the capability of advancing technology coming forth from the latest industrial revolution has the potential to make even bigger and greater improvements on every aspect of our lives changes than the first three industrial revolutions summed together [6].*

One of the notable fruits of Industry 4.0 is AI. Recently, much has been said about the growing application of AI to work processes and its place in the development of tomorrow's workforce [10–12]. Through the lens of workforce development (WfD) pedagogy, the storyline of this chapter will be guided by the following questions: What is AI? What benefits does AI bring to the workplace? What challenges does AI pose to the workforce? What are the WfD pedagogical implications of AI?

2. Defining artificial intelligence

The definition of AI can be elusive, probably because it is still a relatively new concept. Hashimoto et al. defined AI as "the study of algorithms that give machines the ability to reason and perform cognitive functions such as problem solving, object and word recognition, and decision-making" [13]. This definition may be contrasted with Du-Harpur et al.'s definition: "... the ability of a machine to communicate, reason and operate independently in both familiar and novel scenarios in a similar manner to a human" [14].

The two definitions cited above may also be contrasted with the following definition by the University of Notre Dame: "*Artificial intelligence (AI), in its simplest definition, is any technology/machine that can perform complex tasks that are typically associated with human intelligence.* These tasks can include problem-solving, planning, reasoning, and decision-making" [15].

The three randomly selected definitions mentioned above vary slightly. What they have in common is that they all use human ability to reason and solve problems as the yardstick for determining whether a machine's capability qualifies it to be considered as possessing AI. Any technology or machine can be classified as AI if it can perform tasks that would ordinarily require human cognitive abilities, such as reasoning, problem-solving, and decision-making.

Defining AI by making comparisons to human cognitive abilities adds to the challenges of adopting a generally agreed-upon definition of AI, which bases its definiens on human intelligence. As some scholars have pointed out, to define AI by referring to human intelligence calls for a close look at the meaning of intelligence and drawing boundaries between AI and real intelligence [16, 17].

It is, therefore, for reasons of simplicity rather than completeness of the definition, that in this chapter we shall adopt a definition of AI that describes it as the ability of technology or a machine to perform tasks that ordinarily require some human cognitive abilities.

3. Benefits of artificial intelligence to the workforce

The application of AI to work processes is a blessing to the industry because of the several benefits it brings, including enhanced accuracy and efficiency, high productivity, reduction in costs, improved decision-making, and minimal concerns about safety.

3.1 Accuracy and efficiency

AI has made tremendous improvements to accuracy and efficiency in several business sectors, such as healthcare, finance, e-commerce, manufacturing, and marketing [18]. Taking healthcare as an example, AI enhances accuracy and efficiency through its:

> *exceptional ability to detect intricate patterns in medical images, often surpassing human capabilities, which is crucial for diagnosing complex diseases such as cancer and neurological disorders. AI's role in reducing human error, coupled with its rapid processing abilities, not only accelerates diagnostic procedures but also contributes to operational efficiency and cost-effectiveness by minimizing unnecessary repeat scans and the likelihood of misdiagnoses [19].*

3.2 Productivity

Recent research has demonstrated how AI is revolutionizing production processes to enhance productivity. For instance, a study by Gao and Feng [20] concluded that "AI enables firms to improve their productivity, leading to increased main business income and higher added value of products. This results in higher output from all factors of production in the current period compared to the previous period, leading to a significant increase in TFP." This is not surprising because, although human labor has its strengths, it has weaknesses that can affect productivity. For example, human beings can impact productivity through industrial disputes, fatigue, human error, sickness, or resignations. AI is free from these weaknesses. Furthermore, AI is better suited than human workers for repetitive tasks that are characteristic of mass production processes.

3.3 Costs

One of the central motives of business enterprises is to make profits from business operations by minimizing costs. AI comes in handy in this regard. Examples of the contribution of AI to cost reduction include the use of robotics process automation (RPA) to minimize production costs associated with human labor, enhancing predictive analytics through the deployment of machine learning (ML), and applying natural language processing (NLP) technologies to customer services and communication processes [21].

3.4 Decision-making

Effective decision-making is a function of the utilization of data in problem analysis and the selection of problem solution options. AI contributes to the decision-making process by generating and providing the data needed to guide the process. Jarrahi [22] illustrates how AI helps decision-makers when navigating three challenges of decision-making, namely *uncertainty, complexity, and equivocality*. **Table 1** paraphrases Jarrahi's decision-making model.

As **Table 1** shows, AI aids the decision-making process by the speedy provision of necessary information, facilitating quick analysis of data, and analyzing divergent interpretations of a decision domain.

3.5 Safety

The subject of workplace safety and occupational health is critically important. It affects productivity through the loss of man-hours caused by injuries, fatalities, occupational diseases, and accidents [23–25]. The rate of occupational casualties is very high. It is estimated that annually, two million workers worldwide lose their lives due to work-related causes [23], not to mention the number of workers who cannot contribute effectively to productivity because of occupational sickness and injury.

Against this background, the advent of AI-based solutions to occupational health and safety issues is a huge blessing. Examples of solutions include "smart sensors,

Challenge	Uncertainty	Complexity	Equivocality
	(characterized as a lack of information about all the alternatives or their consequences, which makes interpreting a situation and deciding more difficult)	(characterized by an abundance of elements or variables; they demand the processing of masses of information at a speed beyond cognitive capabilities)	(the presence of several simultaneous but divergent interpretations of a decision domain, often due to the conflicting interests of stakeholders, customers, and policymakers.)
Role of human decision-maker	Make swift, intuitive decisions in the face of the unknown	Decide where to seek and gather data, and choose among options with equal data support	Negotiate, build consensus, and rally support
Role of AI	Provide access to "real time" information	Collect, curate, process, and analyze data	Analyze sentiments, and represent diverse interpretations

Table 1.
Use of AI in the navigation of uncertainty, complexity, and equivocality.

drones, and predictive systems" that have proven to significantly reduce accidents and strengthen decision-making capacity in terms of prevention. For example, companies such as Siemens and Bechtel have reported decreases of up to 40% in workplace incidents following the adoption of smart technologies [26]. Prevention is better than cure. Therefore, the greatest contribution of AI-enabled systems to occupational safety is prevention through advanced sensor technologies and other AI-enabled technologies that help in processing huge datasets to identify and warn about health risks and other hazardous situations, and suggest preventative measures in good time [27].

4. Challenges of artificial intelligence

AI poses several challenges, such as ethical issues, fears of the unknown, alienation, worker safety, and job insecurity. As some scholars have perceptively observed, AI in the workplace indeed tends to function as a double-edged sword [28–30].

4.1 Ethical issues

One of the greatest challenges posed by AI is the issue of ethics. Ethics may be defined as "the discipline dealing with right versus wrong, and the moral obligations and duties of entities (e.g., humans, intelligent robots, etc.)" [31]. Concerns that have driven the issue of ethics to prominence include:

> *an increasing number of high-profile cases of harm that has resulted either because of the misuse of the technology (e.g., psychometric voter manipulation, facial recognition surveillance, mass data collection without consent, etc.), or as a result of the technology having design flaws (e.g., bias in cases of recidivism, loan rejection, and medical misdiagnosis, etc.) [32].*

According to Kazim and Koshiyama [32], the major themes in AI ethics are human well-being, safety, privacy, transparency, fairness, and accountability. **Table 2** summarizes the descriptors of these six themes.

4.2 Fear of the unknown

AI is a relatively new phenomenon. Its limits and potential harm are not yet fully understood. The general workforce is not quite sure yet of the possible harm of AI in

Themes	Descriptors	Examples of risks
Human well-being	This theme is grounded in the ethical principle of respect for human dignity and includes psychological, social, and environmental well-being.	Possibility of a system diminishing the deliberative/rational capacity of humans; mental and physical ill-health caused by job stress and anxiety; undesired impact on society and the environment, for example, through pollution caused by poor industrial waste management.
Safety	This theme is based on the ethical principle of preventing harm, where harm is defined in terms of adverse effects on human well-being, that is, the psychological, social, and environmental human well-being.	Data poisoning, model leakage, weaponization of delivery drones; a warehouse robot that causes new safety risks; a malfunction of an AI-powered tool can cause injuries to workers.

Themes	Descriptors	Examples of risks
Privacy	This theme is based on the public and political demand to respect a human's personal information. It relies on a distinction between the private-personal and the public-political/communal sphere, where the former is seen as demanding a higher level of respect for privacy than the latter.	Mass surveillance and personal data are being used to target and engage in recommendation and manipulation (for both political and economic ends); identity theft caused by data breaches or leakages of confidential information.
Transparency	This theme is based on the public and political demand to respect a human's personal information. This relies on a distinction between the private-personal and the public-political/communal sphere, where the former is seen as demanding a higher level of respect for privacy than the latter.	Failure to explain how a system has come to a decision and to make that decision explicable to workers and other stakeholders leads to a loss of trust.
Fairness	This theme is based on the ethical principle of human equality. Fairness falls under debates about justice and is hotly contested. A central question is what definition(s) of fairness/ justice to commit to, that is, there are mutually exclusive theories of fairness, such as corrective, distributive, procedural, substantive, comparative, etc.	Bias in datasets, intentional exploitation of people (e.g., customers/regional pricing), unfair competition, and a hiring tool that ignores qualified female applicants.
Accountability	Ethical AI is a branch of applied ethics and, as such, is inherently concerned with how AI systems impact human beings: how the systems are developed, the processes and logic of decision-making, the allocation of duties with respect to who makes decisions and how, to what extent impacts, risks, and harms are gauged and measured. All this falls under the remit of accountability, and, as the previous sentence indicates, accountability relates to knowing who made decisions, how those decisions were made, and what systems or tools were put in place to measure and track – that is, governance.	Not communicating the capabilities and purposes of the system to those both directly and indirectly impacted; misleading a person to believe that he/she is communicating with a human being by not making it known that they are interacting with a system. A system may be mimicking human subjectivity (e.g., a chatbot).

Table 2.
Descriptors of AI ethical themes based on Kazim and Koshiyama's analysis.

the workplace, which raises fear and anxiety. According to a study by Wang and Xiao [33], "Anxieties induced by AI, such as privacy violation, bias behavior, job replacement, and learning anxiety, are not limited to the field of AI-generated art but are prevalent across various domains. Such concerns may also trigger multifaceted anxieties among individuals and with society at large."

4.3 Alienation

The history of the concept of alienation can be traced to the writings of Karl Marx during the epoch of Industry 1.0 [34]. Work alienation may be defined as "a feeling of the loss of meaningfulness towards the job; this means that an employee will only perceive his job as a way of survival, overlooking factors such as self-fulfillment and satisfaction, which limits the full use of his skills, experiences, and knowledge"

[35]. Apart from the sense of detachment from the job, a worker may also feel detached from himself and other people [36].

Research has shown that workers can easily feel alienated when AI seems to take away their jobs and their sense of ownership of the expertise needed to perform their tasks, creating anxiety and an identity crisis [37]. The consequences of work alienation can be dire:

> *An alienated employee suffers several adverse ramifications. An alienated employee particularly suffers from low quality of work, production, cooperation, motivation, and organizational commitment. He/she would suffer deficiencies in organizational identification and job involvement. He/she would be liable to excessive absenteeism, feeling of purposelessness, disruption of work activities, changes in job designation, more indulgence in immoral activities. He/she would suffer a sense of fatigue along with other health issues; they would feel distant from colleagues, they would have lack of interest in achieving organizational targets and unable to shoulder their responsibilities [38].*

5. Pedagogical implications of artificial intelligence

The increased use of AI in the workplace has some pedagogical implications. Workers need education and reskilling to enable them to maximize the benefits of the ever-changing AI-enabled tools and to minimize, cope with, and survive the impact of the negative effects of AI.

5.1 Knowledge and skills for maximizing the benefits of artificial intelligence

To operate AI-powered machines and to interact well with non-human "workers" (e.g., robots), employees need to acquire appropriate technical as well as soft skills. The term "technical skills," also known as hard skills, refers to skills that enable "employees to compete in tasks, including skills related to scientific knowledge, professional abilities, and technical expertise" [39]. According to a study by Babashahi et al. [40], the technical skills include:

> *proficiency in language processing, machine learning, machine vision, big data analysis, IoT-related technologies, programming, data analytics, cybersecurity, messaging platforms, chatbot development, natural language processing, intent recognition, robotic process automation, assistive robotics skills, human–robot cooperation, exoskeleton robotics, and various scientific and technological proficiency. These technical skills form the backbone of AI integration and technological advancement across industries, driving innovation and efficiency.*

In addition to the technical skills mentioned above, today's workers are expected to have soft skills. There is no commonly agreed-upon definition of soft skills yet. However, the term "soft skills" generally encompasses:

> *personal, social, communication, and self-management behaviors. They cover a wide spectrum of abilities and traits: being self-aware, trustworthiness, conscientiousness, adaptability, critical thinking, attitude, initiative, empathy,*

> *confidence, integrity, self-control, organizational awareness, likability, influence, risk taking, problem solving, leadership, time management, and then some [41].*

5.2 Knowledge and skills for minimizing the impact of the negative effects of artificial intelligence

Not all workplace-related stressors and traumatic situations arising from the unintended consequences of AI-powered technologies can be resolved by education and training. Some responses to psychosocial problems will need non-training-related interventions, such as the formulation of policies for the prevention, detection, and treatment of workplace mental health issues caused by the negative side of Industry 4.0. Other non-training-related interventions include medication or counseling, communication about one's condition, accepting one's condition, and allowing oneself to take time off when needed [42].

That said, there is a place for education and training for supervisors and workers. In their justification of the role of education and training in occupational stress management intervention programs, Richardson and Rothstein observed that:

> *Although it is not possible to eliminate stress entirely, people can learn to manage it. Many organizations have adopted stress management training programs to try and reduce the stress levels of their workforce. A stress management intervention (SMI) is any activity or program initiated by an organization that focuses on reducing the presence of work-related stressors or on assisting individuals to minimize the negative outcomes of exposure to these stressors [43].*

Training content for courses designed to minimize stressors includes soft skills such as communication and conflict resolution, as reported by Molek-Winiarska and Kawka [44]:

> *According to theoretical models, such as the JD–C[1] and HCM[2] presented in the introduction, improving one's skills, such as communication, conflict resolution, teamwork or self-management, is instrumental in the reduction of work-related stress, especially in the circumstances of teamwork. On the basis of the results showing a decrease in perceived stress and an increase in general health, one can support the hypotheses related to the positive influence of soft-skills training on stress reduction.*

5.3 Pedagogical approach for artificial intelligence workforce development

To equip the workforce with both hard skills and soft skills that will enable them to maximize the benefits of AI-enabled tools and minimize the impact of the negative effects of AI, institution-based and workplace-based providers of education and training need to adopt teaching-learning approaches that promote experiential learning (which enhances the acquisition of innovation, deeper learning, lifelong learning, and soft skills), holistic learning, and multidisciplinary learning.

5.3.1 Experiential learning

Experiential learning theory (ELT) is attributed to David Kolb, who defined learning as "a process, in which knowledge is created through transformation of experience" [45]. This approach to learning shifts the burden of knowledge acquisition from the teacher to the learner [46]. Unlike the traditional didactic approach, in which the teacher is the sole provider of knowledge, experiential learning enables the learner to learn by doing and reflect on the process and outcome of learning. The teacher's role is to create a learning environment in which a student constructs knowledge from the practical experience of performing a real-world task. In this approach, learning takes place in the following four-stage cycle [47].

- Concrete experience – the student actively participates in performing a task.
- Reflexive observation – the student consciously reflects on the experience gained from performing the task.
- Abstract conceptualizing – the student tries to conceptualize a theory or model of what they have observed while performing the task.
- Active experimenting – the student tries to plan how to test the model, the theory, or the plan for the forthcoming experience.

School or college classrooms, laboratories, and workshops must be characterized by learner-centered approaches that promote experiential learning, such as project-based learning (PjBL), through which students engage in long-term, real-world projects to apply knowledge and cultivate a sense of responsibility; problem-based learning (PBL), which encourages students to actively solve real-world problems, foster collaboration, and interdisciplinary thinking; and inquiry-based learning (IBL), which fosters active learning involving questioning, exploring, and critical thinking to develop problem-solving skills [48].

Out-of-class learning environments for this approach include work-based learning (WBL) modes such as apprenticeships and internships, which allow learners to acquire AI-related skills by observing experienced workers perform tasks using AI-enabled tools, appreciating the theory behind the tasks, and then performing those tasks under the supervision of their mentors while reflecting on their experiences.

The advantages of experiential learning are many, including the promotion of innovation [49], lifelong learning [50], deeper learning [51], and soft skills [52], all of which are critical Industry 4.0 skills.

5.3.2 Holistic learning

In an AI-dominated working environment, in which workers require not only cognitive and psychomotor skills to get work done but also other skills for coping with psychosocial issues, there is a need for training providers to adopt a holistic learning approach. This approach is "characterised by focus on the 'whole person', and seeking to engage fully all aspects of the learner: cognitive; emotional; and social" [53].

5.3.3 Multidisciplinary learning

Problem-solving, collaboration, and teamwork skills rank among the in-demand Industry 4.0 workforce abilities. Workers are expected to have the ability to collaborate with colleagues specializing in different disciplines to solve real-world production problems. To empower workers with such skills, WfD pedagogies must include multidisciplinary learning approaches. "Multidisciplinarity is understood as interaction between members from varying disciplines in academia, where the disciplinary boundaries are kept intact" [54]. According to Chi et al. [55]:

> *Multidisciplinary technological approaches in higher education are important because they foster interdisciplinary collaboration, develop critical thinking and problem-solving skills, align with the demands of the digital era, and promote innovation and creativity. These approaches aim to work across different disciplines and technologies to enhance learning outcomes, address specific problems, and develop professional skills by drawing on a range of expertise.*

6. Welcome to Industry 5.0

Like all industrial revolutions before it, Industry 4.0 will be remembered for its contribution to increased productivity. It will also be remembered for its unintended negative consequences, such as mental health issues arising from machine-human interactions; anxieties about losing jobs to machines; and social isolation of workers, with its attendant traumatic ramifications for occupational health and well-being. Anxiety about job insecurity or a reduction in wages is not without justifiable reason. It has been estimated that occupations at the highest risk of automation account for about 27% of employment, and according to OECD surveys, about 20% of workers in manufacturing and finance worry about a possible reduction in wages [56]. Technology has always had a negative impact on job security. However, AI appears to have a greater impact on employment retention rates. Studies have shown that the introduction of AI is 26.6% points more likely to be associated with job destruction compared to the introduction of other technology [57].

Industry 4.0's criticisms are not only about concerns for worker well-being; they are also about the centrality of the profit motive in the drive for technological advancements. As Ghobakhloo et al. [58] said, "Industry 4.0 aligns with the long-lasting neoliberal capitalism models that emphasize profitability and shareholder primacy, intensifying some of the prevailing socio-environmental concerns such as regional inequality, environmental degradation, of human-robot collaboration for a smart business ecosystem that promotes resource efficiency and bioeconomy."

The negative impact of Industry 4.0 on workers and the environment has led to the emergence of Industry 5.0. It might be too early for the workforce and environmentalists to sigh with relief upon hearing about the advent of Industry 5.0. However, Industry 5.0 promises to be a game-changer because, as Gamberini and Pluchino [59] have reported:

> *Industry 5.0 represents a pivotal shift from its predecessor, Industry 4.0, by integrating a broader purpose encompassing social and environmental considerations, prioritizing worker well-being, and aiming for a sustainable and resilient industry. This new industrial revolution emerged in response to the increasing automation and removal of human workers from manufacturing processes, which was a hallmark of Industry 4.0. Industry 5.0 seeks to address these shortcomings by*

> *advocating for the reintroduction of human workers for instance alongside robots and collaborative robots . . . thus combining the strengths of humans and machines . . . Industry 5.0 focuses on merging the creativity and expertise of human experts with efficient and intelligent machines.*

In addition to promising industrial practices that place worker well-being at the center of production processes, Industry 5.0 promises to promote technologies that foster environmental sustainability rather than prioritizing profits at the expense of environmental impact [60].

The learner-centered pedagogical approaches discussed in Section 5 (e.g., project-based and holistic learning) resonate with Industry 5.0's goal of placing worker well-being at the center of technological innovations. These teaching-learning approaches prepare workers for a more human-focused industry because they not only target higher-order cognitive skills such as creativity and innovation but also develop socio-cultural competencies (e.g., respect for cultural differences, empathy, and conflict resolution) and emotional intelligence (e.g., the ability to manage one's own emotions and understand the emotions of co-workers, supervisors, and clients).

7. Conclusion

Drawing on literature about Industry 4.0 in general and AI in particular, as well as learning theories associated with technology education and employability skills, this chapter has contributed to the discourse concerning the pedagogical implications of Industry 4.0's increasing use of AI-enabled tools in workplaces. Before proposing pedagogical approaches to the teaching and learning of skills associated with AI, the chapter defined AI and identified its benefits and challenges. The chapter argued that both the benefits and challenges of AI have implications for WfD, as workers need education and re-skilling to enable them to maximize the benefits of the ever-changing AI-enabled tools and minimize the impact of the negative effects of AI.

The chapter has recommended that institutions charged with the responsibility of skills development for the Industry 4.0 workforce should adopt learner-centered pedagogical approaches such as PjBL, PBL, and IBL. These approaches foster experiential learning, prepare learners for lifelong learning, and facilitate the acquisition of soft skills (such as communication, problem-solving, creativity, and critical thinking). The chapter has also recommended the adoption of pedagogical approaches that promote holistic learning (which focuses on the "whole person" – body, mind, and soul) and multidisciplinary learning, which mimics real-world Industry 4.0 environments where workers perform tasks that require multiple disciplines and different talents.

The chapter ends with a word about the advent of Industry 5.0, which is poised to reverse the unintended negative consequences of Industry 4.0, including psychosocial occupational hazards and environmental degradation, by prioritizing employee well-being and environment-friendly technologies.

Conflict of interest

The author declares no conflicts of interest.

Author details

Joseph Mukuni
Virginia Tech University, Blacksburg, VA, USA

*Address all correspondence to: mjoseph7@vt.edu

References

[1] Kuhn SL, Clark AE. Stone tool technology. The Oxford Handbook of the Archaeology and Anthropology of Hunter-Gatherers. Oxford University Press; 2014. Available from: https://academic.oup.com/edited-volume/28290/chapter-abstract/214504217?redirectedFrom=fulltext [Accessed: 2025-October-28]

[2] Wisskirchen G, Biacabe BT, Bormann U, Muntz A, Niehaus G, Soler GJ, von Brauchitsch B.Artificial intelligence and robotics and their impact on the workplace. IBA Global Employment Institute. 2017;**11**(5):49–67

[3] Yin Y, Stecke KE, Li D.The evolution of production systems from Industry 2.0 through Industry 4.0. International Journal of Production Research. 2018;**56**(1-2):848–861

[4] Richmond Vale Academy. The second industrial revolution: The technological revolution. 2022. Available from: https://richmondvale.org/second-industrial-revolution/ [Accessed: 2025-October-28]

[5] Mohajan H. Third industrial revolution brings global development. 2021. Available from: https://mpra.ub.uni-muenchen.de/110972/1/MPRA_paper_110972.pdf [Accessed: 2025-October-28]

[6] Jin BE, Shin DC.The power of 4th industrial revolution in the fashion industry: What, why, and how has the industry changed? Fashion and Textiles. 2021;**8**(1):1–25

[7] McGinnis D. What is the fourth industrial revolution. *Salesforce Blog*. 2023. Available from: https://www.salesforce.com/blog/what-is-the-fourth-industrial-revolution-4ir/ [Accessed: 2025-October-28]

[8] Mukuni Joseph. Behold the fourth industrial revolution and how to keep pace with workplace competencies in an ever-changing world of work! In: El-Farr, Hadi. The Changing Landscape of Workplace and Workforce. IntechOpen; 2023

[9] Xu M, David JM, Kim SH.The fourth industrial revolution: Opportunities and challenges. International Journal of Financial Research. 2018;**9**(2):90–95

[10] AlAli R, Wardat Y.Opportunities and challenges of integrating generative artificial intelligence in education. International Journal of Religion. 2024;**5**(7):784–793

[11] Alier M, García-Peñalvo F, Camba JD. Generative artificial intelligence in education: From deceptive to disruptive International Journal of Interactive Multimedia and Artificial Intelligence. 2024. Available from: https://www.ijimai.org/journal/bibcite/reference/3436: [Accessed: 2025-October-28]

[12] Mao J, Chen B, Liu JC. Generative artificial intelligence in education and its implications for assessment. TechTrends: Linking Research & Practice to Improve Learning. 2024;**68**(1):58–66

[13] Hashimoto DA, Rosman G, Rus D, Meireles OR.Artificial intelligence in surgery: Promises and perils. Annals of Surgery. 2018;**268**(1):70–76

[14] Du-Harpur X, Watt FM, Luscombe NM, Lynch MD.What is

AI? Applications of artificial intelligence to dermatology. British Journal of Dermatology. 2020;**183** (3):423–430

[15] University of Notre Dame. AI overview and definitions. Notre Dame, USA; 2025. Available from: https://learning.nd.edu/resource-library/ai-overview-and-definitions/ [Accessed: 2025-October-28]

[16] Korteling JH, van de Boer-visschedijk GC, Blankendaal RA, Boonekamp RC, and Eikelboom AR. Human-versus artificial intelligence. Frontiers in Artificial Intelligence. 2021;**4**:1–13

[17] González-Calatayud V, Prendes-Espinosa P, Roig-Vila R. Artificial intelligence for student assessment: A systematic review. Applied Sciences. 2021;**11**:5467

[18] Arman M, Lamiyar UR.Exploring the implication of ChatGPT AI for business: Efficiency and challenges. Journal of Innovation Information Technology and Application. 2023;**5** (1):52–64

[19] Khalifa M, Albadawy M. AI in diagnostic imaging: Revolutionising accuracy and efficiency. Computer Methods and Programs in Biomedicine Update. 2024;**5**:1–12

[20] Gao X, Feng H.AI-driven productivity gains: Artificial intelligence and firm productivity. Sustainability. 2023;**15**(11):8934

[21] Ismanov I, Qayumov N, Mukhamadjonova D, Akhmadaliyev B (2024, April). AI and cost management: Strategies for reducing expenses and improving profit margins in business. In *2024 International Conference on Knowledge Engineering and Communication Systems (ICKECS)* (Vol. **1**, pp. 1–7). IEEE

[22] Jarrahi MH.Artificial intelligence and the future of work: Human-AI symbiosis in organizational decision making. Business Horizons. 2018;**61** (4):577–586

[23] Lemos J, de Souza VB, Falcetta FS, de Almeida FK, Lima TM, Gaspar PD. Enhancing workplace safety through personalized environmental risk assessment: An AI-driven approach in industry 5.0. Computers. 2024;**13**(5):120

[24] Savković M, Dasic M, Djapan M, Vukicevic A, Macuzic I, Stefanovic M (2021). Improving workplace safety using advanced Industry 4.0 technologies

[25] Ozobu CO, Adikwu FE, Cynthia OO, Onyeke FO, Nwulu EO. Developing an AI-powered occupational health surveillance system for real-time detection and management of workplace health hazards. World Journal of Innovation and Modern Technology. 2025;**9**(1):156–185

[26] Trujillo HP, Castillo JC, Rangel AD, Patiño OF, Bautista REA, Cárdenas JZC. The global impact of AI on workplace safety, opportunities and challenges for the future of work. Ciencia Latina Revista Científica Multidisciplinar. 2025;**9**(2):7500–7513.

[27] Howard J, Schulte P.Managing workplace AI risks and the future of work. American Journal of Industrial Medicine. 2024;**67**(11):980–993.

[28] Wilkens U.Artificial intelligence in the workplace–A double-edged sword. The International Journal of Information and Learning Technology. 2020;**37**(5):253–265

[29] Chen JJ, Lin JC.Artificial intelligence as a double-edged sword: Wielding the POWER principles to maximize its positive effects and minimize its negative effects. Contemporary Issues in Early Childhood. 2024;**25**(1):146–153

[30] Dong X, Tian Y, He M, Wang T. When knowledge workers meet AI? The double-edged sword effects of AI adoption on innovative work behavior. Journal of Knowledge Management. 2025;**29** (1):113–147

[31] Siau K, Wang W.Artificial intelligence (AI) ethics: Ethics of AI and ethical AI. Journal of Database Management (JDM). 2020;**31**(2):74–87

[32] Kazim E, Koshiyama AS. A high-level overview of AI ethics. Patterns. 2021;**2**(9):1–12

[33] Wang C, Xiao A. Anxiety induced by artificial intelligence (AI) painting: An investigation based on the fear acquisition theory. Psychological Trauma: Theory, Research, Practice, and Policy. 2025;**17**:S214–S224

[34] Marx K. Economic and philosophic manuscripts of 1844. In: Struik, Dirk. Social Theory Re-wired, Routledge, 2023:142–147

[35] García-Contreras R, Muñoz-Chávez JP, Muñoz-Chávez RL, Lezama-León E, Barrios-Quiroz H. Work alienation, deviant workplace behavior and performance in public sector. Sustainability. 2022;**14**(17):1–13

[36] Shah MI. Marx's Concept of Alienation and Its Impacts on Human Life. Al-Hikmat: A Journal of Philosophy. 2015;**35**:43–54

[37] Houssaini Y (2025). *Artificial Intelligence and Professionals' Work Alienation* (Bachelor's thesis, University of Twente). Available from: https://essay.utwente.nl/106836/1/Houssaini_BA_BMS_faculty.pdf [Accessed: 2025-October-29]

[38] Muttar AK, Keir MYA, Mahdi OR, Nassar IA.Antecedents and consequences of work alienation–A critical. Journal of Statistics Applications and Probability. 2019;**8**(3):279–288

[39] Lyu W, Liu J. Soft skills, hard skills: What matters most? Evidence from job postings. Applied Energy. 2021;**300**: 1–12

[40] Babashahi L, Barbosa CE, Lima Y, Lyra A, Salazar H, Argôlo M, Souza JMD.AI in the workplace: A systematic review of skill transformation in the industry. Administrative Sciences. 2024;**14** (6):127

[41] Klaus P, Rohman J, Hamaker M. The Hard Truth about Soft Skills. Bookstore in New York, NY: Harper Collins Publishers; 2007

[42] Hennekam S, Richard S, Grima F. Coping with mental health conditions at work and its impact on self-perceived job performance. Employee Relations: The International Journal. 2020;**42** (3):626–645

[43] Richardson KM, Rothstein HR. Effects of occupational stress management intervention programs: A meta-analysis. Journal of Occupational Health Psychology. 2008;**13**(1):69–93

[44] Molek-Winiarska D, Kawka T. Reducing work-related stress through soft-skills training intervention in the mining industry. Human Factors: The

Journal of the Human Factors and Ergonomics Society. 2024;**66**(5): 1633–1649

[45] Kolb DA. Experiential Learning: Experience as the Source of Learning and Development. Englewood Cliffs, New Jersey: Prentice-Hall; 1984

[46] Bartle E. Experiential Learning: An Overview. *Institute for Teaching and Learning Innovation*. *Australia*: *The university of Queensland*; 2015

[47] Sharlanova V.Experiential learning. Trakia Journal of Sciences. 2004;**2**(4):36–39

[48] Singha R, Singha S. Application of experiential, inquiry-based, problem-based, and project-based learning in sustainable education. In: Lee Goi, Chai Lee. Teaching and Learning for a Sustainable Future: Innovative Strategies and Best Practices. Hershey, Pennsylvania: IGI Global Scientific Publishing, 2024:109–128

[49] Siddique Z, Hardré P, Bradshaw A, Saha M, Mistree F (2010, January). Fostering innovation through experiential learning. In *International Design Engineering Technical Conferences and Computers and Information in Engineering Conference* (Vol. **44144** , p. 791–803)

[50] Passarelli AM, Kolb DA. The learning way: Learning from experience as the path to lifelong learning and development. The Oxford Handbook of Lifelong Learning. **2nd Edn**. 2011. 70–90 p.

[51] Moore C, Boyd BL, Dooley KE.The effects of experiential learning with an emphasis on reflective writing on deep-level processing of leadership students. Journal of Leadership Education. 2010;**9**(1):36–52

[52] Haritha G, Rao R. A holistic approach to professional development: Integrating Kolb's experiential learning theory for soft skills mastery. Journal of Engineering Education Transformations SP. 2024;**37**:415–424

[53] Ene GU, Goulding JS, Adjaie John G, Pour Rahimian Leilabadi F. Holistic Learning: A workforce development paradigm. 2015. Available from: https://clok.uclan.ac.uk/id/eprint/12382/1/Gloria%20et%20al.pdf

[54] Brodin EM, Avery H.Cross-disciplinary collaboration and scholarly Independence in multidisciplinary learning environments at doctoral level and beyond. Minerva. 2020;**58** (3):409–433

[55] Chi C, Yunus MM, Rafiq KRM, Hameed H, Ediyanto E.A systematic review on multidisciplinary technological approaches in higher education. International Journal of Advanced Research in Future Ready Learning and Education. 2024;**36** (1):1–10

[56] Green A. Artificial intelligence and jobs: No signs of slowing labour demand (yet). OECD Employment Outlook. 2023. 102–127 p.

[57] Hunt W, Sarkar S, Warhurst C. Measuring the impact of AI on jobs at the organization level: Lessons from a survey of UK business leaders. Research Policy. 2022;**51**(2):1–12

[58] Ghobakhloo M, Iranmanesh M, Mubarak MF, Mubarik M, Rejeb A, Nilashi M. Identifying industry 5.0

contributions to sustainable development: A strategy roadmap for delivering sustainability values. Sustainable Production and Consumption. 2022;**33**:716–737

[59] Gamberini L, Pluchino P.Industry 5.0: A comprehensive insight into the future of work, social sustainability, sustainable development, and career. Australian Journal of Career Development. 2024;**33**(1):5–14

[60] Rame R, Purwanto P, Sudarno S. Industry 5.0 and sustainability: An overview of emerging trends and challenges for a green future. Innovation and Green Development. 2024;**34**:1–20

Chapter 2

Cognitive Industry 4.0 Reimagined – Human-Centered Intelligence and Organizational Resilience in Post-Industrial Systems

Mitt Nowshade Kabir

Abstract

This chapter introduces Cognitive Industry 4.0 (CI4.0) as a new socio-technical framework that reimagines industrial transformation beyond the traditional metrics of efficiency and speed. Arguing that incumbent Industry 4.0 models are often too rigid for the complexities of post-industrial systems, this work proposes a paradigm centered on co-adaptive intelligence – the symbiotic, real-time learning between humans and machines – as the core mechanism for generating organizational resilience as an emergent property. Through an interdisciplinary synthesis of resilience engineering, complex adaptive systems theory, and critical AI studies, the chapter outlines an integrated architecture of technological pillars, core principles, and measurable outcomes. It details how technologies such as cognitive digital twins, augmented reality, and decentralized governance infrastructures can be operationalized to deliver two key outcomes: human-centered intelligence, which focuses on augmenting human agency and creativity rather than replacing them, and organizational resilience, defined as the capacity to anticipate, adapt, and learn from disruption. Sector-specific case studies in healthcare, education, public administration, and the creative industries illustrate the framework's practical application in both high- and low-resource contexts, providing specific implementation roadmaps and failure mode analyses. The chapter also provides actionable design patterns, governance protocols, and evaluation metrics for implementation. It concludes by addressing the profound ethical dimensions of this transition – including cognitive sovereignty, algorithmic justice, and participatory governance – and proposes a concrete research agenda. Ultimately, CI4.0 is presented as a normative and operational model for designing more humane, adaptable, and equitable socio-technical systems fit for an increasingly uncertain world.

Keywords: Cognitive Industry 4.0, organizational resilience, human-centered intelligence, co-adaptive systems, socio-technical systems, AI governance

1. Introduction: Reimagining Industry 4.0

The prevailing narratives of industrial revolutions have traditionally been framed as stories of mechanization, mass production, and automation – linear progressions optimized for efficiency gains and labor-cost reductions. Industry 4.0, with its intense focus on cyber–physical systems, interconnectivity, and smart automation, marks the latest chapter in this trajectory. However, this framing is increasingly insufficient for the contemporary landscape. Empirical reviews indicate that, while the adoption of Industry 4.0 has indeed improved throughput and efficiency, it has also revealed significant limitations inherent in rigid automation and brittle optimizations. These systems, optimized for stable environments, often fail to address the complex, dynamic challenges of post-industrial contexts, where variability and unpredictability are the norms [1]. This divergence has created a critical gap between technological potential and organizational reality, particularly in advanced knowledge sectors, where adaptability and creative problem-solving hold significantly greater value than static efficiency.

This chapter addresses that gap by proposing Cognitive Industry 4.0 (CI4.0) as a profound socio-technical evolution rather than merely a technological upgrade. We define CI4.0 as a holistic framework in which human and machine agents continuously learn from one another – a process of co-adaptive intelligence – enabling organizational resilience to emerge at the system level. Resilience, in this context, is rigorously defined not as robustness (the ability to resist change) but as the capacity to anticipate, absorb, adapt, and learn under disruption. Distinct from the efficiency-centric Industry 4.0 and the emerging values-centric Industry 5.0, CI4.0 operationalizes resilience via specific mechanisms: decentralized decision rights, human-centered interfaces, and auditable learning loops across the device-edge–cloud continuum. Crucially, the framework is bounded by a rejection of fully autonomous replacement in high-stakes contexts and opposes purely centralized MLOps in favor of distributed, edge-native intelligence.

This focus on the cognitive dimension – where value is generated not from physical goods but from artificial intelligence (AI)-enabled insight, design, and innovation – aligns with theories of cognitive capitalism [2, 3] and the broader economic shift toward an intangible, knowledge-based economy [4]. While the historical arc of industrial progress has been one of offloading human effort to machines, CI4.0 deliberately redirects this trajectory toward augmentation. In this framework, augmentation is defined as the measurable amplification of human agency, judgment, and creativity. For instance, in the creative industries, generative models expand the solution space for a designer, proposing novel starting points that augment rather than direct the creative process. Recent reviews in related domains suggest that digital technology adoption is a key enabler of this capacity, provided it is governed by human-centric principles [5].

To rigorously explore this framework, this chapter addresses several core research questions (RQs) that guide our inquiry. RQ1 examines how different architectural configurations of co-adaptive intelligence impact organizational resilience metrics, specifically "graceful extensibility" – the capacity of a system to sustain adaptive performance as demand outpaces capacity [6]. RQ2 investigates which governance models are most effective at mitigating the risks of cognitive dependency while fostering human-centered intelligence. Finally, RQ3 assesses the extent to which these principles can be applied in knowledge-based sectors to produce measurable gains in adaptability and well-being. By answering these questions, we aim to provide a roadmap for organizations seeking to navigate the transition to a cognitive, post-industrial future.

2. Fundamentals of CI4.0

At its core, CI4.0 represents a fundamental paradigm shift from designing systems for rule-based efficiency to cultivating socio-technical ecosystems that augment human potential. This logic of augmentation, which positions human cognition as the system's creative and adaptive core, is operationalized through an integrated architecture of technological pillars and core principles. These components do not function in isolation but form a cohesive ecosystem designed to support continuous learning and adaptation.

2.1 The technological pillars

The foundational technologies of CI4.0 are grouped into three thematic clusters that span the device-edge–cloud continuum, ensuring intelligence is pervasive yet grounded in physical reality.

The first cluster, "perceptual–cognitive interfaces," mediates the dialog between humans and machines. This layer is led by pervasive AI and machine learning (ML), which are deployed primarily to support human decision-making rather than for autonomous control. In this context, AI agents act as teammates rather than tools. Where partial autonomy exists, it is strictly bounded by invariants, human override capabilities, and post hoc trace/audit mechanisms. A critical element here is explainable AI (XAI). However, for high-stakes control, we prioritize inherently interpretable models [7] over post hoc explanations, acknowledging recent scholarship that highlights the instability of post hoc methods under distribution shifts [8–10]. Complementing this are Immersive Interfaces, including augmented reality (AR), virtual reality (VR), and advanced human–computer interaction (HCI) modalities. These technologies act as cognitive overlays, providing context-sensitive information directly into the user's field of view. To prevent cognitive overload, these interfaces are designed with ergonomic caps based on human-centered design principles and include strict session time limits to mitigate fatigue and simulator sickness.

The second cluster, "physical–cyber integration," grounds the cognitive partnership in the physical world. We ensure seamless interoperability via Open Platform Communications Unified Architecture (OPC UA) [11] and deterministic transport via time-sensitive networking (TSN). This rigorous integration layer is essential for real-time responsiveness. It culminates in the deployment of cognitive digital twins (CDTs), which evolve beyond standard digital twins that merely model physical assets [12]. Baseline digital twin implementations correspond to the manufacturing-oriented framework formalized in ISO 23247, from which CDTs deliberately extend toward organizational and human-process modeling [13].

In this framework, CDTs create dynamic simulations of human and organizational workflows using process-mined traces and agent-based models. These advanced simulations allow organizations to model "soft" factors, such as communication bottlenecks and decision latencies. CDT validity is established via face, construct, and predictive checks. At the same time, worker-level data are strictly aggregated and purpose-limited to prevent the twin from becoming a tool for covert surveillance.

Figure 1 synthesized architecture of CI4.0, illustrating how technological pillars are animated by core principles to produce emergent outcomes.

The third cluster, "Trust & Governance Infrastructure," provides the secure and ethical foundation for the entire framework. To empower users and prevent the centralization of power, it relies on Decentralized Architectures that balance global guardrails with local decision rights. Methodologies like Federated Learning are employed to train models across distributed edges without exposing raw user data, using robust aggregation techniques to mitigate model poisoning attacks. The infrastructure is further secured through confidential computing and strict provenance controls, such as the Software Bill of Materials (SBOM) and the Supply Chain Levels for Software Artifacts (SLSA), to ensure the integrity of the AI supply chain. Auditability is ensured via signed, append-only logs anchored in trusted execution environments (TEEs), providing an immutable record of all human-machine interactions. Operational controls map to established frameworks such as the NIST AI Risk Management Framework (RMF) [14], ISO/IEC 42001 [15], and complementary AI risk guidance articulated in ISO/IEC 23894 [16], ensuring compliance and safety. Industrial cybersecurity controls are aligned with IEC 62443 to address threats across the device-edge–cloud continuum [17].

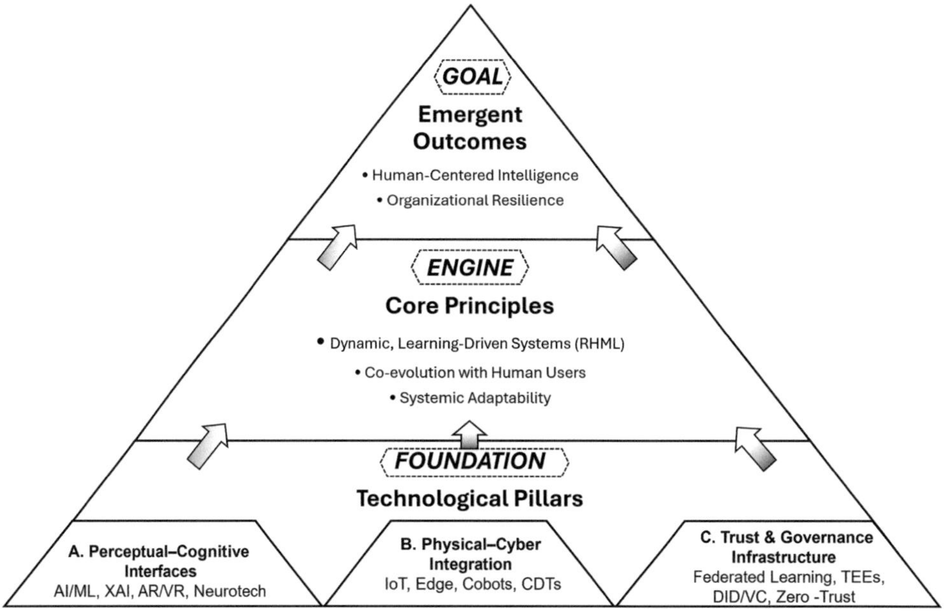

A three-layer pyramid showing how technological pillars (foundation) are animated by core principles (engine) to yield emergent outcomes at the apex.

Figure 1.
The integrated architecture of Cognitive Industry 4.0

2.2 Core operational principles

These pillars are animated by specific operational principles that dictate how the system behaves in practice. First, systems must be dynamic and learning-driven. They embrace reciprocal human–machine learning (RHML), formalized as a continuous control loop: Observe, Propose, Explain, Consent/Override, Act, and Learn. In this loop, the machine observes context, proposes an action, explains its reasoning, waits for human consent or override, executes the action, and then learns from the human's decision. This loop includes safety interlocks based on invariants derived via systems-theoretic process analysis (STPA) – a hazard analysis technique that treats safety as a control problem rather than a reliability problem. Where CI4.0 systems interface with physical automation or collaborative robotics, actuation layers remain constrained by established industrial robot safety standards such as ISO 10218, ensuring that cognitive augmentation cannot bypass embodied safety constraints [18]. Safety-related parts of control systems are further constrained through principles consistent with ISO 13849, ensuring deterministic fallback behavior under fault conditions [19]. These safety invariants are conceptually aligned with functional safety principles articulated in IEC 61508, ensuring that learning-enabled control logic remains bounded by certified safety constraints [20].

Figure 2 shows that the RHML loop operationalizes the co-adaptive cycle.

Second, the systems must support co-evolution with human users. Unlike the static "human-centric" focus of Industry 5.0, which often assumes fixed human roles, a co-adaptive system evolves alongside humans. As the user gains expertise, the system reconfigures its support, fading out basic guidance and introducing

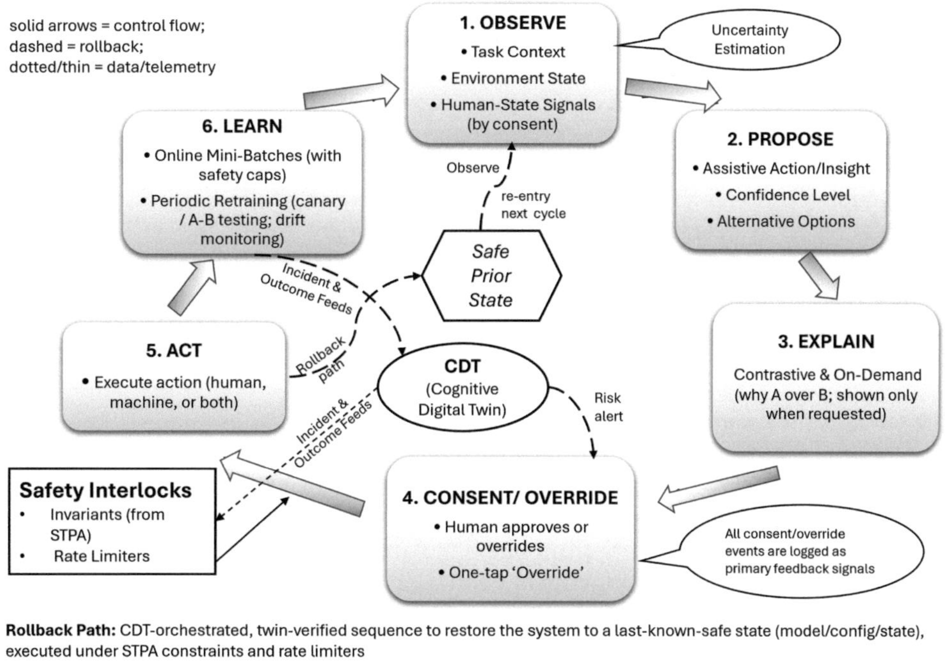

Figure 2.
The reciprocal human–machine learning (RHML) control loop.

advanced diagnostics. Conversely, if the system detects user fatigue or high cognitive load, it can dynamically increase the level of automation to maintain safety. Finally, the principle of systemic adaptability treats resilience as an emergent property cultivated through continuous learning loops. This adaptability is tracked through metrics such as reconfiguration latency (the speed at which the system adapts to a new state) and variance absorption (the ability to handle unexpected fluctuations in input without failing).

3. Human-centered intelligence

Human-centered intelligence is the capability of a socio-technical system to measurably amplify human agency, judgment, and creativity while preserving cognitive sovereignty. It is not merely about "user-friendliness" but about structurally designing systems that respect and enhance human cognitive capabilities. Operationally, this is evidenced by improved decision quality and task performance without increased error or fatigue, calibrated trust, and stable well-being. This approach requires synthesizing insights from cognitive science, ergonomics, and design thinking to address critical psychosocial dimensions that are often overlooked in purely technical deployments.

3.1 Measuring augmentation

To measure augmentation effectively, we must move beyond simple productivity metrics. We track a specific set of key performance indicators (KPIs) normalized to pre-deployment baselines. Decision quality is measured through accuracy, sensitivity/specificity, and decision time. To assess the long-term impact of decisions, we often use offline inverse propensity scoring (IPS) to estimate post-deployment regret. This metric quantifies "what would have happened" under different decision policies. Agency and trust are quantified via the override/accept ratio and trust calibration scales [21]. A healthy system shows a balance of accepts and overrides, indicating that the human is critically evaluating the AI's output. Cognitive load is assessed using the NASA Task Load Index (NASA-TLX) or dwell-time proxies to ensure that the AI is relieving rather than adding to the operator's mental burden. Furthermore, safety and well-being are monitored through incident rates and burnout indices [22], ensuring that efficiency gains do not come at the cost of operator health. Finally, equity and inclusion metrics track performance parity across diverse user groups to detect and mitigate algorithmic bias.

3.2 Governance and design patterns

Realizing the benefits of augmentation requires confronting ethical risks with concrete governance strategies. The risk of algorithmic bias is not uniform; a diagnostic AI trained on data from one population may fail catastrophically on another. Mitigation requires moving beyond technical fixes to embrace participatory design processes and adherence to robust governance frameworks [23]. To counter cognitive dependency – the risk that humans will lose the ability to perform tasks without AI assistance – systems must be designed to scaffold rather than replace human thought.

We implement this through specific human-centered design patterns:

- **Adjustable autonomy:** This pattern allows users to set the assistance level, moving from simple suggestions to previewed changes or full automation for routine tasks. This ensures the user remains in the loop and can adjust the system based on their current context and fatigue level.

- **Explain-on-demand:** Rather than overwhelming the user with constant explanations, this pattern provides contrastive "why A over B" reasoning only when requested, reducing information overload while maintaining accountability.

- **Human-intent checkpoints:** For high-stakes tasks, the system requires explicit confirmation (e.g., "Are you sure you want to shut down the cooling system?"). This prevents the silent auto-application of dangerous actions and forces a moment of reflection.

- **Trust calibration cues:** The system displays confidence ranges and uncertainty badges (e.g., "70% confidence – check manual"). Challenge prompts gently nudge operators to question low-confidence outputs, fostering critical thinking.

- **Accessibility:** Augmentation must be inclusive. We ensure this through low-bandwidth modes for remote areas and multilingual interfaces to support diverse workforces.

3.3 Implementation protocols

A typical deployment follows a phased evaluation protocol to ensure safety and buy-in. It begins with a "shadow phase," during which the system operates in read-only mode. It ingests real-time data and generates predictions, but these are not shown to the operator. This allows the team to establish baselines for decision quality and cognitive load without risk. This is followed by a "guarded assist phase," where recommendations are active but strictly gated by human approval for high-stakes actions. The system acts as a junior partner, learning from the human's corrections. The process culminates in a continuous "Review and Governance Loop," where a diverse stakeholder committee reviews transparent audit summaries of the system's performance to inform updates to models and policies.

4. Organizational resilience in post-industrial systems

Organizational resilience is the second key outcome of the CI4.0 framework. While efficiency optimizes for the expected, resilience optimizes for the unexpected, defined as the capacity to anticipate, adapt, and learn from disruptions [6]. This aligns with Resilience Engineering's emphasis on resilient performance [24] and ISO 22316 principles [25]. In a post-industrial context, where value chains are complex and interdependent, the ability to recover from shocks is often more valuable than marginal gains in steady-state speed.

CI4.0 embeds this adaptive capacity via three falsifiable enablers. The first is decentralized, augmented decision-making. The core mechanism involves pushing perception and computation to the edge, reducing decision latency in safety-critical contexts and supporting operational continuity when central links are impaired. In a centralized system, a network outage paralyzes the entire operation. In a decentralized CI4.0 system, local nodes retain the intelligence to continue operating autonomously. We measure the impact of this enabler through reconfiguration latency (the time elapsed from a shock to stable throughput) and the percentage of decisions resolved locally, which indicates the degree of autonomy from fragile central command structures.

The second enabler is the adoption of flexible, composable infrastructures. This mechanism uses loosely coupled, event-driven services that instantiate dynamic capabilities (sense–seize–transform). Instead of monolithic software stacks that take months to update, CI4.0 relies on microservices that can be swapped or updated in real-time. The efficacy of this infrastructure is measured by variance absorption – the system's ability to maintain performance despite load fluctuations – and graceful extensibility, the capacity to sustain adaptive performance when demands exceed design capacity. Risks such as provenance sprawl (losing track of which service did what) are controlled via SBOM/SLSA protocols and pre-deployment change-impact testing.

The third and most powerful enabler is the use of CDTs for proactive learning. Simulation-based CDTs act as organizational "flight simulators," allowing teams to vet responses before deployment. This mechanism combines process mining with simulation to run "what-if" stress tests (e.g., "What if the supply chain breaks for 3 weeks?"). Success is tracked via learning velocity, the speed at which incident insights are translated into operational updates, and safe-to-fail experiment throughput. A practical protocol involves simulating scenarios, selecting top policies for canary deployment (testing on a small subset), and feeding back real outcomes to recalibrate the twin.

5. Co-Adaptive human–machine systems

Co-adaptive systems embody the engine of the co-evolutionary principle, moving beyond static interactions to create a genuine symbiosis in which humans and machines learn from each other in real time. This symbiosis draws on insights from HCI, affective computing, and neuroergonomics. The goal is to create a coupled system in which the boundaries between human intent and machine execution are fluid, yet distinct enough to maintain accountability.

While technologies like neuroadaptive systems offer the potential to detect cognitive workload and adapt interfaces accordingly (e.g., simplifying a dashboard when the user's electroencephalogram [EEG] shows high stress), fielded evidence remains limited. Neuroergonomics reviews note calibration burdens and signal-to-noise issues, keeping most solutions at early technology readiness levels (TRLs) [26]. The requirement for users to wear headsets or caps can also introduce physical discomfort and resistance, limiting long-term adoption in industrial settings.

Similarly, emotion-aware AI, building on foundational work in affective computing [27], faces significant validity challenges. Large reviews find that facial movements are unreliable signals of emotion across cultures, and the EU AI Act prohibits emotion inference in workplaces and schools due to privacy and reliability concerns. Consequently, the principle of co-adaptation is best applied through less invasive means. For example, in agricultural contexts, a co-adaptive system might refine planting advice based on farmer feedback regarding crop yields and local weather, relying on interaction telemetry (clicks, dwell time, overrides) rather than biosignals. Systems should be instrumented to measure the override/accept ratio and trust calibration without infringing on cognitive sovereignty – the individual's right to control their own mental processes and data [28]. Governance frameworks must be built on data minimalism and revocable consent to protect these rights, ensuring that "co-adaptation" does not become a euphemism for behavioral manipulation.

6. Sector-specific case studies

This section provides a practical demonstration of how the CI4.0 framework can be applied across diverse sectors. By grounding the preceding theoretical discussion in concrete scenarios, we illustrate that the principles of co-adaptation and resilience are not limited to manufacturing but are transferable to any domain where complex systems intersect with human expertise.

6.1 Healthcare: The resilient hospital

In the healthcare sector, systems often operate with brittle workflows that struggle to adapt to demand surges, as evidenced by the systemic failures during recent global pandemics. The traditional response – hiring more staff – is often slow and constrained by labor shortages. CI4.0 addresses this by implementing CDTs that model patient flows, equipment utilization, and real-time staffing levels. Within a Level 1 Trauma Center facing unpredictable surge events, the CDT runs continuous "what-if" simulations rather than relying on static dashboards. For instance, the system might simulate the reconfiguration of a Post-Anesthesia Care Unit into an overflow ICU during a mass casualty event. Implementation typically follows a multi-phase roadmap, beginning with a sensor-fusion phase that aggregates data from electronic health records (EHRs) and RFID tags into a unified data lake. This is followed by a shadow twin phase, where predictions are audited against reality, culminating in an active decision-support phase where the twin pushes "nudge" alerts to unit managers. A critical failure mode in this context is model drift, where the twin optimizes for metrics (throughput) at the expense of unmeasured variables (patient comfort). To mitigate this, a "Human-in-Command" protocol allows staff to flag and override "efficient" but inhumane recommendations, with these overrides fed back to retrain the model.

For clinicians, AI diagnostic tools are deployed to augment expertise rather than replace it. Recent syntheses indicate that AI-assisted readers can significantly improve sensitivity and reduce reading time for medical imaging [29]. In this workflow, the AI acts as a "second pair of eyes," highlighting potential anomalies for the radiologist to review, thereby reducing cognitive load during long shifts.

6.2 Education: The co-adaptive learning environment

Traditional educational models often fail to address diverse learning needs, relying on a "one-size-fits-all" approach that leaves many students behind. In a CI4.0

framework, the classroom is reimagined as a co-adaptive learning environment. Here, an AI platform acts as a metacognitive coach, analyzing learning patterns to provide personalized support. Rather than just delivering content, the system identifies why a student is struggling – distinguishing between a conceptual gap, a lack of prerequisite knowledge, or disengagement. It then prompts the teacher with targeted intervention strategies (e.g., "Student A is stuck on fractions; try the visual 'pizza slice' analogy").

Randomized evaluations have shown that such technology-aided instruction, when designed to augment the teacher rather than replace them, can yield sizable test-score gains [30]. The system handles routine grading and pattern recognition, freeing the teacher to focus on mentorship. The primary failure mode is the "dependency trap," in which teachers stop diagnosing students' issues themselves. To prevent this, the system is designed to "fade out" support as teacher confidence grows.

6.3 Public administration: Adaptive urban governance

Urban governance struggles to manage complex, interconnected systems during crises, often hampered by institutional silos and outdated data. An urban CDT, powered by IoT sensor networks, can model traffic, energy usage, and public services to enhance hotspot prediction and policy targeting.

In the context of a metropolitan government, the twin acts as a unifying layer. During a flood scenario, for example, the twin ingests rainfall data and traffic speeds to simulate rerouting options and predict power outages, allowing emergency services to pre-position assets. A critical risk in this domain is "algorithmic redlining," in which the model prioritizes affluent neighborhoods because of superior sensor coverage. Mitigation requires the deployment of an "Equity Audit" layer that explicitly weights service delivery to underserved areas to counteract historical data biases.

6.4 Creative industries: The augmented studio

In the creative industries, the fear of displacement by Generative AI is palpable. CI4.0 reframes this technology as an "ideation partner" within the "Augmented Studio." Here, generative models expand the solution space, accelerating innovation cycles by proposing novel starting points or variations that a human designer might not have considered. Designers use these models to create hundreds of sketch variants in seconds, shifting their role from execution to curation. Implementation begins with fine-tuning open-source models on the firm's specific design archives, followed by integrating generative plugins into CAD software. Legal governance is paramount; frameworks must be established to tag all AI-generated assets with provenance data to ensure clear authorship trails.

7. Interdisciplinary and cross-scale perspectives

The true power of the CI4.0 framework lies in its inherently interdisciplinary nature, functioning as a meta-framework that bridges siloed domains. Theoretically, we view the organization as a complex adaptive system (CAS), in which individual agents (human and machine) interact to produce emergent global properties, such as resilience [31, 32]. This CAS perspective implies that design leverage comes from shaping local mechanisms – rules, constraints, and safety invariants – rather than top-down control.

It requires a cross-scale mapping in which micro-level interactions (e.g., an operator overriding an AI safety alert) give rise to meso-level coordination (e.g., a shift supervisor reviewing override patterns), which, in turn, produce macro-level outcomes (e.g., organizational policy updates). Feedback channels, such as incident backpropagation, are essential for coupling these scales. Facilitating this requires a synthesis of neuroscience, HCI, and systems engineering, best achieved through socio-technical co-design.

Finally, this perspective demands that we challenge dominant success metrics. We propose shifting the organizational scorecard from efficiency to adaptability and well-being. Organizations should adapt frameworks like the OECD (Organisation for Economic Co-operation and Development) Well-being framework to internal metrics, measuring success by variance absorption during crises rather than just throughput under normal conditions.

8. Technological horizons and emerging opportunities

While grounded in current technologies, the CI4.0 framework offers a lens for evaluating future innovations. In the "near-term," we anticipate the rise of next-generation AI and "world models" (e.g., DeepMind's Genie) that build intuitive understandings of cause and effect. These models can accelerate safe, offline policy learning for CI4.0 pilots. However, they remain brittle under distribution shifts and must be deployed with human-in-command controls.

In the "mid-term," we foresee self-evolving, autonomic infrastructures. Drawing on CAS principles, these systems – manifesting in telecom's "zero-touch" management – will reconfigure at runtime to adapt to changing environments without human intervention. While promising for resilience, high-autonomy reconfiguration raises accountability questions. These systems must be constrained by pre-approved "Safety Cases" and operator kill switches, which can instantly revert the system to a known-good state.

On the "long-term horizon," bio-digital interfaces and advanced neurotechnology promise high-bandwidth connections between biological and computational systems. This could allow for direct cognitive offloading or enhanced perception. However, broader use must embed neurorights – mental privacy, identity, and agency – from the outset [28]. We must prevent a future where employment depends on the willingness to submit to neural monitoring. Policy guardrails for such developments are beginning to emerge, notably through the OECD Neurotechnology Policy Toolkit, which foregrounds mental privacy, agency, and proportionality [33]. Additionally, organizations should begin migrating to post-quantum cryptography (PQC) in line with NIST standards [34] to secure data against future threats.

9. Ethical and societal considerations

The vision of CI4.0 is inseparable from profound ethical challenges. As decision-making becomes increasingly algorithmic, data privacy evolves into the challenge of cognitive sovereignty – the right to self-determination over one's mental life. This defense of sovereignty is critical, as the current trajectory of industrial AI risks converging with the extractive logic described by Zuboff [35] as "surveillance capitalism," where human experience is unilaterally claimed as free raw material for behavioral prediction and modification. In this extractive model, the "behavioral surplus" generated by workers is commodified to optimize systems at the expense of human autonomy. Furthermore, we must confront the material and political realities highlighted by Crawford [36], who argues that AI is not an abstract computation but a physical industry built on resource depletion, labor exploitation, and classification systems that serve as "registries of power." The concentration of cognitive infrastructure power also poses threats to democratic processes [37]. Emerging municipal regulation of automated employment decision tools, such as New York City's Local Law 144, illustrates how governance constraints are beginning to materialize at the operational level [38]. This normative shift is increasingly reflected at the constitutional level, as exemplified by Chile's 2021 constitutional amendment explicitly mandating that scientific and technological development serve people and society at large [39].

To counter these risks, CI4.0 explicitly rejects control architectures. It proposes a mandatory Fundamental Rights Impact Assessment (FRIA) for all high-stakes deployments, modeled on frameworks like Canada's Algorithmic Impact Assessment [40] and requirements in the EU AI Act [41]. This assessment involves stakeholder mapping, rights analysis, and a proportionality test. If risks are identified, a mitigation plan with concrete technical safeguards must be put in place. This governance logic aligns with established frameworks for responsible innovation that emphasize anticipation, reflexivity, inclusion, and responsiveness [42]. By mandating data minimalism and localizing "flight simulator" training to the edge, we aim to decouple organizational resilience from the machinery of surveillance.

10. Conclusion: Toward a cognitive post-industrial society

CI4.0 represents a fundamental re-imagining of the relationship between humans, machines, and the organizations they inhabit. By moving beyond the linear pursuit of efficiency, this framework establishes a viable path toward human-centered intelligence and organizational resilience. Returning to the core inquiries that framed this investigation, the evidence synthesizes into three distinct findings regarding architecture, governance, and sectoral viability.

First, regarding the impact of architectural configurations on resilience (RQ1), our analysis confirms that decentralized, edge-native architectures are essential for achieving "graceful extensibility." We found that pushing perception and compute to the edge – rather than relying on brittle, centralized command structures – significantly reduces reconfiguration latency during disruptions.

Second, regarding the governance models required to mitigate cognitive dependency (RQ2), the findings indicate that technical guardrails alone are insufficient. Effective mitigation requires a "human-in-command" governance model, operationalized through specific design patterns such as Adjustable Autonomy and Human-Intent Checkpoints. By replacing silent automation with active consent loops, organizations can scaffold human judgment rather than eroding it.

Finally, regarding the applicability of these principles in knowledge sectors (RQ3), the case studies demonstrate that CI4.0 yields measurable gains in diverse post-industrial contexts. In healthcare, the application of CDTs has shown clear potential to improve throughput recovery. Similarly, in education, co-adaptive systems have been shown to enhance learning outcomes when deployed as supportive rather than directive tools.

Ultimately, the transition to a cognitive post-industrial society is a choice. While the technical pathways for co-adaptation are now visible, realizing this vision requires a sustained commitment to valuing adaptability over static efficiency. By empirically validating resilience metrics and linking technical design to societal outcomes, we can ensure that cognitive sovereignty and algorithmic justice become mutually reinforcing forces, aligning with the OECD AI Principles [43]. This framework offers a robust starting point, guiding the transformation toward a future that is not only more intelligent but also more humane, resilient, and equitable.

Statement on the use of artificial intelligence

In preparing this chapter, the authors utilized Large Language Model (LLM) tools exclusively for language refinement, copyediting, and reference formatting. All core conceptualization, argumentation, data interpretation, and final verification of sources remain the sole work and responsibility of the authors. This chapter is human-guided; no text was generated without explicit authorial oversight.

Author details

Mitt Nowshade Kabir
Trouvus AI, King, Canada

*Address all correspondence to: nowshade@gmail.com

References

[1] Ahmmed F, Isanaka S, Liou JJ. Promoting synergies to improve manufacturing efficiency in industrial material processing: A systematic review of Industry 4.0 and AI. Journal of Manufacturing Systems. 2024;**73**: 466–485. DOI: 10.1016/j.jmsy.2024.04.004

[2] Moulier-Boutang Y. Cognitive Capitalism. Cambridge (UK): Polity Press; 2011

[3] Vercellone C. From formal subsumption to general intellect: Elements for a Marxist reading of the thesis of cognitive capitalism. Historical Materialism. 2007;**15**(1): 13–36

[4] Haskel J, Westlake S. Capitalism without Capital: The Rise of the Intangible Economy. Princeton (NJ): Princeton University Press; 2017

[5] Baharuddin K, Omar SS. A systematic review of organizational resilience through digital technology adoption: Trends and insights in a decade (2014–2024). Cogent Business & Management. 2024;**11** (1):2359482

[6] Woods DD. Four concepts for resilience and the implications for the future of resilience engineering. Reliability Engineering & System Safety. 2015;**141**:5–9

[7] Rudin C. Stop explaining black box machine learning models for high stakes decisions and use interpretable models instead. Nature Machine Intelligence. 2019;**1**(5):206–215

[8] Lipton ZC. The mythos of model interpretability. arXiv. 2016. DOI: 10.48550/arXiv.1606.03490

[9] Lundberg SM, Lee SI. A unified approach to interpreting model predictions. Advances in Neural Information Processing Systems. 2017;**30**:4765–4774

[10] Ribeiro Mt, Singh S, Guestrin C. "Why should I trust you?" Explaining the predictions of any classifier. In: Proceedings of the 22nd ACM SIGKDD International Conference on Knowledge Discovery and Data Mining; 2016. p. 1135–1144

[11] International Electrotechnical Commission. OPC Unified Architecture (IEC 62541). 2020

[12] Tao F, Qi Q, Liu A, Kusiak A. Data-driven smart manufacturing. Journal of Manufacturing Systems. 2018;**48**:157–169. DOI: 10.1016/j.jmsy.2018.01.006

[13] International Organization for Standardization. Automation Systems and Integration — Digital Twin Framework for Manufacturing (ISO 23247). Geneva (Switzerland): International Organization for Standardization; 2021

[14] National Institute of Standards and Technology. Artificial Intelligence Risk Management Framework (AI RMF 1.0). Gaithersburg (MD): U.S. Department of Commerce; 2023

[15] International Organization for Standardization & International

Electrotechnical Commission. Information Technology — Artificial Intelligence — Management System (ISO/IEC 42001). 2023

[16] International Organization for Standardization & International Electrotechnical Commission. Information Technology — Artificial Intelligence — Guidance on Risk Management (ISO/IEC 23894). 2023

[17] International Electrotechnical Commission. Security for Industrial Automation and Control Systems IEC 62443. 2018

[18] International Organization for Standardization. Robots and Robotic Devices — Safety Requirements for Industrial Robots (ISO 10218). 2011

[19] International Organization for Standardization. Safety of Machinery — Safety-related Parts of Control Systems (ISO 13849). 2015

[20] International Electrotechnical Commission. Functional Safety of Electrical/electronic/programmable Electronic Safety-related Systems (IEC 61508). 2010

[21] Lee Jd, See KA. Trust in automation: Designing for appropriate reliance. Human Factors: The Journal of the Human Factors and Ergonomics Society. 2004;**46**(1):50–80

[22] Maslach C, Jackson SE, Leiter MP. Maslach Burnout Inventory Manual. Palo Alto (CA): 3rd ed. Consulting Psychologists Press; 1996

[23] Batool F, Zowghi D, Bano M. Responsible AI governance: A systematic literature review. arXiv. 2023. DOI: 10.48550/arXiv.2305.12782

[24] Hollnagel E. Safety-I and Safety-II: The past and Future of Safety Management. Boca Raton (FL): CRC Press; 2014

[25] International Organization for Standardization. Security and Resilience — Organizational Resilience — Principles and Attributes (ISO 22316). 2017

[26] Johnson A, Proctor RW. Neuroergonomics: A Cognitive Neuroscience Approach to Human Factors and Ergonomics. Hoboken (NJ): Palgrave Macmillan; 2013. DOI: 10.1057/9781137347305

[27] Picard RW. Affective computing. Cambridge (MA): MIT Press; 1997

[28] Farahany NA. The Battle for Your Brain: Defending the Right to Think Freely in the Age of Neurotechnology. New York (NY): St. Martin's Press; 2023

[29] Topol EJ. High-performance medicine: The convergence of human and artificial intelligence. Nature Medicine. 2019;**25**(1):44–56. DOI: 10.1038/s41591-018-0300-7

[30] Muralidharan K, Singh A, Ganimian AJ. Disrupting education? Experimental evidence on technology-aided instruction in India. American Economic Review. 2019;**109**(4): 1426–1496

[31] Holland Jh. Studying complex adaptive systems. Journal of Systems Science and Complexity. 2006;**19**(1):1–8

[32] Gunderson Lh, Holling CS, editors. Panarchy: Understanding Transformations in Human and Natural Systems. Washington (DC): Island Press; 2002

[33] OECD. Neurotechnology Policy Toolkit. Organisation for Economic Co-operation and Development. 2019

[34] National Institute of Standards and Technology. Module-Lattice-Based Key-Encapsulation Mechanism Standard (FIPS 203). Gaithersburg (MD): U.S. Department of Commerce; 2024

[35] Zuboff S. The Age of Surveillance Capitalism: The Fight for a Human Future at the New Frontier of Power. New York (NY): PublicAffairs; 2019

[36] Crawford K. Atlas of AI: Power, politics, and the Planetary Costs of Artificial Intelligence. New Haven (CT): Yale University Press; 2021

[37] Coeckelbergh M. Why AI Undermines Democracy and What to Do about It. Cambridge (UK): Polity Press; 2024

[38] New York City Council. Local Law 144 of 2021: Automated Employment Decision Tools. New York (NY): New York City Council; 2021

[39] Chile. Law No. 21.383: Modifies the Political Constitution of the Republic to establish scientific and technological development at the service of people. 2021

[40] Canada Treasury Board Secretariat. Algorithmic Impact Assessment (AIA). Ottawa (ON): Government of Canada; 2019

[41] European Union. Regulation (EU) 2024/1689 of the European Parliament and of the Council laying down harmonised rules on artificial intelligence (Artificial Intelligence Act). Official Journal of the European Union. 2024

[42] Stilgoe J, Owen R, Macnaghten P. Developing a framework for responsible innovation. Research Policy. 2013;**42** (9):1568–1580

[43] OECD. Recommendation of the Council on Artificial Intelligence. OECD/LEGAL/0449; 2019

Chapter 3

Perspective Chapter: The Imperative of Proactive Workforce Management – A Methodology for a Human Capital Digital Twin

Romain Stevens and Elsa Negre

Abstract

The accelerating transformation of Industry 4.0 has provided organizations with advanced technologies to monitor and optimize their machines and processes. Yet, approaches to managing human capital are still largely reactive and fragmented. This paradox highlights a critical strategy-to-execution gap. Traditional Knowledge Management and Human Resources systems remain mostly descriptive and disconnected from strategic processes. This chapter addresses this limitation by introducing the Human Capital Digital Twin (HCDT), a conceptual framework designed to transform static organizational knowledge into proactive and actionable insights for workforce management. The HCDT is designed around three integrated components: a workforce ontology acting as the semantic backbone by modeling resources, competencies, external market, and objectives; an Artificial Intelligence-based foresight engine that predicts skill evolution and scenario outcomes; and the Organizational Actions Recommender System that provides explainable and actionable strategic recommendations. The system's ability to operationalize business objectives into concrete and executable plans is demonstrated through an illustrative case on market diversification. It focuses on balancing trade-offs between cost, expertise readiness, risk, and operational continuity. The results show the framework's potential to enhance operational efficiency and strategic agility, closing the strategy-to-execution gap. Finally, a multi-dimensional evaluation methodology that assesses forecasting accuracy, recommendation diversity, and organizational impact is proposed, establishing the HCDT as a technically robust and strategically disruptive decision-support system that addresses the human capital challenges in Industry 4.0.

Keywords: Human Capital Digital Twin, Industry 4.0, ontology, recommender system, workforce management

1. Introduction

The Industry 4.0 era reveals a profound paradox at its core. While organizations leverage predictive maintenance, exploiting digital twins and cyber-physical systems to simulate machine states and anticipate breakdowns in real time, no equivalent exists to support the management of their most crucial asset: human capital [1, 2]. While being mostly reactive in their workforce planning, using descriptive analytics and historical data, organizations are left vulnerable, unable to simulate strategic scenarios as they do with supply chains [3]. This breach results in a "strategy-to-execution" gap, exposing organizations in a context of rapid technological advances and intensified competition [4].

The rapid and continuous technological evolution, as well as the shortening of skills half-life and non-linear career paths, have redefined the competitive landscape [5, 6]. The traditional Knowledge Management (KM) and Human Resources (HR) systems provide efficient data repositories but fall short in converting knowledge into actionable strategic levers. Designed for a stable context, they lack the semantic structure that enables the foundation of predictive algorithms.

This chapter underscores a critical gap in the literature: current KM systems are mostly passive and not adapted to the Industry 4.0 context. HR systems, still administrative and operationally focused, lack contextualization and analytics. Foresight methods are slow and detached from real-time data. As a result, organizations suffer from the absence of a "Human Capital Digital Twin" (HCDT) to connect strategy and operational workforce management. This model would serve as a dynamic and interconnected framework of the organization's resources, enabling scenario planning, anticipating future workforce needs, and recommending actionable strategies aligned with organizations' objectives.

This chapter seeks to investigate the following research question: How can organizations integrate a dynamic and proactive framework, based on an ontological knowledge representation and Artificial Intelligence (AI) analytics, that functions as a HCDT and closes the strategy-to-execution gap in workforce management?

This chapter is presented as an exploratory work, providing a conceptual foundation and a methodological roadmap for implementing proactive workforce management systems. It is intended for HR leaders, KM operators, and system architects who seek to align strategic decision-making with organizational resources in an Industry 4.0 context. By first revealing the limitations of traditional KM, HR systems, and foresight approaches, this work bridges the identified literature gap by introducing a framework methodology composed of three core components:

(1) A workforce ontology modeling internal and external organizational knowledge;

(2) An AI mechanism that transforms this static knowledge representation into a dynamic predictive engine;

(3) An Organizational Actions Recommender System (OARS) that converts insights into actionable strategic recommendations.

2. Theoretical foundations

2.1 Knowledge Management in the Industry 4.0 context

Knowledge Management refers to practices and systems aimed at acquiring, structuring, sharing, and applying knowledge to improve organizational performance. In the context of the knowledge economy [7], KM allows valorization of employees' technical and business expertise. This knowledge capitalization strengthens internal capabilities, supports employee upskilling, and enhances service quality and innovation [8].

Knowledge Management is based on the hierarchical Data-Information-Knowledge-Wisdom (DIKW) pyramid, which illustrates the transformation of raw data into actionable decisions [9]. Data are unprocessed facts, which become information once meaning is attributed to them through contextualization. Knowledge is a collection of interpreted information exploited for decision-making. Finally, wisdom is the ability to use this knowledge ethically and strategically to make decisions. KM relies on the transfer of explicit and tacit knowledge [10]:

- Explicit knowledge is formalized, codified, logically articulated, and can be transmitted (e.g., in the form of documents);
- Tacit knowledge is non-formalized, based mostly on personal experience and practical skills.

This distinction influences knowledge transfer strategies because tacit knowledge, even though hardly shareable, is a sustainable competitive source [10]. KM is characterized by the existence of three types of main factors [11]:

- Generators collect and assemble knowledge, motivated by individual benefits such as personal development and/or the improvement of their own skills;
- Transformers facilitate knowledge transfer and sharing, motivated by organizational performance;
- Users exploit and benefit from this knowledge, motivated by individual and organizational benefits.

Geisler [11] presents a managerial KM strategy composed of four cyclic major steps:

- Generation corresponds to the conversion of explicable tacit knowledge into exploitable and codified formats;
- Transfer is done through reports, audits, and meetings to make knowledge accessible;
- Implementation deals with the application of knowledge within the organization's services;

- Absorption corresponds to the integration of new knowledge into the culture and processes of the organization, making them integral parts of the organization.

This cycle is part of the continuous learning process, allowing adaptation and innovation, and must be taken into account when implementing new KM processes in an organization.

The emergence of Industry 4.0 has fundamentally altered the environment in which organizations grow and, therefore, rendered past KM processes obsolete. This new context, seen as turbulent and discontinuous [12, 13] is characterized by the VUCA (Volatility, Uncertainty, Complexity, and Ambiguity) paradigm [14]:

- Volatility shows the acceleration of technological evolution affecting markets;
- Uncertainty is linked to this same evolution, appearing to be difficult to predict;
- Complexity highlights the deep interconnections between the socio-technical systems;
- Ambiguity refers to the challenging interpretation of events and the anticipation of causalities in such environments.

Traditional knowledge transfer processes are no longer enough for organizations to remain competitive. Relying only on past experiences and using incremental knowledge appears to be outdated. It is no longer sufficient to capture "what is"; organizations must focus on anticipating "what will be." Therefore, new KM dynamics must integrate proactivity and foresight to address the short lifecycle of skills as well as the nonlinear nature of career paths [5].

However, traditional KM systems face limitations in fulfilling these new demands. North and Kumta [15] highlight the system's reactive nature, showing that they are mainly conceived toward past knowledge and not predictive intelligence.

This analysis reveals that current KM systems are rendered obsolete in the Industry 4.0 context. It highlights the foundational necessity of moving beyond static repositories toward dynamic systems designed to anticipate future needs based on internal and external organizational knowledge.

Bridging these gaps requires technological enablers capable of embedding semantics, predictive reasoning, and actionable insights into KM platforms. This opens the way for the integration of emerging technologies as core components of the next-generation KM architecture.

2.2 Technology integration: Enablers for a proactive KM

Technological enablers such as AI, data-driven predictive analytics, big data, and ontologies are an opportunity to reshape KM systems from static repositories to dynamic systems. KM systems are no longer limited to storage but can integrate simulation, prediction, and reasoning perspectives. For example, Dawson et al. [16] developed a Machine Learning (ML)-based predictive model to correlate skills demand with the labor market, and Margherita [17] focused on identifying competency gaps and actions to resolve them, such as reskilling plans.

While AI and big data ensure computational processing, ontologies play a key role in structuring data to support these new applications. An ontology is an explicit specification of a conceptualization, which contains the objects, concepts, and other entities that are supposed to exist in a field of interest and the relationships that bind them [18]. In sum, ontology consists of a formal structure that defines terms and concepts representing the meaning of a domain and a set of instances. The key role of ontologies is, therefore, to specify a modeling representation of information at a high level of abstraction, so that data can be exported, translated, queried, and unified for independently developed systems.

An applied ontology or knowledge graph (KG) is intended as a dynamic structure, going beyond a simple relationship of entities. It exploits a reasoning engine to generate new knowledge. Ehrlinger et al. [19] provide the following definition: "A knowledge graph acquires and integrates information into an ontology and applies reasoning to derive new knowledge." For example, structured representations of, for example, skills, roles, and resources enable reasoning to produce scenarios aligned with organizational strategy. Dorn et al. [20] worked on ontology-based systems applied to the HR domain, illustrating their potential to align workforce capabilities with business requirements.

Organizational tacit knowledge, for example, project documentation, employee or client feedback, is usually not exploitable for traditional KM systems because of its unstructured nature. AI techniques, and particularly Natural Language Processing (NLP), can extract, structure, and understand this unstructured knowledge, making it accessible to KM systems. Therefore, tacit knowledge becomes an actionable insight for decision-making processes. A recent study [21] showed the potential of NLP algorithms to convert tacit knowledge into explicit knowledge following a defined ontology and allowing it to be operationalized.

Despite presenting potential in proactive KM, the presented enablers lack a strategical focus. While ML can predict skill evolution [22], identify resource gaps in the organization [17], or extract tacit knowledge [21], the enablers are often isolated and remain purely technical rather than interconnected strategic systems. This chapter points out the importance of a semantically connected framework enabling predictions to translate into actionable organizational workforce planning.

Such technological advances open the possibility of reimagining KM systems not as administrative but as knowledge-driven strategic systems. By anticipating needs and revealing organizational tacit expertise, they can support HR and workforce planning through strategic decision-making support platforms. This shift marks HR's transition from operational to strategical in an Industry 4.0 context.

2.3 The failure of current HR systems and foresight approaches

2.3.1 The limits of the HR system: Operational rather than strategically focused

Digital HR systems, such as Human Resource Information Systems (HRIS), Talent Management Systems (TMS), or Learning Management Systems (LMS), have evolved from basic administrative tools to integrated systems that centralize data of the same type and enable the automation of complex processes, such as descriptive analytics. However, despite the operational efficiency of these solutions, they remain reactive and purely operational rather than proactive and strategic. This limitation, further emphasized by the Industry 4.0 era, highlights the lack of capabilities to align workforce strategy and needs with VUCA business objectives, thereby justifying a transition from descriptive to predictive systems.

Furthermore, these HR digital platforms are fragmented systems divided for their specific tasks, for example, recruitment, performance, and learning management, leading to siloed data. This fragmentation prevents HR management from reaching a holistic view of organizational resources and capabilities. Recent studies highlight the integration of HR analytics into strategic frameworks as mandatory to drive HR actions toward organizational goals. Minbaeva [22] explains the role of HR analytics in predicting workforce needs, employee management, and talent retention. The author integrates technological enablers such as AI and ML to emphasize the importance of data-driven approaches in HR management practices. Despite the vast data collection, HR digital systems fail to contextualize data within specific organizational contexts or strategies. Therefore, even when analytic systems exist, they fall short in semantic depth and contextual links, which are mandatory for strategic scenario planning.

In summary, HR digital systems have proven their administrative and operational efficiency. However, they still face multiple challenges, such as a lack of analytics, fragmentation, and contextualization. They fail to meet HR strategic needs, remaining focused on descriptive and diagnostic insights. To address this issue, organizations have adopted foresight methods such as scenario planning and weak signals analysis to drive future evolutions. Yet, these methods were primarily designed for long-term strategy and not for the rapidly changing and ever-evolving Industry 4.0 context. In the face of Industry 4.0 evolutions, organizations are left vulnerable, making decisions based on outdated or incomplete data about their workforce resources.

2.3.2 The shortcoming of foresight approaches: Strategic but not operationally integrated

Given the rapid technological evolution and the highly competitive environment in Industry 4.0, integrating foresight into workforce planning has become a key concern. The World Economic Forum (2020) [23] even stated that five years is now the half-life duration of skills (even less for technical competencies), meaning that today's acquired competencies will rapidly become obsolete. It added that 39% of current skills will change by 2030 [23]. These predicted changes must be considered alongside the technological disruption possibilities arising from the unpredictability of technological appearances or evolutions [24]. Moreover, career paths have become mostly nonlinear, following markets with project-based employment, cross-sector mobility, and the exponential growth of multidisciplinary jobs [6]. As we discussed earlier, traditional workforce planning strategies are obsolete in addressing these new constraints. Therefore, organizations have turned to various foresight methodologies present in the scientific literature, each having its own limitations when applied to workforce planning in a dynamic and uncertain context.

The Delphi method [25, 26], viewed as a traditional strategic forecasting approach, emphasizes communication among experts to achieve a consensus on future skill needs. While being a strong choice for long-term and high-level strategic support, the Delphi method comes with subjective analysis, cost, and slow speed, making it unsuitable for the rapid workforce adaptation needed in the Industry 4.0 context. Although scientists have tried to create new methods like Real-time Delphi [27] to optimize cycle times, they still struggle with the VUCA context and live data integration.

As a systemic approach, a combination of scenario planning and war-gaming solutions allows organizations to simulate alternative futures and stress-test strategic options to evaluate organizational resilience through uncertainties [28]. Despite being useful for breaking conventional thinking, these approaches are manual, costly in resources, and detached from operational data. The insights remain vague and theoretical, rarely influencing real-time workforce management. This disconnect is often seen as a common strategic planning failure, highlighting the lack of integrated tools and widening the reactive-to-proactive gap [29].

Moving toward data-driven solutions, organizations started to adopt predictive statistical approaches, offering scalability and objective anticipation using field data. However, because of their simple and single-variable analysis, they fail to capture the multi-dimensional, complex context of workforce real-life scenarios [30]. The predictions are mostly narrow and fail to consider the impact of a decision about skills, for example, on projects, roles, or the market.

While statistical forecasting and scenario planning solutions have been studied, other approaches have emerged by leveraging recommender systems (RS). Using personalized information filtering, RS offer elements likely to interest a user. In the HR context, RS have primarily been employed at the individual level for personalized job and training recommendations or career path suggestions [31, 32]. Despite showing huge potential for data-driven personalization, they are still employee-focused and dissociated from the organizational level and strategy. They often fail to bridge workforce development with an organization's objectives. This opens a window to widen the use of RS from a micro-level to a macro-level decision-making framework to recommend strategic actions (e.g., hire, upskill, and partner).

Finally, these methods showed a common persistent limit: a "strategy-to-execution gap" [4]. Although they can anticipate workforce needs using reports and exercises, they lack computational mechanisms to translate insights into concrete, actionable, and multi-dimensional workforce decisions. As a result, workforce planning remains a periodic and standalone exercise instead of an embedded and continuous organizational mechanism.

2.4 Synthesis: The Human Capital Digital Twin needs

This chapter exposes a fundamental paradox at the core of Industry 4.0. From one perspective, organizations have made critical advances in operational technology through digital twins and cyber-physical systems that simulate machine health and anticipate failures in real time. On the other hand, the management of their crucial asset – their workforce – remains reactive, descriptive, and fragmented. This chapter's literature reveals that the breach is not only technological but also conceptual and architectural, with the need for the HCDT.

The literature review highlighted powerful but isolated solutions, each bringing their own brick into a bigger framework foundation. **Table 1** summarizes this disconnect:

- KM systems enable static structure but are passive, retrospective, and limited in tacit knowledge;
- Technological Enablers are capable of providing predictive and semantic insights but are still used fragmentarily and not strategically focused;
- HR systems demonstrate data centralization and prove their operational efficiency, but they remain descriptive and fragmented;
- Traditional foresight solutions perform well in strategic planification but remain slow, costly, and detached from operational executions;
- Data-driven foresight methods provide personalized, objective insights, but they are still disconnected from the organizational perspective.

This architectural disconnect deepens the "strategy-to-execution gap." Knowledge cannot be converted from operational data to strategic insight and then injected back into operational decisions. Rather than being integrated and continuous, workforce management is still periodic and mostly theoretical.

This chapter underscores the capable but disjointed components that must be integrated into a new architectural framework. This computational model would serve as a dynamic representation of the organization's ecosystem, reflecting its workforce resources in a virtual environment that allows prediction, scenario simulation, and optimization.

Approach	Advantages	Limits	Needs
KM in VUCA context	Structures and helps capitalize on knowledge; supports organizational learning and innovation.	Reactive, passive, static; struggling with tacit knowledge and lacking predictive and contextual tools.	Proactive knowledge processes.
Technological enablers (AI, Big Data, NLP, ontologies)	Allow prediction, machine-interpretable structuring, and structuring tacit knowledge.	Lack strategic integration; fragmented.	Architectural integration.
Digital HR systems	Administrative efficiency, centralized systems, simple analytics.	Reactive, mostly descriptive, lacks semantics, still fragmented.	Strategy, focus, and predictive analytics.
Traditional foresight solutions	Objective predictions, scenario planning, personalization, prediction.	Cost, slow speed, detached from operational data.	Operational focus and real-time integration.
Data-driven foresight methods	Individual personalized recommendations, data filtering.	Disconnected from the organizational strategy.	Strategic prescription.

Table 1.
Comparative analysis of KM, HR systems and foresight approaches in the Industry 4.0 context.

To bridge this gap, this chapter proposes a strategy-to-execution framework designed as a digital twin and composed of three core components:

(1) A structure layer, based on an organizational-level workforce ontology, provides a structured and semantic representation of organizational knowledge and relationships between, for example, activities, competencies, strategic objectives, and resources;

(2) A predictive layer using an AI analytic mechanism converts the static model into a dynamic system, enabling forecasting and scenario simulations;

(3) A prescriptive layer, the OARS, that converts insights into actionable and explainable strategic recommendations.

By overcoming **Table 1**'s limitations and integrating its advantages, this framework is envisioned as a digital twin that contributes to the alignment of workforce needs with the Industry 4.0 evolutions. The next section outlines the structure and mechanisms of this system.

3. Propositions: Introducing the Human Capital Digital Twin

3.1 The architecture of the Human Capital Digital Twin

To address the strategy-to-execution research gap, we introduce an integrated methodology to transform organizational data into strategic workforce levers. This framework surpasses traditional reactive KM systems and standalone foresight solutions by proposing an integrated, continuous decision-support platform.

Designed to ingest, structure, process, and operationalize organizational knowledge, the foundation of the HCDT is based on a six-core pipeline, as presented in **Figure 1**:

(1) Data ingestion: Collected and pre-processed heterogeneous data from multiple internal systems (e.g., Learning & Talent Management System (LTMS) and Human Resources Information System (HRIS)), external sources mostly related to labor markets, and finally business plans, which provide the strategic context.

(2) Knowledge graph (KG): Used as a taxonomic structuration, standardizing relationships between key entities (skills, roles, employees, projects, clients, context data) to enable learning and reasoning processes.

(3) Foresight engine: Using predictive solutions in the ontology, forecasting skill evolution, and extracting tacit knowledge enable future scenario simulations and ontology enrichment.

(4) OARS: Based on the ontology, the OARS will generate contextual, actionable, strategic recommendations for workforce management.

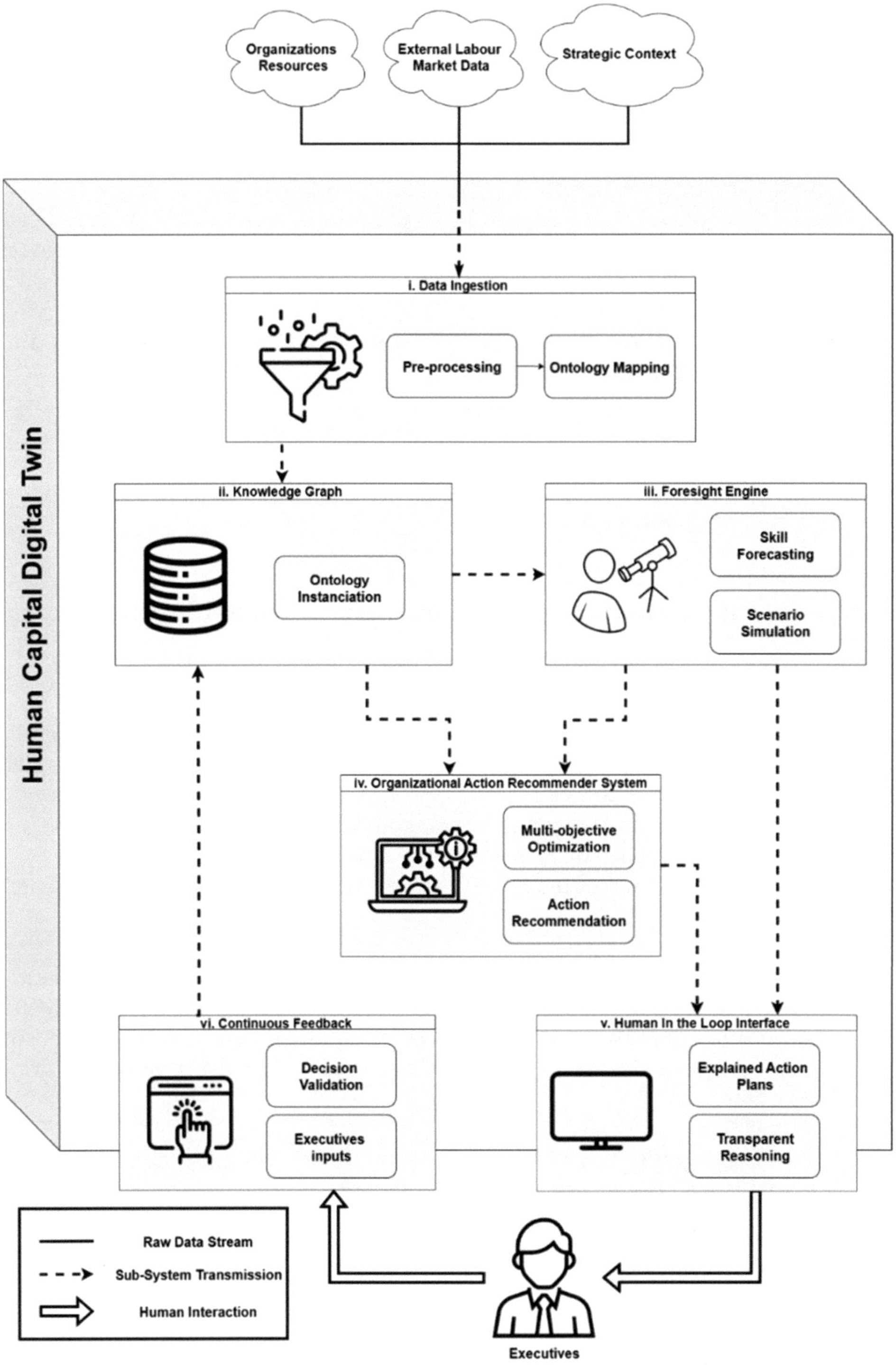

Figure 1.
Architecture of the Human Capital Digital Twin framework.

(5) Human-in-the-Loop interface: Ensures that executives maintain a central position in decision-making by providing transparent and explainable recommendations, which can be adjusted and reviewed by experts.

(6) Continuous feedback: Integrates expert inputs and action outcomes to refine the KG. It ensures organizational alignment and continuous learning with reality.

This architecture enables dynamic simulation and modeling of human capital aligned with strategic objectives. The next sections detail the mechanisms of each integrated sub-system by illustrating how they transform static organizational data into proactive, actionable, explainable, and strategically aligned recommendations.

3.2 Ontology development: The semantic backbone

The ingestion and structuration of raw data sources is the first pipeline stage (cf. Data Ingestion in **Figure 1**) which serves as the raw material to build the semantic foundation of the HCDT. This second section also presents how data is transformed into an organizational KG or applied ontology (cf. Knowledge Graph in **Figure 1**). As defined in the literature, the ontology is not just a data model but the semantic backbone that gives meaning to data. This unified and consistent framework represents the organization's resources in a machine-understandable model, overcoming the limitations of siloed systems.

The ontology is based on three complementary data dimensions:

- Static knowledge: related to perennial organizational resources such as competencies, roles, collaborators, learning modules, and projects;

- Dynamic context: linked to situational and temporal factors such as strategic objectives, market trends, clients' needs impacting skills evolution;

- Action scope: reference decision-making levers that can be used by decision-makers, for example, up-skilling, internal mobility, recruitment, outsourcing, and partnerships.

Developed using a specific ontology engineering methodology [33], the simplified ontology, as shown in **Figure 2**, links Agents (e.g., people, organizations, teams), available Resources (e.g., human, financial, material, technological), organizations' Objectives (e.g., strategic or operational), Actions (e.g., training, allocating, recruiting, outsourcing), mobilizable Competences, and the External Context, which can influence Objectives. This structure serves as the bridge between workforce knowledge and organizational priorities.

The ingestion and processing of heterogeneous data sources is the foundation for building the KG. Internal systems, including, for example, HRIS (employee profiles, competencies, and projects) and LTMS (learning pathways and modules), as well as organizational structures, are associated with external sources such as market intelligence (skill and technological evolutions, competitor strategies). Datasets are then extracted, cleaned, and mapped following the ontology representation and instantiated into a KG. Tacit

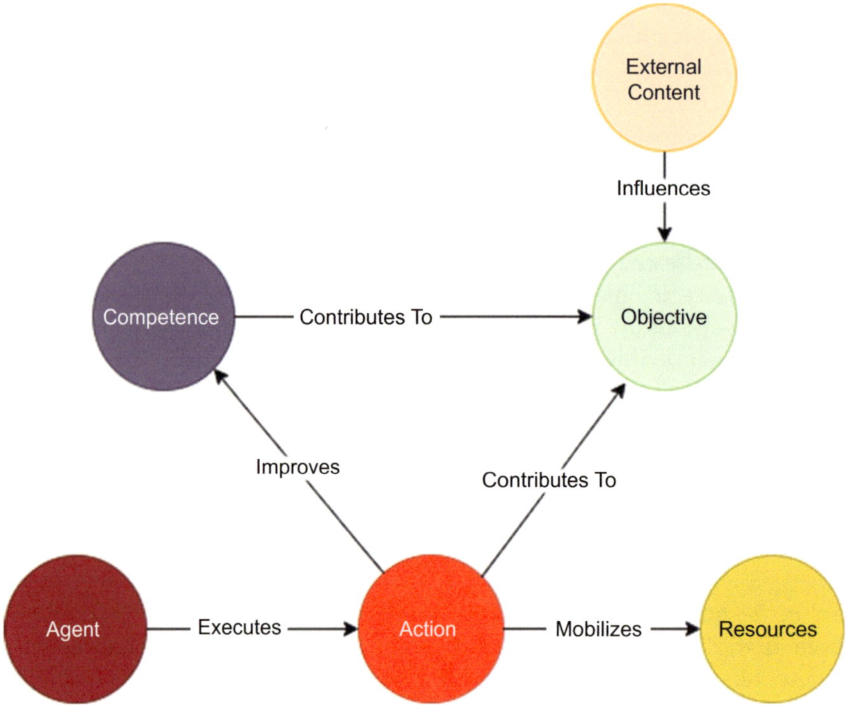

Figure 2.
Extract of the core components of the workforce ontology.

knowledge is also extracted using supervised and unsupervised Natural Language Processing solutions, transforming unstructured documents (e.g., reports, feedback, and learning documentation) into explicit knowledge that enriches the organizational KG [21]. The continuous data ingestion process keeps the ontology as a living model, updated from organizational and external data flows.

Existing standards and ontologies (e.g., European Skills/Competences Taxonomy[1], Occupational In-formation Network[2] , Human Resource Management Ontology[3] , Friend of a friend[4], and Simple Knowledge Organization System[5]) are used to ensure interoperability and alignment with international reference models.

The ontology is not just a static representation but the foundation of a reasoning and prediction engine. It bridges both workforce knowledge and organizational priorities, providing context and objectives for analytics and recommendations. The ontology sets the foundation to connect raw data and decision-support systems, establishing the semantic backbone of Machine Learning algorithms to generate

[1] https://esco.ec.europa.eu/en/classification

[2] https://www.onetonline.org/

[3] https://interoperable-europe.ec.europa.eu/collection/eu-semantic-interoperability-catalogue/solution/human-resources-management-ontology

[4] https://en.wikipedia.org/wiki/FOAF

[5] https://en.wikipedia.org/wiki/Simple_Knowledge_Organization_System

actionable insights. This semantic backbone parallels digital twin systems in Industry 4.0, using standardized ontologies to integrate diverse data sources and cyber-physical processes. However, it shifts from using sensors and machine processes to organizational workforce resources and external context to create an HCDT that supports strategic decision-making.

Providing a formal representation, the ontology allows knowledge to be expressed in measurable terms. As an example, the available supply S and demand D of competence i at a time t can be represented as the variables $S_i(t)$ and $D_i(t)$ with $i \in N$. Their difference $D_i(t) - S_i(t)$ highlights the skill gap that the organization must address. The strategic actions A (e.g., training, recruiting, partnership, outsourcing) can be modeled as the possible interventions that would influence the evolution of competencies at $t + 1$.

By integrating data with the ingestion and structuration mechanisms, the KG becomes the semantic interface between knowledge and decision-support systems. The next section details the use of AI to transform the KG into a foresight engine, allowing scenario simulations.

3.3 AI integration: The foresight engine

The formalized ontology models the organizational knowledge through a structured and machine-interpretable representation. However, it persists as a "passive" framework. The AI integration leverages this static representation into a dynamic and predictive engine for strategic decision support (cf. Foresight Engine in **Figure 1**). This transformation is accomplished using a structured analytical framework:

- Pattern identification (What is): The KG enables the identification of relevant patterns, such as "Skill Q and Skill J are usually linked to Role R in the market context C." It can also identify clusters by grouping similar entities and, therefore, revealing workforce distribution [34].

- Predictive foresight (What if and what will be): The KM can integrate temporal modeling of entities, making it aware of competencies' supply and demand evolution. As an example, imagine the future supply of competence i expressed with:

$$S_i(t+1) = S_i(t) + \sum_{a \in A} \left(b_{a,i} - \delta_i(t)\right)$$

  where actions $a \in A$ (e.g., upskilling, recruiting) increase supply using $b_{a,i}$ the contribution of action a to competence i and $\delta_i(t)$ its natural attrition.

This mechanism mirrors predictive maintenance in Industry 4.0 systems: while sensor and machine states are analyzed to anticipate failures, here the skill supply and demand allow to anticipate skill gaps and future bottlenecks in the organization. This competence evolution at $t + 1$ could be instantiated through a predictive function that would, for example, combine organizational context and market dynamics. This function could be instantiated with time series model or even advanced Graph Neural Network that would simulate the impact of external changes across the organizational infrastructure [35].

- Optimization (What to do): Here, the framework introduces the workforce planning problem as an optimization decision problem. The system evaluates multiple actions to determine the best subset that would minimize the skill gap while respecting organizational priorities and constraints (duration, budget, objectives). This optimization problem establishes the foundation of the OARS (*cf.* OARS in **Figure 1**).

The AI integration fundamentally changes the role of the ontology: from a descriptive framework to a proactive decision-support system able to anticipate workforce evolutions and react to it. This foresight capability sets the foundation for the OARS detailed in Section 3.4.

3.4 Organizational Actions Recommender System

While AI enables predictive insights, the recommender system converts these insights into concrete and actionable strategies for decision-makers. Unlike traditional recommenders designed for individuals [36], OARS operates at the macro-organizational level, directly supporting executives in aligning workforce resources with business objectives, making OARS the decision-making layer of the HCDT. The recommendations are made under resolutions, ensuring they are feasible in real-world scenario.

The organization is seen as a user having a portfolio of available actions $A = \{a1,\ a2, \dots,\ an\}$ with $n \in N$. Each action is associated with attributes (e.g., cost, duration, addressing a specific competency). Therefore, choosing to take one or multiple actions would influence the evolution of skills supply by reducing or increasing its gap. For example, recruiting specific profiles might increase instantly the availability of competencies. However, upskilling takes time but strengthens internal talent over a period.

In its simplest form, the recommender identifies the best subset $\mathbf{A}^*$ of actions A that minimizes the anticipated skill gaps:

$$\min O(x) = \sum_i w_i G_i(t+1)$$

where $G_i(t+1)$ is the skill gap for a competency i at time t+1 and w_i the contribution of each competence to objectives.

However, realistic organizations' decision-making rarely optimizes a single criterion. Since an organization comprises multiple stakeholders, the system must consider multi-preferences, for example, executives want alignment with their long-term objectives; HR wants to close skill gaps quickly by upskilling or recruiting employees; and Finance requires controlled and low cost.

Optimizing a single criterion, such as cost, would lead the engine to recommend "do nothing" or "always go for the cheapest solution." To reflect real-world decision-making, OARS considers the problem as a multi-objective optimization problem:

$$\min O(x) = \alpha \sum_i w_i G_i(t+1) + \beta \sum_a c_a x_a + \gamma R(x)$$

where:

- α, β,γ represents the weight variables defined by executives;
- $G_i(t+1)$: skill gap for competence i at time t + 1;

- $c_a x_a$ is the cost of an action taken;

- $R(x)$ is the disruption of actions' impact (e.g., conduct of change resistance, execution feasibility, risk). This variable is crucial since an action that seems optimal in terms of cost or duration might be encountering strong organizational resistance, operational disruption, or even a cultural misalignment.

OARS delivers the final output through a ranked set of actions according to the organization's objectives. For example: $\mathbf{A}^*$ = {"Reallocate Team 1 to Project P", "Upskill Team 1 with Python Training", "Hire 5 Data Scientists in Team 1"}.

To guide decision-makers, each recommendation proposed by the OARS comes with:

- A predicted impact on skill gaps;

- A confidence score computed with predictive models;

- An explanation justifying the choice of an action.

3.5 Organizational fit: The Human-in-the-Loop principle

As represented in **Figure 1**, the OARS is designed to operate with a Human-in-the -Loop Interface and a Continuous Feedback Loop [37]. These mechanisms ensure that executives remain central to the decision-making process. The HCDT provides output modules (e.g., *Explained Action Plan*, *Transparent Reasoning*), allowing users to evaluate and interpret recommendations, as well as input modules (*Decision Validation* and *Executive Inputs)* through which validated decisions and outcomes are reinjected into the system.

A recommendation is valuable if the user can understand, verify, and adjust it [38]. Therefore, explainability in the HCDT is mandatory to guarantee trust, organizational alignment, and ensure a meaningful action's impact. The multi-objective function integrates customizable weight variables. Leaders can prioritize strategic objectives while understanding the choice of the system over an action:

- $\alpha \sum_i w_i G_i(t+1)$: Setting α high will make the system close skill gaps faster, despite higher cost;

- $\beta \sum_a c_a x_a$: If β is high, system rejects expensive actions focusing on cheaper ones;

- $\gamma R(x)$: High γ focuses on stability, very useful for risk sectors.

Additional terms can be added to the optimization function to consider shifting strategic priorities without altering the system, ensuring the scalability and adaptability of OARS. To build transparency and clarity, the HCDT provides an explainability framework at every reasoning stage:

- Traceability: Each recommendation is grounded in the KG. The reasoning path can be displayed through subgraphs such as:

Objective S –[requires] → Competence i –[heldBy] → Team 1 –[improvedBy] → Action

- Predictive interpretability: Predictions are followed by interpretable signals concerning counterfactual reasoning (consequences of not taking an action), feature importance (most influential factor in a prediction), and scenario simulation (outcomes under different conditions).

- Recommendation explainability: Recommended actions include confidence scores and trade-offs, justifying the selection of one action over another.

- Organizational-level communication: Explanations must be translated into vocabulary that can be understood by executives. The objective is not to present mathematical terms but actionable and decision-ready narratives such as: "Upskilling Team A reduces the AI skills gap by 30% within six months at medium cost, with low disruption risk."

Consequently, explainability is a critical aspect to consider. AI-driven recommendations must not only be optimal from an efficiency perspective but also trustworthy. Just as predictive maintenance digital twins are required to explain machine predictions and suggested interventions, the HCDT must justify organizations' predictions, suggested actions, and their contribution to strategic objectives. The objective is a collaborative system in which AI recommends and humans decide. To illustrate the shift from abstract principles to practical application, this section concludes with an illustrative case demonstrating the role of the HCDT in supporting executives in a real strategic challenge.

3.6 Illustrative case: Strategic market entry

3.6.1 Context and problem

This illustrative case highlights the impact of the HCDT framework in a real-case scenario: market diversification. Our considered company operates in the railway sector through five main technical domains: signaling, rolling stock, infrastructure, and high voltage. Executives identify an opportunity in the energy sector through the multiplication of client requests to integrate energy expertise into railway projects. The board defines a new strategic objective: enter the energy market in the next 24 months. This complex challenge brings a series of questions to be addressed:

- Should engineers be upskilled, and which groups would be the most cost-effective to retrain?

- Should the company hire externally for immediate capacity, despite the higher costs?

- Should partnerships or outsourcing be used to mitigate risk?

- How do these choices affect ongoing, high-priority railway projects that cannot be delayed?

Traditional HR and planning systems struggle to answer these interdependent questions in an integrated and quantifiable way. The HCDT addresses this gap.

3.6.2 Ontology instantiation and state formalism

The HCDT begins by instantiating the company's state into the workforce ontology, providing a dynamic representation of resources, competencies, objectives, and constraints (cf. Knowledge Graph in **Figure 1**).

- Objective: Enter the energy market within 24 months.
- Resources: 2,000 engineers, €10 M budget, average training time of ~12 months for new competencies.
- Competencies: Railway skills (signaling, rolling stock, high/low voltage, infrastructure) vs. energy skills (grid management, renewable energy systems).
- Ongoing Activities:
    - o High-speed rail upgrade (deadline: 12 months, heavy demand for high-voltage engineers, €50k penalty per month of delay).
    - o Bridge renewal (18 months, demand for infrastructure engineers, €30k penalty per month).
    - o Regular maintenance (medium priority; uses high-voltage engineers).

Using the ontology, the system evaluates competence overlaps to estimate the ease of transition:

- High-voltage engineers share 70% of competencies with energy, making them relatively cheap to retrain.
- Infrastructure engineers share 50%.
- Signaling and rolling stock engineers share almost none, making retraining costly and time-consuming.

Recruitment of new energy experts can be considered but would be more expensive (e.g., ~€120k per engineer). Outsourcing or even partnerships can fill the gaps quickly but include long-term expenses. This formalized state captures the hidden leverage of cross-domain competence reuse, while also mapping constraints such as budget, deadlines, and penalties for disrupting existing projects.

3.6.3 The activity-aware optimization problem

The OARS formulates the decision as a multi-objective optimization: how to reach the required level of energy competence within 24 months, under a €10 M budget, while minimizing disruption to ongoing railway projects and risks. It formalizes the problem as:

$$\min O(x) = \alpha \sum_i w_i G_i(t+1) + \beta \sum_a c_a x_a + \gamma R(x) + \delta U(x)$$

where:

- α, β, γ represents the weight variables defined by executives;
- $G_i(t+1)$: skill gap for competence i at time t+1;
- $c_a x_a$: cost of an action taken;
- $R(x)$: disruption impact of actions (e.g., conduct of change resistance, execution feasibility);
- $U(x)$: risk exposure (e.g., Market instability, Project delays, Uncertainty in hiring success).

Constraints ensure that:

- Total cost stays within budget.
- Energy competencies are sufficient at the 24-month horizon.
- Railway projects are not fatally compromised.

This framing highlights the complexity of balancing strategic objectives, cost-optimization, risk impact, and operational stability. Each action (upskill, recruit, outsource, reallocate) has cascading effects on both future readiness and current commitments.

3.6.4 Scenario-based recommendations with explainability

OARS proposes alternative scenarios by varying weights (α, β, γ, δ) and setting different decision stance (cf. **Table 2**).

The OARS outputs are summarized in **Table 3**.

All the scenarios are generated through the Action Recommendation module of OARS, while executives validate and adjust them using the Human-in-the-Loop Interface and Continuous Feedback (cf. **Figure 1**).

Decision stance	**Description**	α	β	γ	δ
Aggressive	Prioritize speed and competence buildup.	High	Low	Low	Low
Balanced	Seek equilibrium between growth and stability.	Medium	Medium	Medium	Medium
Conservative	Protect current operations and minimize risk.	Low	Medium	High	High

Table 2.
Organizational decision stance.

Scenario	Main actions	Cost	Energy readiness	Operational Disruption	Risk level	Explainability insights
Aggressive	Upskill 40 high-voltage engineers. Recruit 20 external energy specialists.	7 M€	100% in 18 months	Minor delays (around 3 M €3M penalty).	High	Traceability; counterfactual reasoning; trade-off visualization.
Balanced	Upskill 20 engineers. Outsource initial energy projects to partners.	4 M€	50% in 24 months	No major disruption.	Medium	Scenario simulation; feature importance; partner reliance clarity.
Conservative	Recruit a small cohort of external energy experts only.	9 M€	30% in 24 months	No disruption.	Low	Trade-off visualization; human-in-the-loop validation.

Table 3.
Comparative evaluation of strategic scenarios under the HCDT framework.

3.6.5 Outcome and contribution

This illustrative case demonstrates the impact of the HCDT in transforming an abstract strategy into actionable, concrete, and explainable plans. It starts by structuring organizational knowledge within its integrated ontology. The cascading impacts of decisions are then simulated on real ongoing activities, costs, and strategic objectives. The OARS delivers ranked recommendations supported by transparent reasoning, visualization, and scenario simulations. Executives are actively involved in the validation and configuration of decisions. Therefore, organizations can make confident, data-driven decisions aligned with organizational constraints, cultures, and priorities, making the strategy-to-execution gap quantifiable, apparent, and manageable.

4. Conclusion and future work

This journal chapter has explored a critical paradox at the core of Industry 4.0: despite having mastered predictive maintenance and the utilization of digital twins, organizations lack an equivalent for their most vital asset: human capital. The chapter reviewed the shortcomings of traditional reactive KM systems, operational HR systems, and isolated foresight methodologies. Organizations face a strategy-to-execution gap in their workforce management due to the passiveness of KM systems, the fragmented and descriptive HR systems, as well as the operationally disconnected foresight methods.

In response, this chapter introduces a conceptual framework for an HCDT. Seen as an integrated system, it is designed to transform passive organizational data into dynamic and actionable insights. This interconnected framework is built around three core components:

(1) A dynamic workforce ontology acting as the semantic backbone, which provides a unified, machine-readable model of the organization's resources, objectives, and external context.

(2) An AI predictive engine that transforms the static ontology into a dynamic system, capable of forecasting skill evolution, simulating scenarios, and identifying future gaps.

(3) An OARS, which translates predictive insights into actionable, explainable, contextual, and optimized recommendations for workforce planning.

The illustrative case demonstrated the HCDT's potential through a real example of strategic market entry. Transitioning from abstract business objectives to a concrete and executable plan, the OARS bridged the strategic-to-execution gap by providing guidance and linking workforce decisions to strategic outcomes. The Human-In-The-Loop layer is embedded within the HCDT to ensure explainability and transparency throughout the framework. Its objective is not to replace but to enhance executive decision-making with trust and organizational alignment. It establishes a foundational step to address a crucial vulnerability in Industry 4.0 strategic management.

While the HCDT was conceived to serve data-rich and complex organizations, its logic is also valuable for Small and Medium-sized Enterprises (SMEs). A practicable implementation would take the concept of a Lightweight HCDT, centered on a minimal set of data inputs such as key resources, core competencies, and organizational objectives. The system could be maintained by a defined stakeholder using simple tools such as structured spreadsheets and prioritizing semantic clarity rather than data automation. Proportionate data governance practices (e.g., data minimization, controlled access, and consistency checks) would ensure that even this minimal data setup remains secure and reliable. Even in its lightweight form, the framework would function as a strategic compass by converting implicit knowledge into explicit knowledge and revealing critical dependencies between workforce capabilities and strategic objectives. This approach provides a scalable foundation, offering immediate strategic value as well as laying the groundwork for future advanced capabilities in case the organization's data maturity evolves. This adaptation for low-data contexts underscores the practical relevance of the HCDT and sets the stage for empirical validation and refinement.

This chapter can be seen as an exploratory foundation to reflect on future research pathways and practical implementations to develop the HCDT. Therefore, the next logical step is to move from theory to empirical validation. Future work will focus on structuring a functional prototype, followed by a robust evaluation methodology to ensure real-world efficiency and impact.

It is crucial that the evaluation strategy moves beyond traditional technical metrics to capture the true value of the HCDT as a strategic decision-support system. The assessment needs to be multi-faceted to address three critical dimensions. First, predictive accuracy ensures the reliability of forecasts and recommendations. Traditional metrics such as Mean Absolute Error for skill gap predictions, as well as Precision and Recall for action suggestions, will set a solid operational foundation [39]. Secondly, Recommendations Diversity will assess the system's strategic creativity to generate varied and non-evident action plans to avoid proposing only conventional options [40]. Finally, organizational impact and utility will represent the last validation layer by combining expert evaluations, decision-making quality metrics (e.g., reduced planning cycles, better trade-off understanding), and simulated impact analyses.

Table 4 synthesizes this multi-dimensional evaluation framework by outlining the corresponding metrics, observation windows, and their strategic intent.

Evaluation dimension	Metric(s)	Sampling/ observation windows	Intent
Predictive accuracy	Mean Absolute Error, Precision, Recall.	Each forecasting or recommendation generation cycle.	Validates forecasts' and recommendations' reliability; establishes a quantitative baseline for operational robustness.
Recommendation diversity	Shannon Index, Intra-list diversity [40].	Each recommendation generation cycle.	Evaluates the variety and non-redundancy of strategic recommendations, supports the system's creativity, and avoids convergence on conventional decisions.
Organizational impact and utility	Expert utility score, simulated scenario impact.	Semi-annual organizational review.	Measures the real strategic contribution across organizational efficiency, decision-making quality, and alignment between workforce capabilities and strategic objectives.

Table 4.
Operational evaluation framework for the HCDT.

These three dimensions-based evaluations will determine that the HCDT is not only technically robust but also strategically disruptive. It aims to prove that the HCDT can be considered as a reliable and value-generating decision-support tool, empowering organizations to navigate the complexity of Industry 4.0 and transform their human capital into a source of competitive advantage.

Author contributions

First author (Romain Stevens): conceptualization, methodology, writing – original draft; second author: (Elsa Negre): methodology, supervision, writing – review and editing.

Funding

This work is carried out as part of a CIFRE PhD, an industrial research program funded by the French National Association for Research and Technology (ANRT).

Ethical statement

Not applicable.

Data availability statement

Data are available from the corresponding author upon request.

Conflict of Interest (mandatory section)

The authors declare no conflict of interest.

Author details

Romain Stevens[1,2]* and Elsa Negre[1]

1 Paris-Dauphine University, PSL Research University, Paris, France

2 IKOS Consulting, IKOS Lab, Levallois-Perret, France

*Address all correspondence to: romain.stevens@lamsade.dauphine.fr

References

[1] Delery JE, Roumpi D. Strategic human resource management, human capital and competitive advantage: Is the field going in circles? Human Resource Management Journal. 2017;**27** (1):1–21. DOI: 10.1111/1748-8583.12137

[2] Polzer JT. The rise of people analytics and the future of organizational research. Research in Organizational Behavior. 2023;**43**:100181. DOI: 10.1016/j.riob.2023.100181

[3] Marler JH, Boudreau JW. An evidence-based review of HR Analytics. The International Journal of Human Resource Management. 2017;**28**(1):3–26. DOI: 10.1080/09585192.2016.1244699

[4] Rohrbeck R, Kum ME. Corporate foresight and its impact on firm performance: A longitudinal analysis. Technological Forecasting and Social Change. 2018;**129**:105–116. DOI: 10.1016/j.techfore.2017.12.013

[5] World Economic Forum. The Future of Jobs Report 2020. Geneva: World Economic Forum; 2020. Available from: https://www.weforum.org/reports/the-future-of-jobs-report-2020 [Accessed: 2025-October-13]

[6] Hirschi A. The Fourth Industrial Revolution: Issues and implications for career research and practice. The Career Development Quarterly. 2018;**66** (3):192–204. DOI: 10.1002/cdq.12142

[7] Powell WW, Snellman K. The knowledge economy. Annual Review of Sociology. 2004;**30**:199–220. DOI: 10.1146/annurev.soc.29.010202.100037

[8] Xue CTS. A literature review on knowledge management in organizations. Research in Business and Management. 2017;**4**(1):30–41. DOI: 10.5296/rbm.v4i1.10856

[9] Bellinger G, Castro D, Mills A. Data, information, knowledge, and wisdom [Internet]. 2004 Available from: https://www.systems-thinking.org/dikw/dikw.htm [Accessed: 2025-October-03]

[10] Addis M. Tacit and explicit knowledge in construction management. Construction Management and Economics. 2016;**34** (7–8):439–445. DOI: 10.1080/01446193.2016.1180416

[11] Geisler E. A typology of knowledge management: Strategic groups and role behavior in organizations. Journal of Knowledge Management. 2007;**11** (1):84–96. DOI: 10.1108/13673270710728259

[12] Gupta A, Singh RK, Kamble S, Mishra R. Knowledge management in Industry 4.0 environment for sustainable competitive advantage: A strategic framework. Knowledge Management Research & Practice. 2022;**20**(6):878–892. DOI: 10.1080/14778238.2021.1933097

[13] Popova N, Shynkarenko V, Kryvoruchko O, Zéman Z. Enterprise management in VUCA conditions. Annual Economics-XXI. 2018;**170**(3–4):27–31. DOI: 10.21003/ea.V170-05

[14] Bennett N, Lemoine GJ. What a difference a word makes: Understanding threats to performance in a VUCA world. Business Horizons.

2014;**57**(3):311–317. DOI: 10.1016/j.bushor.2014.01.001

[15] Akhavan P, Ale Ebrahim N, Fetrati MA, Pezeshkan A. Major trends in knowledge management research: A bibliometric study. Scientometrics. 2016;**107**(3):1249–1264. DOI: 10.1007/s11192-016-1938-x

[16] Dawson N, Rizoiu M-A, Johnston B, Williams M-A. Predicting skill shortages in labour markets: A machine learning approach. In: 2020 IEEE International Conference on Big Data. 2020;3052–3061. DOI: 10.1109/BigData50022.2020.9377773

[17] Margherita A. Human resources analytics: A systematization of research topics and directions for future research. Human Resource Management Review. 2022;**32**(2):100795. DOI: 10.1016/j.hrmr.2020.100795

[18] Guarino N. Understanding, building, and using ontologies. International Journal of Human-Computer Studies. 1997;**46**(2–3):293–310. DOI: 10.1006/ijhc.1996.0091

[19] Ehrlinger L, Wöß W. Towards a definition of knowledge graphs. SEMANTiCS (Posters, Demos, SuCCESS). 2016; p. 48. Available from: http://ceur-ws.org/Vol-1695/paper4.pdf [Accessed: 2025-January-25]

[20] Dorn J, Naz T, Pichlmair M. Ontology development for human resource management. Knowledge Management: Innovation, Technology and Cultures. 2007;109–120. DOI: 10.1142/9789812770592_0010

[21] Zaoui Seghroucheni O, Lazaar M, Al Achhab M. Using AI and NLP for Tacit knowledge conversion in knowledge management systems: A comparative analysis. Technologies. 2025;**13**(2):87. DOI: 10.3390/technologies13020087

[22] Minbaeva DB. Building credible human capital analytics for organizational competitive advantage. Human Resource Management. 2018;**57**(3):701–713. DOI: 10.1002/hrm.21848

[23] World Economic Forum. Future of Jobs Report 2023. Geneva: World Economic Forum; 2023. Available from: https://www.weforum.org/reports/future-of-jobs-report-2023 [Accessed: 2025-October-13]

[24] Frey CB, Osborne MA. The future of employment: How susceptible are jobs to computerisation? Technological Forecasting and Social Change. 2017;**114**:254–280. DOI: 10.1016/j.techfore.2016.08.019

[25] Linstone HA. The Delphi Technique. In: Covello VT, Mumpower JL, Stallen PJM, Uppuluri VRR, editors. Environmental Impact Assessment, Technology Assessment, and Risk Analysis. Berlin: Springer; 1985. DOI: 10.1007/978-3-642-70634-9_22

[26] von der Gracht HA. Consensus measurement in Delphi studies: Review and implications for future quality assurance. Technological Forecasting and Social Change. 2012;**79**(8):1525–1536. DOI: 10.1016/j.techfore.2012.04.013

[27] Beiderbeck D, Frevel N, von der Gracht HA, Schmidt SL, Schweitzer VM. Preparing, conducting, and analyzing Delphi surveys: Cross-disciplinary practices, new directions, and advancements. MethodsX. 2021;**8**:101401. DOI: 10.1016/j.mex.2021.101401

[28] Bradfield R, Wright G, Burt G, Cairns G, van der Heijden K. The origins and evolution of scenario techniques in long range business planning. Futures. 2005;**37**(8):795–812. DOI: 10.1016/j.futures.2005.01.003

[29] Vecchiato R. Creating value through foresight: First mover advantages and strategic agility. Technological Forecasting and Social Change. 2015;**101**:25–36. DOI: 10.1016/j.techfore.2014.08.016

[30] McCartney S, Fu N. Bridging the gap: Why, how and when HR analytics can impact organizational performance. Management Decision. 2022;**60**(13):25–47. DOI: 10.1108/MD-12-2020-1581

[31] Zhou Q, Liao F, Chen C, Ge L. Job recommendation algorithm for graduates based on personalized preference. CCF Transactions on Pervasive Computing and Interaction. 2019;**1**:260–274. DOI: 10.1007/s42486-019-00022-1

[32] Huang X, Yang F, Zheng J, Feng C, Zhang L. Personalized human resource management via HR analytics and artificial intelligence: Theory and implications. Asia Pacific Management Review. 2023;**28**(4):598–610. DOI: 10.1016/j.apmrv.2023.04.004

[33] Noy NF, McGuinness DL. Ontology development 101: A guide to creating your first ontology [Internet]. Stanford: Stanford Knowledge Systems Laboratory; 2001. Available from: https://protege.stanford.edu/publications/ontology_development/ontology101.pdf [Accessed: 2025-October-13]

[34] Ji S, Pan S, Cambria E, et al. A survey on knowledge graphs: Representation, acquisition, and applications. IEEE Transactions on Neural Networks and Learning Systems. 2021;**33**(2):494–514. DOI: 10.1109/TNNLS.2021.3070843

[35] Olaniyi FG, Olaniyi OO, Adigwe CS, et al. Harnessing predictive analytics for strategic foresight: A comprehensive review of techniques and applications in transforming raw data to actionable insights [Internet]. SSRN Preprint 4635189; 2023. Available from: https://ssrn.com/abstract=4635189 [Accessed: 2025-October-13]

[36] Roy D, Dutta M. A systematic review and research perspective on recommender systems. Journal of Big Data. 2022;**9**:59. DOI: 10.1186/s40537-022-00592-5

[37] Asad U, Khan M, Khalid A, et al. Human-centric digital twins in industry: A comprehensive review of enabling technologies and implementation strategies. Sensors. 2023;**23**(8):3938. DOI: 10.3390/s23083938

[38] Charles V, Emrouznejad A, Kunz WH. Advancements in Artificial Intelligence-based prescriptive and cognitive analytics for business performance: A special issue editorial. Journal of Business Research. 2025;**200**:115576. DOI: 10.1016/j.jbusres.2025.115576

[39] Nègre E. Information and Recommender Systems. New Jersey: John Wiley & Sons; 2015. DOI: 10.1002/9781119102779

[40] Parapar J, Radlinski F. Towards unified metrics for accuracy and diversity for recommender systems. In: Proceedings of the 15th ACM Conference on Recommender Systems; 2021. p. 75–84. DOI: 10.1145/3460231.3474234

Chapter 4

Perspective Chapter: The Role of Artificial Intelligence in Workforce Training and Talent Acquisition under Industry 4.0

Mohammed Gronfula and Mohammed Alshammari

Abstract

This chapter critically examines the transformative influence of artificial intelligence (AI) on workforce development and talent acquisition paradigms within the context of Industry 4.0. Given the widespread digital transformation across industries, AI technologies are becoming fundamental for conceptualizing and implementing personalized, scalable, and adaptive learning frameworks essential for enhancing digital competencies. The exposition highlights how AI-driven platforms not only facilitate intelligent recruitment processes through automated candidate screening and predictive workforce analytics but also enable continuous professional development and skill realignment via empirically informed training programs. Illustrative case studies and contemporary trends will be presented to exemplify AI's strategic utility in bolstering human resource initiatives across diverse sectors. Furthermore, the discussion will address inherent challenges, including ethical dilemmas, algorithmic biases, and data privacy implications. Ultimately, this discourse emphasizes the symbiotic relationship between technological innovation and human capital, offering strategic guidance for industry leaders, educational stakeholders, and policymakers committed to aligning talent cultivation with the dynamic demands of the evolving digital economy.

Keywords: artificial intelligence, Industry 4.0, workforce training, talent acquisition, digital skills development

1. Introduction

The Fourth Industrial Revolution, characterized by the fusion of digital, physical, and biological spheres, is profoundly reshaping industries and economies worldwide. At its core, artificial intelligence (AI) stands out as a pivotal technology, driving unprecedented transformations in how organizations manage human capital – from talent acquisition to continuous workforce development. This paradigm shift, often termed Industry 4.0, necessitates a fundamental reevaluation of traditional employment models, skill prerequisites, and learning methodologies. As AI capabilities rapidly advance, they offer both immense opportunities for optimizing human potential and significant challenges related to ethics, bias, and privacy. This chapter critically examines the multifaceted influence of AI on workforce training and talent acquisition, exploring its role as a catalyst for personalized learning, intelligent recruitment, and continuous professional development (CPD). By delving into the symbiotic relationship between AI and human capabilities, analyzing current applications and emerging trends, and addressing inherent challenges, this discourse aims to provide strategic guidance for navigating the complex landscape of the evolving digital economy and fostering a future-ready workforce.

2. AI as a catalyst in workforce transformation under Industry 4.0

The integration of AI fundamentally alters employment patterns by automating routine tasks, creating new job roles, and redefining existing ones, thereby necessitating a paradigm shift in workforce training [1, 2]. This transformation underscores the urgent need for reskilling and upskilling initiatives to bridge emerging skill gaps and improve workforce adaptability in an AI-driven landscape [3]. This necessitates a comprehensive transformation strategy within educational paradigms to equip learners with both technical competencies and critical soft skills to navigate an AI-infused era [4]. As AI continues to reshape industries and job markets, defining AI literacy for lifelong learning becomes increasingly critical to ensuring individuals can effectively collaborate with AI systems and leverage their capabilities (see **Figure 1**). This involves not only understanding the technical aspects of AI but also its ethical implications and societal impacts, ensuring a responsible integration of AI into professional practices [5].

Figure 1.
Integration of AI in modern work environments: illustrating how AI reshapes employment patterns and enhances workforce capabilities in Industry 4.0 [6].

2.1 Defining workforce transformation in the digital age

This section elaborates on the multifaceted shifts occurring in labor markets due to pervasive digitalization and automation, necessitating a reevaluation of traditional employment models and skill prerequisites. The rapid advancements in AI necessitate a proactive approach to reskilling and upskilling the workforce to bridge emerging skill gaps and enhance adaptability [7]. This includes adapting to the ethical and socioeconomic considerations of AI integration, which affect diverse regions and economies [8]. Moreover, the acceleration of digitalization trends, partly spurred by events such as the COVID-19 pandemic, further highlights the criticality of employee resilience and well-being in adapting to job and technological disruptions [9]. The shift toward advanced technologies, such as smart technology, cloud computing, and the Internet of Things, fundamentally alters the nature of work, creating an urgent demand for new skills and competencies [9].

2.2 Core AI technologies shaping human capital management

This includes a detailed examination of machine learning algorithms for predictive analytics, natural language processing for enhanced communication and content generation, and robotic process automation for streamlining repetitive tasks within HR functions [10]. These technologies collectively enable more efficient talent acquisition, personalized employee development, and data-driven strategic workforce planning, thereby optimizing human capital management [11, 12]. The application of AI in HR also extends to sophisticated areas such as affective computing for nuanced candidate assessment and adaptive learning platforms that tailor training modules to individual employee needs, promoting continuous growth and performance enhancement [13, 14]. The pervasive influence of AI is poised to significantly impact organizational income by increasing workforce productivity through the automation of routine tasks and the expansion of employee skill sets, thereby elevating the overall value of their contributions [15].

2.3 The symbiotic relationship between AI and human capabilities

AI should be regarded as a pivotal factor for future industries, enhancing human capabilities and enabling a symbiosis between technology and the workforce [16]. This collaboration emphasizes complementarity rather than competition, fostering improved productivity and organizational growth through the strategic integration of technological and cognitive proficiencies [11, 17]. This perspective posits that AI acts as an enabler, augmenting human decision-making and operational efficiency, rather than serving as a direct replacement for human ingenuity [18]. This cooperative framework aims to leverage AI's analytical prowess to automate mundane tasks and provide data-driven insights, thereby freeing human employees to focus on complex problem-solving, creativity, and interpersonal interactions [19]. The human–AI synergy facilitates more nuanced and informed decisions, particularly in complex fields such as health care and finance, by combining AI's data processing capabilities with human intuition and emotional intelligence [20].

3. Transitioning to AI-enhanced personalized and adaptive learning

The integration of AI fundamentally alters traditional training methodologies by creating individualized learning paths that adapt to diverse learner preferences, paces, and expertise levels [21]. This personalized approach, driven by AI algorithms, optimizes educational outcomes by providing tailored content and real-time feedback, thereby fostering a more engaging and effective learning environment [22, 23]. Moreover, AI-powered adaptive learning systems can dynamically adjust the curriculum based on a learner's performance and engagement, ensuring that content remains challenging yet achievable, which is crucial for sustained motivation and skill acquisition [24, 25]. This iterative adjustment ensures that the learning experience remains highly relevant and optimizes knowledge retention [26]. At its core, personalized learning leverages AI to cater to the unique needs, learning styles, and pace of individual learners, moving beyond a "one-size-fits-all" educational model (see **Figure 2**) [23]. Recent studies confirm the efficacy of adaptive learning, showing improved retention and faster skill acquisition in AI-driven platforms [26–28]

3.1 AI algorithms for individualized skill pathways

These algorithms leverage sophisticated machine learning models to analyze performance data, identify knowledge gaps, and recommend highly targeted learning modules, thereby optimizing the acquisition of specific competencies. Such systems utilize techniques such as Bayesian inference and reinforcement learning to continuously refine personalized educational trajectories, ensuring that learners receive instruction optimally aligned with their evolving needs and cognitive profiles [30, 31]. This granular personalization, supported by AI, significantly enhances the efficiency and efficacy of upskilling and reskilling initiatives, directly addressing the dynamic demands of Industry 4.0 [28, 32]. The strategic deployment of AI in this context also includes leveraging neural networks and fuzzy logic to model learning behaviors and manage the inherent

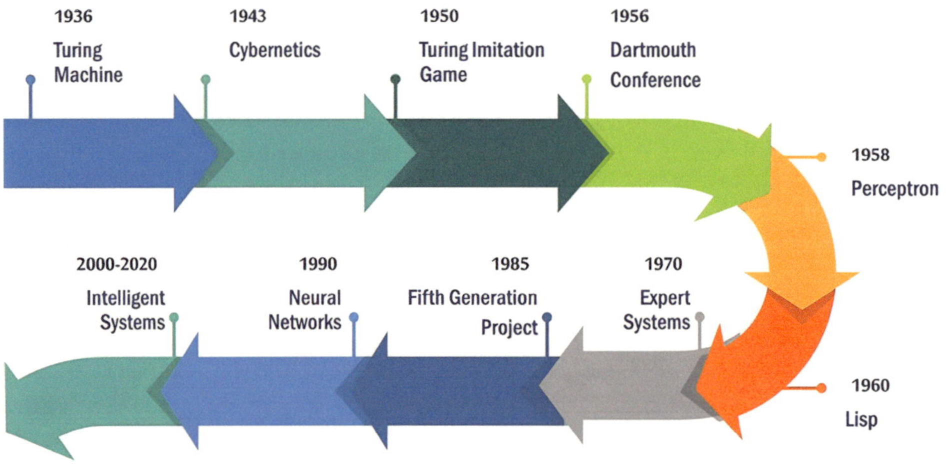

Figure 2.
Timeline of workforce transformation: from traditional models to AI-driven digital ecosystems, highlighting critical inflection points in labor markets.

uncertainties in human cognitive processes, thereby further refining recommendations for learning activities [33].

3.2 Adaptive learning platforms for workforce competency building

These platforms dynamically adjust content difficulty and presentation based on real-time learner interactions, ensuring a more efficient and effective upskilling process for critical roles within evolving industrial landscapes [34]. These advanced systems leverage AI to provide unparalleled levels of personalization, efficiency, and scalability in learning, thereby revolutionizing traditional pedagogical paradigms by individualizing instruction to the specific needs of each learner [25]. This approach goes beyond conventional methods by utilizing analytical tools and algorithms to create automated, adaptive learning activities and materials, ensuring that training is specifically tailored to individual requirements and learning styles [35]. Furthermore, these AI-driven platforms often incorporate intelligent tutoring systems that offer immediate feedback and remedial support, optimizing knowledge retention and application in practical scenarios [25]. The capacity of AI to analyze vast datasets of learner interactions enables the identification of intricate patterns in comprehension and engagement, allowing for highly targeted interventions and curriculum adjustments [36].

3.3 Scaling training programs through intelligent automation

By automating administrative tasks and content delivery, AI enables organizations to deploy training initiatives at scale, reaching a larger workforce more efficiently and consistently [31]. This scalability is crucial for rapidly disseminating essential skills across diverse geographical locations and organizational departments, fostering a uniformly skilled workforce capable of adapting to technological shifts. The implementation of intelligent automation within training frameworks also facilitates continuous improvement by collecting and analyzing vast amounts of performance data, leading to iterative refinements in curriculum design and delivery [37]. This data-driven approach allows for the identification of areas where training programs can be optimized, ensuring that resources are allocated effectively and that learning outcomes are consistently met. Further, these systems can identify at-risk learners, providing targeted interventions and supplementary resources to ensure comprehensive skill acquisition across the workforce [38]. Moreover, AI-powered tools such as facial recognition systems and predictive analytics can further assist educators by analyzing student attitudes and behaviors during the learning process, thereby enabling more targeted instructional adjustments [39].

4. AI-powered talent acquisition: Automation, screening, and predictive analytics

The utilization of AI in talent acquisition transforms the traditional recruitment life cycle by introducing advanced automation, rigorous screening mechanisms, and sophisticated predictive analytics, which collectively optimize the identification and selection of suitable candidates. This integration streamlines processes, reduces human bias, and enhances the accuracy of hiring decisions, thereby improving organizational efficiency and talent quality [40]. These AI applications not only accelerate candidate sourcing and evaluation but also provide strategic foresight into future workforce requirements, allowing organizations to proactively address talent gaps [41] The integration of AI technologies, such as machine learning, deep learning, and natural language processing, offers powerful computational capabilities for precise data analysis and intelligent learning algorithms [42]. Specifically, AI tools are increasingly employed to automate various aspects of the employment process, from initial screening to final selection, with an estimated 99% of Fortune 500 companies already utilizing some form of AI assistance in their hiring decisions to improve recruitment quality and efficiency [43, 44] (see **Figure 3**). Explainable AI approaches are increasingly advocated to ensure transparency and trust in candidate screening [27].

4.1 Automated candidate sourcing and screening processes

AI-driven tools facilitate rapid and extensive data analysis, surpassing human capacities in evaluating large datasets to identify optimal candidates and provide robust decision recommendations [46]. This enhanced capability allows for the efficient sifting through numerous applications, identifying patterns and predictors of success that might be overlooked by human recruiters [47]. Furthermore, these systems can analyze resumes, cover letters, and even social media profiles to extract relevant keywords and phrases, matching them against job requirements and desired competencies [48]. This automation significantly reduces the time and effort traditionally associated with manual screening, allowing human resource professionals to focus on more strategic aspects of talent management [49]. Beyond mere keyword matching, advanced AI algorithms can perform semantic analysis to understand the contextual relevance of applicant data, leading to a more nuanced assessment of qualifications and potential [50]. However, recent evidence shows that many AI systems still exhibit methodological and quality limitations—including inadequate validation and transparency—that may affect the reliability of their

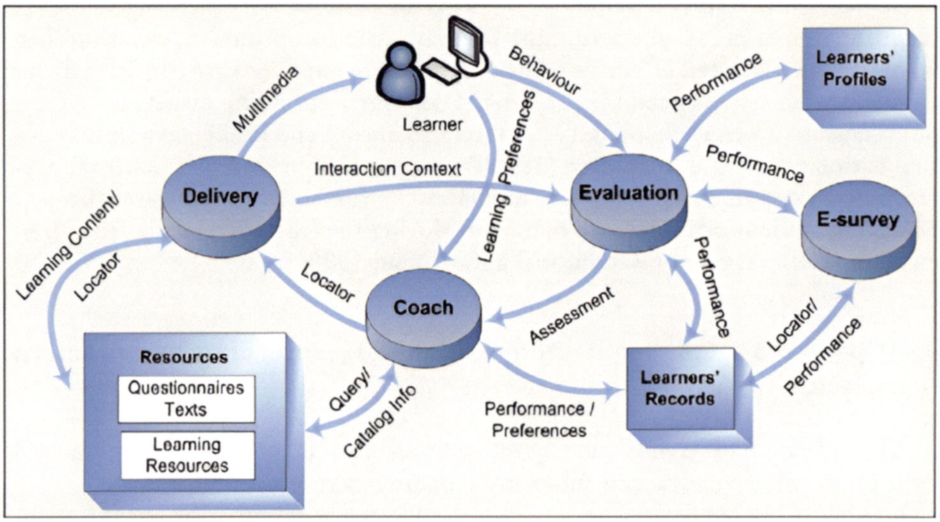

Figure 3.
AI-based individualized skill plan generator: Learner profiles dynamically guide adaptive training resources, improving workforce competency as proposed by Tzouveli et al. [29].

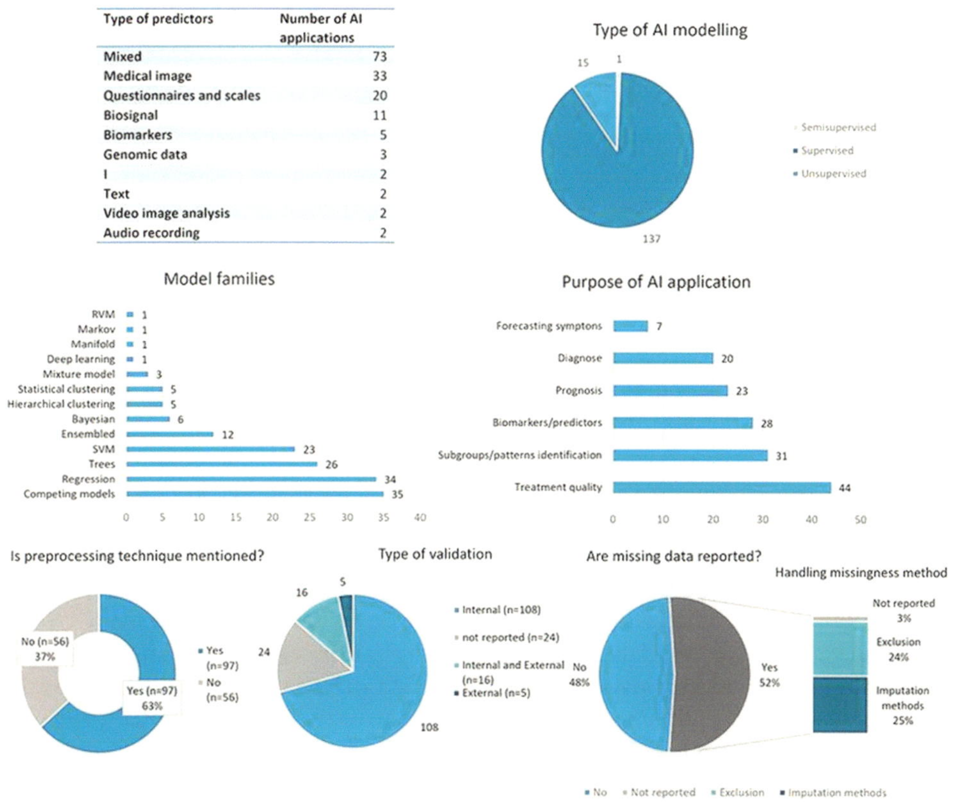

Figure 4.
Overview of the AI model-development workflow and key preprocessing steps, including components such as electronic health records (EHR), relevance vector machines (RVM), and support vector machines (SVM) [45].

outputs, as illustrated in **Figure 4** showing the AI development and preprocessing pipeline (AI: artificial intelligence; EHR: electronic health record; RVM: relevance vector machine; SVM: support vector machine) [45].

4.2 Predictive models for workforce demand forecasting

These models leverage historical data and external economic indicators to forecast future skill requirements and staffing needs, enabling proactive talent pipeline development and strategic workforce planning [51]. By anticipating future demands, organizations can develop targeted recruitment strategies and invest in upskilling initiatives, ensuring a continuous supply of qualified talent. Moreover, predictive analytics can identify potential talent shortages or surpluses, guiding strategic decisions on recruitment campaigns and training investments. This allows companies to better navigate the global war for talent, which often sees traditional recruiting methods fail to cope with intense competition [52]. Furthermore, AI can analyze complex datasets to identify optimal sourcing channels and engagement strategies, further refining the efficiency of talent acquisition processes.

4.3 Reducing recruitment cycles through intelligent matching

AI algorithms can rapidly process vast applicant pools, matching candidate profiles with job requirements to significantly shorten time-to-hire while simultaneously improving the quality of hire [53]. This accelerated matching process, facilitated by machine learning and natural language processing, streamlines the initial screening phases, allowing human recruiters to focus on qualitative assessments and candidate engagement [49]. Moreover, AI tools can help mitigate unconscious biases in the initial screening process, leading to a more diverse and equitable candidate pool [54]. This is particularly relevant given that traditional recruitment methods often yield low success rates in identifying ideal candidates, with some estimates suggesting a mere 16% effectiveness [55]. This efficiency gain is crucial in competitive markets, where swift and accurate hiring decisions can provide a significant strategic advantage.

5. CPD through data-driven insights

By analyzing comprehensive performance metrics and learning behaviors, AI systems generate actionable insights that facilitate targeted interventions and personalized learning recommendations, ensuring continuous skill enhancement and career progression for employees [23, 32]. This data-driven approach fosters a culture of perpetual learning, aligning individual development with organizational strategic objectives. These insights enable organizations to proactively address skill gaps, tailor training programs, and predict future competency needs, thereby optimizing workforce agility and long-term organizational resilience [56]. This comprehensive data analysis further allows for the dynamic adaptation of learning pathways, ensuring that educational content remains relevant and engaging for individual learners. This continuous adaptation ensures that training programs remain effective in addressing evolving industry demands and individual career aspirations. CPD would be used in the following investigation.

5.1 Real-time performance monitoring and feedback loops

This enables dynamic adjustments to training programs and immediate constructive criticism, fostering an agile learning environment [57]. These systems can identify skill gaps in real time, providing personalized learning pathways that address specific deficiencies and capitalize on individual strengths, thereby optimizing professional development [53]. Moreover, AI-powered feedback loops can also mitigate unconscious biases in performance evaluations by focusing on objective metrics and behaviors, ensuring fairer and more accurate assessments of employee capabilities. Additionally, AI can be leveraged to analyze communication patterns and team dynamics, providing insights into collaborative effectiveness and areas for improvement in interpersonal skills. This continuous monitoring allows organizations to identify and nurture high-potential employees, ensuring that talent is developed and retained strategically within the enterprise. Such systems also empower educators with real-time data visualization, which enhances their capacity to make informed decisions regarding instructional strategies and resource allocation, ultimately leading to more personalized and responsive learning experiences for students [58].

5.2 AI-guided career pathways and upskilling initiatives

These initiatives leverage AI to map individual competencies against future industry demands, providing tailored recommendations for reskilling and professional growth to maintain workforce relevance and competitiveness. This proactive approach not only benefits individual employees by enhancing their marketability but also secures organizational resilience in the face of rapid technological advancements and evolving economic landscapes. Furthermore, AI can identify potential skill obsolescence within the workforce, enabling organizations to proactively implement targeted training programs that address emerging needs and ensure long-term adaptability [59]. This proactive identification of future skill requirements through AI modeling enhances organizational agility and responsiveness to market changes [60]. Such AI-driven career guidance systems can also incorporate predictive analytics to anticipate disruptions in specific job roles, allowing for timely transitions into new, high-demand areas.

5.3 Integrating learning analytics into workforce planning

This integration allows organizations to strategically align learning investments with long-term business objectives, ensuring a skilled workforce capable of addressing emerging challenges and opportunities. By leveraging AI to analyze vast datasets pertaining to employee performance, learning efficacy, and market trends, organizations can proactively identify critical skill gaps and forecast future talent requirements with remarkable precision. This enables the creation of personalized learning paths that adapt to individual needs and organizational shifts, fostering a culture of continuous learning and development [7, 22]. Such sophisticated analytics allow for the optimization of training resource allocation, directly contributing to enhanced organizational performance and competitiveness in the dynamic Industry 4.0 landscape. Beyond optimizing resource allocation, AI-driven learning analytics also provide organizations with the capability to measure the return on investment of training programs, thus justifying further investments in human capital development [61]. This systematic approach to workforce planning, underpinned by HR analytics, is crucial for mitigating risks, reducing attrition rates, and building a value-added training culture within the organization [62].

5.4 The role of AI in optimizing employee performance and engagement

AI-driven HR analytics provide actionable insights into employee performance, satisfaction, and engagement, enabling organizations to implement personalized strategies that foster a productive and motivated workforce [63]. By utilizing AI to reduce employee workload and provide tailored training, organizations can significantly enhance employee engagement and overall company performance [64]. This analytical capability also allows for the identification of potential risks to employee retention and well-being, enabling proactive interventions to mitigate such issues [65, 66]. Moreover, AI can forecast future workforce needs by analyzing internal performance data and external market trends, allowing for proactive talent acquisition and development strategies [12]. This includes leveraging AI for assessing potential and identifying talent to maintain a competitive advantage [67]. The key performance indicators (KPIs) for AI-enhanced workforce strategies are listed in **Table 1.**

KPI	Definition	Example benchmark/outcome
Time-to-hire	Average days from requisition to candidate acceptance	Reduced from 45 → 20 days via AI screening
Offer acceptance rate	% of offers accepted vs. extended	Increase from 70% → 88% through personalized matching
Skill-gap closure time	Time to upskill/reskill employees to competency	Cut from 12 months → 6 months with adaptive learning
Rework/FPY	First-pass yield (FPY) or % error-free task completions	FPY improved by 15% with AI-assisted inspections
MTTR (maintenance)	Mean time to repair (MTTR) for equipment	Reduced by 30% using AR/simulator training
Incidents/safety	# of workplace incidents per year	20% reduction with predictive dashboards
Retention rate	% employees retained over 1–3 years	Improved by 10% via career-path guidance
Training cost per outcome	Cost per employee competency achieved	25% lower using AI-powered adaptive modules

Table 1.
Key performance indicators (KPIs) for AI-enhanced workforce strategies.

6. Case studies and emerging trends in AI-enabled workforce strategies

This section will delve into specific instances where AI has been successfully deployed to revolutionize human resource practices, examining both their measurable impacts and the foundational principles enabling their efficacy. These case studies will illustrate how AI optimizes recruitment, enhances training effectiveness, and supports CPD, providing concrete examples of AI's transformative potential in HR. It will also explore emerging trends such as the integration of AI with virtual and augmented reality for immersive training experiences and the use of explainable AI to ensure fairness and transparency in HR decision-making processes. The academic literature increasingly acknowledges AI as a megatrend significantly impacting human resource management, yet there remains a notable gap in comprehensive studies on the full scope of these changes [68]. Despite widespread adoption in various HR functions, the practical implementation of AI in talent management faces significant challenges, primarily related to high deployment costs for many organizations [69]. Furthermore, the ethical considerations and potential for algorithmic bias inherent in AI systems necessitate careful attention to ensure equitable and transparent HR practices [70, 71]. Predictive maintenance is very important product data management (PdM).

6.1 Illustrative industrial vignettes

To contextualize the measurable outcomes of AI-enabled workforce strategies, short role-specific vignettes are presented:

- Operators with AR work instructions

  In a manufacturing setting, operators used AR glasses overlaying real-time AI-driven assembly guidance. The outcome was a 40% reduction in time-to-competency for new hires, enabling them to achieve operational proficiency in weeks instead of months.

- Maintenance technicians using simulators
  Technicians trained with AI-powered simulators replicating complex fault scenarios. Compared with traditional manuals, simulator-trained staff achieved a 30% decrease in MTTR while also reducing secondary errors during live interventions.

- Quality inspectors in human-in-the-loop vision
  Quality inspectors paired with AI vision systems to flag potential defects, maintaining human oversight in decision-making. This human–AI collaboration yielded a 15% increase in FPY and reduced rework costs significantly.

- Supervisors with predictive dashboards
  Supervisors accessed AI-driven dashboards integrating safety, performance, and training analytics. This visibility enabled preemptive interventions, leading to a 20% decline in workplace safety incidents within a year.

6.2 Sector-specific AI adoption in workforce development

Examples include the utilization of AI in health care for automated talent matching with specialized medical roles and in manufacturing for predicting skill gaps in advanced robotics operations. Similarly, in the financial sector, AI assists in identifying and nurturing talent with expertise in algorithmic trading and cybersecurity, while in education, it helps tailor learning content to individual student needs for future workforce readiness. Furthermore, the travel industry has seen AI adoption in streamlining tasks such as booking management and customer service, allowing human staff to focus on complex problem-solving and personalized interactions. This human-centric approach ensures that AI enhances, rather than replaces, human expertise, leading to improved employee performance and customer experience. In professional services, AI-powered platforms are transforming consulting and legal fields by automating research and document analysis, freeing up professionals for higher-value strategic tasks [72].

6.3 Global best practices in AI-enhanced talent management

These practices often involve leveraging AI to foster collaborative human–machine dynamics and cultivate an engaged workforce aligned with Industry 4.0 demands [73]. For instance, leading organizations are deploying AI platforms that not only automate routine HR tasks but also provide predictive analytics for workforce planning, enabling proactive talent acquisition and retention strategies based on real-time data and market intelligence. Furthermore, ethical considerations surrounding algorithmic bias and data privacy are increasingly being integrated into global best practices, ensuring that AI-driven talent management systems are not only efficient but also equitable and transparent. This global perspective underscores the critical need for organizations to adopt robust governance frameworks and ethical guidelines when implementing AI in human capital management, ensuring compliance with diverse international regulations and cultural norms. This holistic approach ensures sustainable service development and enhances organizational resilience by fostering a dynamic human–AI collaboration [74]. The careful calibration of AI's role alongside human discretion is crucial for optimizing organizational decision-making and minimizing potential disruptions from technological updates [75].

6.4 Innovations shaping the future of AI in human resources

This includes advancements in explainable AI for fairer hiring practices and sophisticated natural language processing models for nuanced sentiment analysis in employee feedback [76]. These innovations highlight the ongoing evolution of AI from a tool for process automation to a strategic partner in fostering organizational growth and employee well-being. Moreover, the integration of AI is increasingly transforming decision-making processes in HR, especially concerning employee involvement and ethical challenges [71]. This evolution necessitates a critical examination of how AI influences human perceptions and behaviors within the workplace, particularly concerning trust, fairness, and the acceptance of AI-driven assessments in personnel selection [77]. Furthermore, addressing these ethical challenges requires interdisciplinary collaboration between I–O psychologists, computer scientists, and legal scholars to develop and implement fair and unbiased AI applications in organizational contexts [78]. Such collaborations are essential to ensure that AI technologies enhance human capabilities and decision-making, rather than replacing them, while upholding principles of transparency and accountability in the employment life cycle [71, 76].

7. Navigating challenges: Ethics, bias, and data privacy in AI applications

The rapid integration of AI into human resource management necessitates a thorough examination of the inherent challenges, particularly regarding ethical considerations, algorithmic bias, and the imperative of data privacy [14, 79]. These challenges underscore the need for robust regulatory frameworks and ethical guidelines to ensure that AI applications in HR are implemented responsibly and equitably [80, 81]. Addressing these concerns is paramount to fostering trust in AI systems and maximizing their beneficial impact on workforce development and talent management [82]. Specifically, the development and deployment of AI in HR must account for potential human rights impacts, ensuring that these technologies do not inadvertently conflict with human autonomy or perpetuate discrimination [83]. This calls for a proactive approach in designing AI algorithms that prioritize fairness, transparency, and accountability, thereby mitigating the risks associated with biased outputs and ensuring equitable opportunities for all individuals [78, 84]. Emerging literature emphasizes systematic frameworks for fairness and bias mitigation in AI hiring [85].

7.1 Identifying and mitigating algorithmic bias

Algorithmic bias, which can perpetuate or even amplify existing societal prejudices, is a critical concern that requires robust methodologies for identification and mitigation to ensure equitable outcomes in AI-driven HR processes [86]. This involves rigorous data validation, the use of diverse training datasets, and continuous auditing of AI models to detect and correct discriminatory patterns [87]. Furthermore, the development of explainable AI techniques is crucial for understanding the decision-making processes of algorithms, thereby enhancing transparency and accountability in HR applications [88, 89]. This is particularly important in areas such as recruitment and performance evaluation, where biased algorithms can lead to unfair or discriminatory outcomes against certain demographic groups [90]. The opacity of many machine learning algorithms, often referred to as "black box" models, exacerbates this issue by making it difficult to ascertain the exact reasons for a particular output, thus complicating bias detection and remediation [13]. Consequently, the legal implications of algorithmic bias are substantial, as existing anti-discrimination laws may apply to AI-driven decisions, necessitating compliance measures and a thorough understanding of how AI systems can inadvertently produce discriminatory results [66].

7.2 Ensuring transparency and accountability in AI systems

This involves designing AI models with interpretable decision-making processes and establishing clear frameworks for oversight and redress, thereby building trust and ensuring responsible AI deployment in HR [71]. This also necessitates the development of clear accountability mechanisms, defining who is responsible when AI systems produce unintended or biased outcomes. Such mechanisms must encompass the entire life cycle of AI deployment, from design and training to continuous monitoring and iterative refinement [91]. Furthermore, continuous monitoring and auditing of AI systems are crucial to identifying and reducing biases, thereby ensuring that AI algorithms align with ethical principles throughout their operational lifespan [50, 92]. Moreover, the ethical imperative extends to incorporating diverse and inclusive teams in the development and validation processes of AI systems to minimize the potential for biased outcomes and foster a comprehensive array of viewpoints [83]. This includes adopting methodologies such as discrimination-aware machine learning, which quantifies nondiscrimination regulations into machine-interpretable constraints to systematically remove bias from potentially skewed datasets [14].

7.3 Balancing data utility with privacy protection

This delicate equilibrium necessitates the implementation of stringent data governance policies, anonymization techniques, and secure data architectures to safeguard sensitive employee information while still harnessing its potential for insightful analytics and improved HR strategies [93]. Moreover, compliance with evolving regulatory frameworks, such as GDPR and CCPA, is paramount to prevent legal repercussions and maintain organizational integrity [94]. The inherent opacity of some AI systems also raises concerns regarding human control and ethical decision-making, necessitating a deeper understanding of the ethical implications of AI in HRM [14, 70]. A primary challenge resides in ensuring transparency and accountability within AI decision-making processes, as the "black box" nature of many AI systems impedes understanding and explaining their outputs, potentially undermining trust and compliance [95]. Addressing these ethical dilemmas requires a multifaceted approach, encompassing robust technical solutions for explainability, clear ethical guidelines for AI development, and comprehensive regulatory frameworks to govern AI's application in sensitive domains such as HR [96–99]. Given that AI often operates as a sociotechnical system, its fairness and ethical alignment are intrinsically linked to the conditions under which it is designed, developed, and deployed, necessitating a comprehensive understanding of both computational and human factors.

8. Strategic pathways for aligning human capital with the digital economy

This section outlines actionable strategies for organizations, educators, and policymakers to bridge the talent gap and foster a workforce capable of thriving in an increasingly AI-driven landscape.

This involves developing comprehensive frameworks that integrate human considerations into AI design and deployment within human resource management, thereby fostering trust and ensuring the ethical alignment of technological advancements with human capital strategies [76, 84]. The integration of AI into HR functions also necessitates a reevaluation of traditional HR paradigms, emphasizing the importance of employee involvement in AI-driven HR decision-making processes [71]. A critical aspect of this reevaluation involves anticipating future trends and challenges at an organizational level through well-thought-out strategic management approaches [100]. This anticipatory approach should include proactive investment in upskilling and reskilling initiatives, aligning workforce capabilities with emerging technological demands to maintain a competitive advantage [75]. Furthermore, collaboration between industry, academia, and government is essential to develop robust educational pipelines that equip individuals with the requisite digital competencies and analytical skills for the future workforce [15, 97].

8.1 Policy recommendations for AI-driven workforce ecosystems

These recommendations should prioritize the development of ethical guidelines for AI adoption, investment in digital infrastructure, and the creation of public–private partnerships to facilitate skill development and talent mobility. Furthermore, fostering collaboration between academia and industry can ensure that educational curricula are continually updated to meet the evolving demands of AI-driven roles, thereby preparing a future-ready workforce [76]. Additionally, policy frameworks must address the implications of AI on employment patterns, promoting reskilling initiatives and social safety nets to support workers through transitional periods. Moreover, national AI strategies often include components addressing the societal, economic, and ethical implications of AI, alongside recommendations for investing in AI workforce development and research [101]. Such policies should also consider the potential for AI to create new job categories and enhance human capabilities rather than solely focusing on job displacement [11, 16].

8.2 Collaborations between industry, academia, and government

Such collaborations are crucial for establishing comprehensive training programs, funding research into ethical AI applications, and formulating policies that support a resilient and adaptable workforce capable of navigating the complexities of the digital economy. These multistakeholder partnerships are essential for developing robust talent management strategies that leverage AI's potential while mitigating associated risks, ultimately contributing to sustained economic growth and societal well-being [67]. These strategic alignments between various stakeholders are indispensable for shaping an AI ecosystem that is both innovative and ethically sound, ensuring that the benefits of technological advancement are equitably distributed across the workforce [102]. A critical component of this collaboration involves establishing shared platforms for knowledge exchange and resource pooling, enabling the development of standardized best practices for AI integration in human capital management [103]. These concerted efforts can foster a dynamic learning environment where continuous upskilling and reskilling are the norm, preparing the workforce for evolving job roles and technological advancements [7, 104].

8.3 Future skills roadmaps for sustainable competitiveness

These roadmaps must anticipate future demands by identifying emerging skill sets, such as advanced data literacy and human–AI collaboration, and integrate them into continuous learning pathways to ensure long-term workforce relevance and economic vitality. This proactive approach ensures that educational institutions and professional development programs can adapt swiftly to technological shifts, preparing individuals for roles that may not yet exist but will be critical in the evolving digital landscape [5, 75]. To fully capitalize on the opportunities presented by AI, companies must actively develop comprehensive AI strategies that are both human-centric and ethically responsible, fostering innovation and positive feedback [103]. This involves proactively identifying and nurturing critical skills such as critical thinking, complex problem-solving, and adaptive learning, which are increasingly vital in an AI-augmented work environment [105]. Furthermore, nurturing socioemotional intelligence and creativity becomes paramount as AI handles routine tasks, demanding a workforce capable of nuanced human interaction and innovative problem-solving [4]. Additionally, fostering a culture of lifelong learning is imperative to ensure that individuals remain adaptable and proficient amid rapid technological advancements, enabling them to continuously acquire new competencies [106].

8.4 Resilience and change-management pathways

Successful AI adoption in workforce development requires a structured change-management approach:

- Adoption pathway: Organizations should follow a phased approach:

    1. Sponsor – Executive leadership provides strategic commitment and resources.

    2. Pilot – deploy AI tools in controlled, low-risk environments to demonstrate value.

    3. Scale – expand proven solutions organization-wide, integrating feedback into system refinement.

- Stakeholder roles:
  - o Executives – ensure funding and policy alignment.
  - o Managers – champion local adoption and provide feedback.
  - o Employees – participate actively in training and evaluation loops.
- Trust and acceptance measures: Transparency in AI decision-making (e.g., explainable AI in candidate screening), coupled with continuous feedback mechanisms, enhances user confidence and reduces resistance.
- Monitoring indicators: Adoption rates, training completion, employee satisfaction, ethical compliance audits, and performance outcomes (e.g., FPY, MTTR, retention) serve as leading metrics of resilience.

9. Conclusion – Forging the future workforce: A symbiotic path forward with AI

The transformative journey detailed throughout this chapter underscores AI's indispensable role in redefining workforce training and talent acquisition within Industry 4.0. We have observed how AI transcends conventional methods by enabling highly personalized and adaptive learning pathways, optimizing recruitment through intelligent automation and predictive analytics, and fostering CPD fueled by data-driven insights. From algorithmic precision in skill mapping to real-time performance feedback and AI-guided career progression, the integration of AI demonstrably enhances organizational efficiency and individual competency. Crucially, this evolution is not merely about technological adoption but about cultivating a symbiotic relationship between advanced AI systems and human ingenuity, where AI augments human capabilities, allowing for greater focus on complex problem-solving, creativity, and interpersonal interactions.

However, the path forward is not without its complexities. The imperative to address ethical considerations, mitigate algorithmic bias, and ensure robust data privacy remains paramount for responsible AI deployment. These challenges necessitate transparent AI systems, clear accountability frameworks, and a delicate balance between data utility and individual protection. Ultimately, aligning human capital with the demands of the digital economy requires strategic, multistakeholder collaboration among industry, academia, and government. By developing forward-looking policy recommendations, fostering cross-sector partnerships, and crafting dynamic future skills roadmaps, organizations can proactively bridge skill gaps and cultivate a resilient, adaptable, and ethically empowered workforce. As we move deeper into the digital age, embracing AI as a strategic partner, while diligently managing its implications, will be fundamental to achieving sustainable competitiveness and fostering a thriving human–AI ecosystem.

Author details

Mohammed Gronfula* and Mohammed Alshammari
Alasala Colleges, Dammam, Saudi Arabia

*Address all correspondence to: mohammed.gronfula@alasala.edu.sa

References

[1] Du J. AI and your job what's changing and what's next. Frontiers in Science and Engineering. 2024;**4**(7):116–124

[2] Tabbassum A, et al. The impact of AI on future employment patterns. International Journal of Global Innovations and Solutions (IJGIS). 2024:**1**(1):9

[3] Patil D. Impact of Artificial Intelligence on Employment and Workforce Development: Risks, Opportunities, and Socioeconomic Implications. Amsterdam, Netherlands: Elsevier BV; 2024

[4] Tomaskinova J, et al. Shaping The Future: Navigating the AI era with an innovative training model, EDULEARN Proceedings. 2024, IATED. p. 6295–6303

[5] Ozer M, et al. Adapting to the AI disruption: Reshaping the IT landscape and educational paradigms. In: Communications in Computer and Information Science. Cham, Switzerland: Springer Nature Switzerland; 2025. p. 366–374

[6] Macpherson RJS. Moral Agency in the Age of AI: A Meta-Ethical Framework for Educative Leadership. Wilmington, DE: CMJ Publishers; 2025

[7] Morandini S, et al. The impact of artificial intelligence on workers' skills: Upskilling and reskilling in organisations. Informing science. The International Journal of an Emerging Transdiscipline. 2023;**26**:039–068

[8] Pavashe AS, et al. The impact of artificial intelligence on employment and workforce trends in the post-pandemic era. International Journal for Research in Applied Science and Engineering Technology 2023;**11**:154–157

[9] Trenerry B, et al. Preparing workplaces for digital transformation: An integrative review and framework of multi-level factors. Frontiers in Psychology. 2021;**12**:12

[10] Lozie DR, et al. Examining the impact of generative artificial intelligence on work dynamics. Human Resources Management and Services. 2024;**6**:3420

[11] Bobitan N, et al. Shaping tomorrow: Anticipating skills requirements based on the integration of artificial intelligence in business organizations—A foresight analysis using the scenario method. Electronics. 2024;**13**:2198

[12] Meharunisa S, et al. The impact of artificial intelligence on women's empowerment, and work-life balance in Saudi educational institutions. Frontiers in Psychology. 2024;**15**:1384471. DOI: 10.3389/fpsyg.2024.1384471

[13] Cheong I, Huh YE, Puntoni S. Lay beliefs about AI assessment of interpersonal skills in personnel selection. Scientific Reports. 2025;**15**(1):25317. DOI: 10.1038/s41598-025-10358-7

[14] Nyathani R. AI-Powered recruitment the future of HR digital transformation. Journal of Artificial Intelligence & Cloud Computing. 2022;**1**(4):1–5. Available from: https://www.onlinescientificresearch.com/articles/ai powered-recruitment-the-future-of-hr-digital-transformation.pdf

[15] Purwaamijaya BM, Prasetyo Y. The effect of artificial intelligence (AI) on human capital management in Indonesia. Jurnal Manajemen Dan Kewirausahaan. 2022;**10**(2):168–174

[16] Ibrahim F, et al. The Technology Acceptance Model and Adopter Type Analysis in the Context of Artificial Intelligence. Lausanne, Switzerland: Frontiers Media SA; 2024. DOI: 10.3389/frai.2024.1496518

[17] Ei-Aini HA, Mohamed K, Mohammed YH. Effect of mold types and cooling rate on mechanical properties of Al alloy 6061 within ceramic additives. The 2nd International Conference on Energy Engineering ICEE-2. 2010

[18] Abdellah MY, et al. Mechanical, thermal, and acoustic properties of natural fibre-reinforced polyester. Proceedings of the Institution of Mechanical Engineers, Part E: Journal of Process Mechanical Engineering 2024;**238**:1436–1448

[19] Fernández-Martínez C, Fernández A. AI and recruiting software: Ethical and legal implications. Paladyn. Journal of Behavioral Robotics. 2020;**11**(1):199–216

[20] Akinnagbe OB. Human-AI collaboration: Enhancing productivity and decision-making. International Journal of Education, Management, and Technology. 2024;**2**(3):387–417

[21] Tariq MU. The role of AI in skilling, upskilling, and reskilling the workforce. In: Advances in Educational Technologies and Instructional Design (AETID) Book Series, Hershey, PA: IGI Global; 2024. p. 408–420

[22] Bayly-Castaneda K, Ramirez-Montoya MS, Morita-Alexander A, Crafting personalized learning paths with AI for lifelong learning: A systematic literature review. Frontiers in Education, 2024. **9**

[23] Jian MJKO. Personalized learning through AI. Advances in Engineering Innovation. 2023;**5**(1):16–19

[24] Ward B, et al. Analyzing the impact of AI tools on student study habits and academic performance, 2025 IEEE 15th Annual Computing and Communication Workshop and Conference (CCWC). 2025, IEEE. p. 00434–00440

[25] Jose B, et al. The cognitive paradox of AI in education: Between enhancement and erosion. Frontiers in Psychology. 2025;**16**:1550621. DOI: 10.3389/fpsyg.2025.1550621

[26] Kamalov F, Calonge DS, Gurrib I. New era of artificial intelligence in education: Towards a sustainable multifaceted revolution. Sustainability. 2023;**15**(16):12451. DOI: 10.3390/su151612451

[27] Choung H, David P, Ross A. Trust in AI and its role in the acceptance of AI technologies. International Journal of Human–Computer Interaction. 2023;**39**(9):1727–1739

[28] Mahmoud CF, Sørensen JT. Artificial intelligence in personalized learning with a focus on current developments and future prospects. Research and Advances in Education. 2024;**3**(8):25–31

[29] Tzouveli P, Mylonas P, Kollias S. An intelligent e-learning system based on learner profiling and learning resources adaptation.

Computers & Education. 2008;**51**(1):224–238

[30] Dabingaya M. Analyzing the effectiveness of AI-Powered adaptive learning platforms in mathematics education. Interdisciplinary Journal Papier Human Review. 2022;**3**(1):1–7

[31] Ramachandran KK, et al. Developing AI-powered training programs for employee upskilling and reskilling. Journal of Informatics Education and Research. 2024;**4**(2):1186. Available from: http://jier.org/index.php/journal/article/view/903

[32] Maghsudi S, et al. Personalized education in the artificial intelligence era: What to expect next. IEEE Signal Processing Magazine. 2021;**38**:37–50

[33] Troussas C, et al. A fuzzy-neural model for personalized learning recommendations grounded in experiential learning theory. Information. 2025;**16**:339

[34] Abdellah MY, et al. Stress distribution and fracture toughness of underground reinforced plastic pipe composite. Polymers. 2021;**13**:2194

[35] Hashim S, et al. Trends on technologies and artificial intelligence in education for personalized learning: systematic literature review. International Journal of Academic Research in Progressive Education and Development. 2022;**11**(1):884–903. DOI: 10.6007/IJARPED/V11-I1/12230

[36] Opesemowo OAG, Adewuyi HO. A systematic review of artificial intelligence in mathematics education: The emergence of 4IR. Eurasia Journal of Mathematics, Science and Technology Education. 2024;**20**(7):em2478

[37] Abdellah MY, et al. Experimental evaluation of mechanical and tribological properties of segregated Al-Mg-Si alloy filled with alumina and silicon carbide through different types of casting molds. Metals. 2023;**13**:316

[38] GESR Editorial Board. Global Educational Studies Review. United Kingdom: Humanity Publications (HumaPub); 2020

[39] Saputra I, et al. Integration of artificial intelligence in education: Opportunities, challenges, threats and obstacles. A Literature Review. The Indonesian Journal of Computer Science. 2023;**12**(4):1590–1605. Available from: http://ijcs.net/ijcs/index.php/ijcs/article/view/3266

[40] Ouakili OE. The impact of artificial intelligence (AI) on recruitment process. Open Journal of Business and Management. 2025;**13**(02):749–762

[41] Karakuş N, Gedik K, Kazazoğlu S. Ethical decision-making in education: A comparative study of teachers and artificial intelligence in ethical dilemmas. Behavioral Sciences. 2025;**15**(4):469

[42] Yu X, Fan L, Yu Y. Artificial intelligence and corporate ESG performance: A mechanism analysis based on corporate efficiency and external environment. Sustainability. 2025;**17**(9):3819

[43] Wilson K, Caliskan A. Gender, race, and intersectional bias in resume screening via language model retrieval. Proceedings of the AAAI/ACM Conference on AI, Ethics, and Society, 2024. 7: p. 1578–1590

[44] Tahir M, Hassan FD, Shagoo MR. Role of artificial intelligence in education: A conceptual review. World

Journal of Advanced Research and Reviews. 2024;**22**(1):1469–1475

[45] Tornero-Costa R, et al. Methodological and quality flaws in the use of artificial intelligence in mental health research: Systematic review. JMIR Mental Health. 2023;**10**:e42045. DOI: 10.2196/42045

[46] Chen Z. Ethics and discrimination in artificial intelligence-enabled recruitment practices. Humanities and Social Sciences Communications. 2023;**10**(1):567. DOI: 10.1057/s41599-023-02079-x

[47] Boussioux L, et al. The crowdless future? Generative AI and creative problem-solving. Organization Science. 2024;**35**:1589–1607

[48] Albassam WA. The power of artificial intelligence in recruitment: An analytical review of current AI-based recruitment strategies. International Journal of Professional Business Review. 2023;**8**(6):e02089

[49] Upadhyay AK, Khandelwal K. Applying artificial intelligence: Implications for recruitment. Strategic HR Review. 2018;**17**(5):255–258

[50] Kondra S, et al. AI and diversity, equity, and inclusion (DEI): Examining the potential for AI to mitigate bias and promote inclusive communication. Journal of Artificial Intelligence and Machine Learning. 2025;**3**(1)

[51] Silva G, Godwin G, Jayanagara O. The impact of AI on personalized learning and educational analytics. International Transactions on Education Technology (ITEE). 2024;**3**(1):36–46

[52] Chen Z. Collaboration among recruiters and artificial intelligence: Removing human prejudices in employment. Cognition, Technology & Work. 2022;**25**(1):135–149

[53] Vogler A, et al. Towards a QBLM-Based qualification-management methodology supporting human-resource management and development. Information. 2024;**15**:600

[54] Oman N, Siddiqua N, Noorain N. Artificial Intelligence and its ability to reduce recruitment bias. World Journal of Advanced Research and Reviews. 2024;**24**(1):551–564

[55] Meshram R. The role of artificial intelligence (AI) in recruitment and selection of employees in the organisation. Russian Law Journal. 2023;**11**(9s):1628–1653. DOI: 10.5840/ruslj20231195

[56] Reena DR, K MMA, Jayakrishnan SS. Emerging trends in human resource analytics in upcoming decade. International Journal of Engineering Applied Sciences and Technology. 2019;**04**(08):260–264

[57] Guo S, Halim HBA, Saad MRBM. Leveraging AI-enabled mobile learning platforms to enhance the effectiveness of English teaching in universities. Scientific Reports. 2025;**15**(1):15873. DOI: 10.1038/s41598-025-00801-0

[58] Egunjobi D, Adeyeye OJ. Revolutionizing learning: The impact of augmented reality (AR) and artificial intelligence (AI) on education. International Journal of Research Publication and Reviews. 2024;**5**(10):1157–1170

[59] Ruel H, Njoku E. AI redefining the hospitality industry. Journal of Tourism Futures. 2020;7(1):53–66

[60] Shafiabady N, et al. Using Artificial Intelligence (AI) to predict organizational agility. PLOS ONE. 2023;**18**:e0283066

[61] AlGosaibi AA, et al. Developing an intelligent framework for improving the quality of service in the government organizations in the kingdom of Saudi Arabia. International Journal of Advanced Computer Science and Applications. 2020;**11**(12)

[62] Momin WYM, Mishra K. HR analytics as a strategic workforce planning. International Journal of Applied Research. 2015;**1**(4):258–260

[63] Ajiga D, Ayanponle L, Okatta C. AI-powered HR analytics: Transforming workforce optimization and decision-making. International Journal of Science and Research Archive. 2022;**5**(2):338–346

[64] Rožman M, Oreški D, Tominc P. Artificial-Intelligence-Supported reduction of employees' workload to increase the company's performance in today's VUCA environment. Sustainability. 2023;**15**(6):5019

[65] Wijayati DT, et al. A study of artificial intelligence on employee performance and work engagement: The moderating role of change leadership. International Journal of Manpower. 2022;**43**:486–512

[66] Kim PT. Addressing algorithmic discrimination. Communications of the ACM. 2021;**65**(1):25–27

[67] França TJF, et al. Artificial intelligence applied to potential assessment and talent identification in an organisational context. Heliyon. 2023;**9**(4):e14694. DOI: 10.1016/j.heliyon.2023.e14694

[68] Dima J, et al. The effects of artificial intelligence on human resource activities and the roles of the human resource triad: Opportunities and challenges. Frontiers in Psychology. 2024;**15**:1360401. DOI: 10.3389/fpsyg.2024.1360401

[69] Kalra I. Artificial intelligence and its role in human resource management. International Journal for Modern Trends in Science and Technology. 2020;**10**(8):310–317

[70] Rodgers W, et al. An artificial intelligence algorithmic approach to ethical decision-making in human resource management processes. Human Resource Management Review. 2023;**33**(1):100925. DOI: 10.1016/j.hrmr.2022.100925

[71] Taslim WS, Rosnani T, Fauzan R. Employee involvement in AI-driven HR decision-making: A systematic review. SA Journal of Human Resource Management. 2025;**23**:a2856. DOI: 10.4102/sajhrm.v23i0.2856

[72] Frenette J, The human-centric approach to AI in the travel industry. 2022, White paper

[73] Krishnan L, AI structuring work practices and fuelling employee outcomes-manufacturing industry. 2024

[74] Xu C, Cho S-E. Factors affecting human–AI collaboration performances in financial sector: Sustainable service development perspective. Sustainability. 2025;**17**(10):4335

[75] Zahidi F, Kaluvilla BB, Mulla T. Embracing the new era: Artificial

intelligence and its multifaceted impact on the hospitality industry. Journal of Open Innovation: Technology, Market, and Complexity. 2024;**10**(4):100184. DOI: 10.1016/j.joitmc.2024.100184

[76] Fenwick A, Molnar G, Frangos P. Revisiting the role of HR in the age of AI: Bringing humans and machines closer together in the workplace. Frontiers in Artificial Intelligence. 2024;**6**:1272823. DOI: 10.3389/frai.2023.1272823

[77] Choung H, David P, Ross A. Trust in AI and its role in the acceptance of AI technologies. International Journal of Human–Computer Interaction. 2022;**39**(9):1727–1739

[78] Gonzalez M, et al. "Where's the I-O?" Artificial intelligence and machine learning in talent management systems. Personnel Assessment and Decisions. 2019;**5**(3):5. DOI: 10.2501/pad.v5i3.5

[79] Nyberg AJ, et al. A brave new world of human resources research: Navigating perils and identifying grand challenges of the GenAI revolution. Journal of Management 2025;**51**: 2677–2718

[80] Fabris A, et al. Fairness and bias in algorithmic hiring: A multidisciplinary survey. ACM Transactions on Intelligent Systems and Technology. 2025;**16**:1–54

[81] Gupta P, Fernandes SF, Jain M. Automation in recruitment: A new frontier. Journal of Information Technology Teaching Cases. 2018;**8**(2):118–125

[82] Lepri B, Oliver N, Pentland A. Ethical machines: The human-centric use of artificial intelligence. iScience. 2021;**24**(3):102209. DOI: 10.1016/j.isci.2021.102209

[83] Kriebitz A, Lütge C. Artificial intelligence and human rights: A business ethical assessment. Business and Human Rights Journal. 2020;**5**(1):84–104

[84] Prasad DBV. The impact of technology on human resource management: Trends and challenges. Educational Administration: Theory and Practice. 2024;**30**(5):9746–9752. DOI: 10.51576/kuey.30.5.4635

[85] Fabris A, et al. Fairness and bias in algorithmic hiring: A multidisciplinary survey. ACM Transactions on Intelligent Systems and Technology. 2025;**16**:Article16

[86] Madanchian M. From recruitment to retention: AI tools for human resource decision-making. Applied Sciences. 2024;**14**(24):11750. DOI: 10.3390/app142411750

[87] Shams RA, Zowghi D, Bano M. AI for all: Identifying AI incidents related to diversity and inclusion. Journal of Artificial Intelligence Research. 2025;**83**:1–28. DOI: 10.1613/jair.1.15350

[88] Machucho R, Ortiz D. The impacts of artificial intelligence on business innovation: A comprehensive review of applications, organizational challenges, and ethical considerations. Systems. 2025;**13**(4):264

[89] Livieri G, et al. Ethical issues arising from the use of AI in drug discovery. Journal of Politics and Ethics in New Technologies and AI. 2024;**3**:e37093

[90] Osasona F, et al. Reviewing the ethical implications of AI in decision making processes. International Journal

of Management & Entrepreneurship Research. 2024;**6**:322–335

[91] Dhopte A, Bagde H. Smart smile: Revolutionizing dentistry with artificial intelligence. Cureus. 2023;**15**(6): e41227. DOI: 10.7759/cureus.41227

[92] Kandasamy UC. Ethical leadership in the age of AI challenges, opportunities and framework for ethical leadership. arXiv. 2024;arXiv 2410.18095. preprint

[93] Sucipto H. The impact of artificial intelligence (AI) on human resource management practices. Management Studies and Business Journal (PRODUCTIVITY). 2024;**1**(1):138–145

[94] Li S, Li K, Lu H. National origin discrimination in deep-learning-powered automated resume screening. arXiv. 2023;arXiv:2307.08624. preprint

[95] Adelakun BO, Majekodunmi TG, Akintoye OS. AI and ethical accounting: Navigating challenges and opportunities. International Journal of Advanced Economics. 2024;**6** (6):224–241

[96] Abujaber AA, Nashwan AJ. Ethical framework for artificial intelligence in healthcare research: A path to integrity. World Journal of Methodology. 2024;**14** (3):94071. DOI: 10.5662/wjm.v14.i3.94071

[97] Papagiannidis E, et al. Toward AI governance: Identifying best practices and potential barriers and outcomes. Information Systems Frontiers. 2022;**25**:123–141

[98] Weiner EB, et al. Ethical challenges and evolving strategies in the integration of artificial intelligence into clinical practice. PLOS Digital Health. 2025;**4**:e0000810

[99] Gerlich M. Public anxieties about AI: Implications for corporate strategy and societal impact. Administrative Sciences. 2024;**14**(11):288

[100] Artene AE, Domil AE, Ivascu L. Unlocking business value: Integrating AI-driven decision-making in financial reporting systems. Electronics. 2024;**13** (15):3069

[101] Tobin S, et al. A brief historical overview of artificial intelligence research. Information Services and Use. 2019;**39**:291–296

[102] Hernández EG. Towards an ethical and inclusive implementation of artificial intelligence in organizations: A multidimensional framework. arXiv. 2024;arXiv 2405.01697. preprint

[103] Shabbir J, Anwer T. Artificial intelligence and its role in near future. arXiv. 2018;arXiv:1804.01396. preprint

[104] Rosales MA, et al. Artificial intelligence: The technology adoption and impact in the Philippines, 2020 IEEE 12th International Conference on Humanoid, Nanotechnology, Information Technology, Communication and Control, Environment, and Management (HNICEM). 2020, IEEE. p. 1–6

[105] Dumitru D, Halpern DF. Critical thinking: Creating job-proof skills for the future of work. Journal of Intelligence. 2023;**11**(10):194

[106] Romero M. Lifelong learning challenges in the era of artificial intelligence: A computational thinking perspective. arXiv. 2024; arXiv:2405.19837. preprint

Chapter 5

Knowledge Management in Production: Human-Centered and AI-Supported Approaches for the Shop Floor

Jannis Pfister, Daniela Schmidt, Matthias Jenner, Till Günther and Rainer Müller

Abstract

Knowledge management is a critical enabler for competitiveness in industrial production, especially under demographic change, workforce fluctuation, and digital transformation. Small- and medium-sized enterprises face particular risks, as experiential knowledge is often bound to individuals and easily lost when employees leave. This paper emphasizes the need for systematic approaches to knowledge retention and transfer. While established frameworks such as the knowledge ladder, SECI model, and knowledge balance provide foundations, they remain limited in capturing implicit, action-related expertise essential for production. To address this gap, a model-based approach is proposed that combines cognitive task analysis, methods-time measurement, and AI-assisted tracking technologies, including motion sensors and video analysis, to externalize implicit knowledge. The transformation of human actions and decision-making into structured knowledge objects enables integration into digital assistance and learning systems. The study outlines a research agenda for scalable, technology-enhanced knowledge management systems to safeguard experiential knowledge, reduce training times, and strengthen innovation.

Keywords: knowledge management for SME, human-centered implementation strategies, implicit knowledge, knowledge retention, experiential knowledge, digital assistance systems, best practices, knowledge capture, knowledge preservation, knowledge externalization, artificial intelligence

1. Introduction

In global industry, companies face the challenge of systematically securing experiential knowledge and ensuring its sustainable transfer. Small- and medium-sized enterprises[1] (SMEs) are especially affected, as this knowledge is often bound to individuals and quickly lost when employees leave. In a knowledge-based economy, such loss reduces quality, prolongs training, increases error rates, and weakens innovation capacity. In this chapter, the focus is on experiential knowledge, which in the literature is commonly referred to as implicit knowledge. Both terms are used synonymously, describing practice-based expertise acquired through long-term experience, routines, and situational decision-making.

OECD data show a steady decline in average job tenure. In Germany, it is around 11 years [1]; in the United States, only 4.1 years [2]. In Chinese industrial centers such as Shenzhen and Guangzhou, employees often remain with a company for less than five years, highlighting high labor market mobility [3, 4]. Even moderate attrition on the shop floor can accumulate to significant knowledge loss over time. At the same time, episodes of unusually high turnover demonstrate how disruptive such dynamics can become. A prominent example is the so-called Great Resignation in 2021, when more than 47 million employees voluntarily left their jobs in the United States [5]. In German manufacturing, turnover is lower and more stable. IAB data indicate exporting firms show 2–4% points less turnover than nonexporting ones. Even moderate attrition on the shop floor can erode implicit expertise when knowledge remains undocumented [6].

Younger employees often leave before their knowledge is transferred, while older employees tend to stay longer but keep much of their expertise within established routines. Without deliberate structures for internal exchange, experiential knowledge remains personal rather than organizational. Companies must, therefore, create systematic mechanisms for knowledge sharing and handover.

The economic effects are well documented. A ZEW study shows that a 1% increase in turnover leads to a 0.074% decline in value added. Losing one-third of the workforce results in an average productivity loss of €22,000 per employee, not fully offset even after two years [7]. Replacing a single employee can cost up to €45,529, including lost productivity, training, and recruitment. For an SME with 30 employees and 20% annual turnover, this equates to more than €314,000 annually [8]. According to Wharton, replacing experienced skilled workers costs 100–150% of their salary [9]. The loss is especially harmful when no systematic knowledge management exists, particularly in specialized tasks or roles with customer relationships [10].

When experienced employees leave, training times increase, quality decreases, error rates rise, and project delays occur. Barriers to knowledge transfer include time constraints, knowledge gaps, biases, and missing incentives. Digital tools remain ineffective if not integrated into processes and culture. Existing approaches, such as structured handovers, tandem models, or knowledge maps, illustrate that methods to mitigate knowledge loss are available, yet they are often insufficiently applied or integrated [11].

Compared to large firms, SMEs face particular challenges in digitalization and knowledge management. They are characterized by flat hierarchies, limited resources, and niche expertise [12]. A major barrier to formal structures is the lack of

[1] The EU defines SMEs as companies with fewer than 250 employees and either revenue below €50 million or a balance sheet below €43 million [81, 82]. SMEs account for nine out of 10 companies and create two out of three new jobs [83]. They drive entrepreneurship and innovation and are thus vital to competitiveness.

IT expertise [13]. Specialists are scarce or expensive, preventing sustainable internalization. This leads to inconsistent data and weak database infrastructures. Limited resources and low Research and Development capacity further restrict digitalization and standardization [12–14].

Skills' shortages and expert departures are especially critical for SMEs, as knowledge is highly person-specific [14]. Employees often perform cross-functional tasks requiring broad project knowledge. Expertise is rarely documented; sometimes even process knowledge remains implicit. In many cases, critical know-how is concentrated in a single individual, creating immediate knowledge gaps when that person leaves [15]. Beyond the technical challenge, building a knowledge base requires cultural and organizational change. SMEs' strong face-to-face culture accelerates information flow but reinforces reliance on implicit knowledge. Employees often hesitate to share knowledge if they see little benefit [14].

Modern technologies can help retain knowledge. AI-based tools, semantic models [16], and eye-tracking methods [17] are promising. IT-based measures, such as digital knowledge management systems, mitigate the risk of knowledge loss [18].

With rising process complexity, shorter innovation cycles, and global competition, employee expertise becomes strategically critical. SMEs, in particular, risk losses in quality, efficiency, and costs without knowledge structures. This chapter, therefore, systematically examines the causes, consequences, and countermeasures to experiential knowledge loss. At the core are the externalization, structuring, and dissemination of knowledge across organizations. Technological solutions, digital platforms, and cultural frameworks are discussed. A human-centered perspective emphasizes employees as active knowledge carriers and codesigners of adaptive structures. The goal is to establish pragmatic and cost-effective methods for capturing and applying experiential knowledge.

2. State of the art

This chapter establishes the conceptual and theoretical foundations for a systematic approach to knowledge management in an industrial context. It clarifies the distinctions between data, information, and knowledge, outlines various forms of knowledge, and discusses theoretical concepts relevant to the collection, storage, and utilization of experiential knowledge. These explanations provide the basis for the subsequent analysis of technological, organizational, and cultural design approaches.

2.1 Theoretical and normative foundations

In modern organizations, knowledge is considered a strategic resource and a core element of intangible corporate capital. It is central to innovation, quality assurance, and continuous improvement. In knowledge-intensive production environments, performance depends on the availability, accessibility, and usability of relevant knowledge. Companies can thus be seen as decision-making factories in which information and knowledge are central production factors [19].

A frequently used classification model is the knowledge pyramid, which describes the gradual transition from data to information to knowledge. Data are context-free symbols; information emerges when data are structured and assigned meaning;

knowledge results from context-related interpretation, evaluation, and application [20–22]. North and Maier emphasize that knowledge arises from the active processing of information, manifested in cognitive processes such as thinking, decision-making, and action. These processes are always embedded in individual or social contexts [23].

The pyramid can also be interpreted as a transformation process: Knowledge forms the basis for skills, which are developed through repeated application and experience. The competent application of these skills in specific situations leads to proficiency. Probst et al. argue that value creation depends on translating knowledge into concrete action [24]. According to DIN ISO 30401 and 10015, a distinction is made between knowledge, skill, ability, and competence as different levels of application-oriented utilization. DIN ISO 30401 defines a skill as the learned ability to perform a task according to specified requirements [25]. Competence is the capacity to apply knowledge and skills to achieve desired outcomes. DIN ISO 10015 introduces ability as a prerequisite in cognitive, social, or motor terms, which can be developed through training and experience.

From a theoretical and normative perspective, knowledge is, therefore, understood as an overarching concept that encompasses both content and applicability. It forms the foundation for individual and organizational action, particularly in the digital recording and use of experience-based production knowledge [21, 26, 27].

A detailed understanding of different shapes of knowledge is essential for the design of knowledge management systems. The basic distinction is between implicit and explicit knowledge. As Polanyi already noted in 1966, a significant part of knowledge is not verbalizable [28]. Implicit knowledge is experience-based, action-oriented, and context-sensitive, and is expressed primarily in action. Explicit knowledge, in contrast, can be articulated, documented, and transferred comparatively easily [24, 28]. This distinction remains central in contemporary knowledge management, especially in the identification, transfer, and digitalization of experiential knowledge in production or maintenance.

From a cognitive psychology perspective, knowledge can also be divided into declarative knowledge (knowing that) and procedural knowledge (knowing how). Declarative knowledge refers to consciously retrievable facts and contexts, while procedural knowledge manifests itself in routines, automated skills, and strategies for action. Procedural knowledge is closely linked to implicit components [22]. The dimensions of articulability and proximity to action overlap in practice and cannot be strictly separated.

The anchoring of knowledge carriers is another important aspect. Knowledge may reside at the individual level (in persons), the collective level (in groups), or the organizational level (in processes, routines, systems) [24]. This structure strongly influences how knowledge can be identified, stored, and disseminated, and determines the extent of potential knowledge loss due to demographic change or staff turnover.

Context-specific knowledge forms an additional cross-sectional dimension. It refers to content that can only be clearly interpreted and applied during action. In production, knowledge is effective only when activated in the relevant context. This situational dependency makes both documentation and technological recording difficult, posing a major challenge for digital knowledge management and AI-based solutions [29]. Formalized approaches, such as enterprise models and configurable knowledge services, can mitigate this challenge by dynamically adapting knowledge provision to production roles, product variants, and process stages. This increases the relevance and effectiveness of knowledge usage in distributed manufacturing networks [30].

For targeted implementation in production, it is necessary to differentiate knowledge shapes. According to DIN SPEC 91443 [31], product knowledge, process or action knowledge, technical knowledge, and experiential knowledge are central categories. These shapes differ in structuring, ownership, and digitizability, and thus provide a practical basis for designing knowledge-based systems [22]. Emphasis on these four forms highlights two points: the key role of implicit, action-oriented knowledge in production, and the need to systematically document explicit content and ensure its accessibility across the organization.

These theoretical and normative foundations can be synthesized into an integrated framework. **Figure 1** illustrates the progression from data to information, knowledge, skills, and competences, and links these dimensions with key management models and standards. It thereby provides a visual synthesis of the section and prepares the ground for the following discussion of organizational knowledge management frameworks.

Building on these foundations, several frameworks have become particularly influential. North's knowledge ladder conceptualizes knowledge as part of a transformative process from data to competence, highlighting the interplay of knowledge, motivation, and action. Probst's building block model structures knowledge management into eight interdependent fields, providing a systemic basis for strategic design. Nonaka and Takeuchi's SECI model emphasizes dynamic knowledge conversion between implicit and explicit forms, while other systemic approaches, such as the Munich model, complement this spectrum.

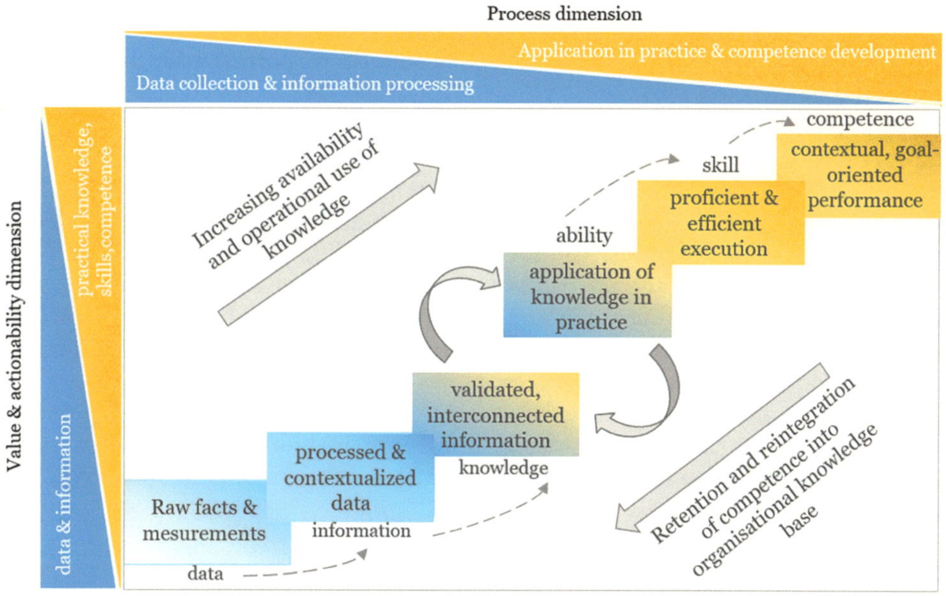

Figure 1.
Integrated framework of knowledge and competence models – own illustration based on Klaus North's knowledge ladder [23].

The knowledge ladder views knowledge as part of a transformative process from data to information, then to skills and competence. Central to this process is the interplay of knowledge, motivation, and the capacity for action, through which individual knowledge evolves into organizational competence [32]. The building block model by Probst et al. conceptualizes knowledge management as the interaction of eight fields: goal definition, knowledge identification, acquisition, development, distribution, use, preservation, and evaluation [24]. This systemic, process-oriented framework provides a basis for the strategic design of knowledge processes.

The SECI model emphasizes the interaction between implicit and explicit knowledge. It distinguishes four modes of knowledge conversion: socialization (implicit to implicit), externalization (implicit to explicit), combination (explicit to explicit), and internalization (explicit to implicit). This cycle illustrates how organizations generate and expand knowledge through interaction, reflection, and application [33].

The figure serves as a link between abstract theoretical models and the subsequent discussion of knowledge management as a socio-technical system.

Beyond these frameworks, knowledge management must be understood as a socio-technical system because its implementation depends on the interplay of people, organization, and technology. While models such as SECI or the building blocks describe knowledge flows, their effectiveness in practice is determined by human-centered design and employee involvement.

Knowledge management rests on three pillars: people, organization, and technology. People are the most important, as they are the knowledge carriers whose expertise and competencies are developed through HR management and training. Organizations provide the framework for the efficient handling of knowledge. Technology supplies tools such as databases that present required information transparently or generate it with AI support. However, a large part of knowledge transfer still occurs through personal communication. This shows that knowledge management is not solely based on databases but also on participation and organizational culture [34].

By applying work-scientific design criteria within a human-centered approach, acceptance and participation can be strengthened [35]. This includes participation, competence development, autonomy, transparency, and sustainable measures. The International Labour Organization addressed these aspects in its concept of "decent work," known in Germany as "Gute Arbeit" [36]. It emphasizes long-term, sustainable work design and integrates technical and social assessments to ensure transparency and traceability. The aim is an iterative process that evaluates safety, health, data protection, job security, and training while considering legal regulations, work-scientific findings, and employee interests [15].

Building on these principles, various technological approaches now support the acquisition, structuring, and transfer of knowledge, including natural language processing (NLP), large language models (LLMs), data science, computer vision, and semantic knowledge representation.

2.2 Digital and AI-driven technologies for knowledge management

Natural language processing comprises methods that enable machines to interact with, understand (natural language understanding, NLU), and generate (natural language generation, NLG) human language. Positioned at the intersection of computer science, AI, and linguistics, NLP includes functions such as translation, speech recognition, summarization, segmentation, classification, and information extraction [37]. For knowledge management, textual information is key, as much of organizational knowledge is stored in unstructured form, for instance, in reports, technical documentation, or internal protocols. NLP techniques preprocess, normalize, and structure such data, improving accessibility and transfer. Typical applications include automatic classification, extraction of key information, and the summarization of extensive technical documents into concise overviews [38–40]. From a process perspective, NLP follows three fundamental steps: text recognition (e.g., from images, scanned documents, or audio), text understanding (semantic and structural analysis), and text generation (creating or rewriting content). Together, these steps underpin intelligent management of text-based knowledge and form the basis of modern systems for decision support, workplace learning, and systematic knowledge transfer [41, 42].

Large language models are advanced AI systems designed to process, understand, and generate human language [43, 44]. They analyze and comprehend text and generate context-appropriate responses. Unlike systems based on predefined modules, LLMs interact through natural written language in full complexity [45]. Technically, they are neural networks with self-attention mechanisms that assign weights to input elements, enabling contextual interpretation and prediction of subsequent content. Through deep learning, this capability extends to generating entire sections of text [46, 47]. In industrial and business settings, LLMs are increasingly used for conversational agents, automated text generation, machine translation, literature analysis, content summarization, and sentiment analysis of customer feedback [44, 48, 49]. Retrieval-augmented generation (RAG) further extends their functionality by supplying external knowledge sources at query time. This reduces hallucinations and improves factual reliability [50–52]. Compared to fine-tuning, RAG is more flexible, though fine-tuning remains useful for style adaptation and structured output, especially when domain-specific or output format constraints are stringent [53, 54].

In environments with strict data protection requirements, deployment architecture is crucial: Local/offline deployment of LLMs with RAG-based pipelines can offer maximal administrative control and minimize reliance on external providers, while still enabling retrieval of up-to-date domain data [55–57]. Recent advances in hardware efficiency allow large models to run on standard- or high-performance workstations. Open-source LLMs provide transparency, adaptability, and independence [46, 53]. Self-hosted models enable tailored solutions, including fine-tuning for specialized tasks. From an engineering and knowledge management perspective, the combination of LLM and RAG offers a robust architecture for integrating human expertise with computational processing, supporting structured preservation, dissemination, and targeted application of organizational knowledge [44, 51, 52].

Data science and machine learning represent another core pillar of knowledge-oriented systems. Data are among the most valuable assets of organizations, as it enables evidence-based decisions that improve efficiency. Data science aims to identify patterns and extract meaning from large datasets, whether structured or unstructured, by means of analytics, AI, machine, and deep learning. Its aim is to identify patterns and extract meaning from large datasets, whether structured or unstructured [58, 59]. Methods draw on mathematics, statistics, programming, data engineering, pattern recognition, visualization, and high-performance computing [38]. Machine learning focuses on systems that learn from examples rather than explicit programming, using approaches such as supervised, unsupervised, reinforcement, and self-supervised learning [60, 61]. Applied to knowledge management, data science and machine learning support knowledge capture and analysis, the development of intelligent training and guidance systems, the creation of structured repositories, and the prediction or prevention of errors during training. These applications significantly optimize knowledge transfer in industrial environments [62, 63].

Computer vision is another central field. Inspired by human vision, it processes and analyzes images and videos to extract information for actions or recommendations. Methods include image processing, pattern recognition, and geometric modeling, aiming at machine-based image understanding [64]. For knowledge capture, two- and three-dimensional images and videos are essential. Computer vision tools support image segmentation, object detection, classification, human activity recognition, quality inspection, robotic perception, and immersive AR/VR applications [65]. Advanced models such as YOLOv12 enable data collection, classification, and workflow analysis, turning raw data into structured information for training, documentation, and process improvement [66].

Knowledge representation and modeling are prerequisites for structuring, transferring, and systematically applying organizational knowledge. Semantic technologies provide declarative descriptions and formal relationships between concepts [67]. Glossaries, taxonomies, and thesauri represent increasingly detailed structures, while ontologies provide comprehensive, formally defined models. Ontologies enable machine-readable representations that support logical reasoning and automated processing [67]. Knowledge graphs (KGs) build on this principle by representing entities and relations as nodes and edges, integrating heterogeneous data sources, providing context-sensitive access, and forming the basis for reasoning.

Combining KGs with AI is a major research focus. Ontologies supply the semantic backbone, KGs contain the structure of the knowledge, and LLMs provide natural language access. This interplay enhances contextual relevance and reliability by grounding AI outputs in structured and validated knowledge bases. For SMEs, structuring and modeling knowledge is especially challenging due to missing governance, reluctance to adopt new technologies, and difficulties in formalizing implicit or procedural expertise [68]. KGs can mitigate these challenges by consolidating diverse knowledge, improving efficiency, reducing risks, and facilitating systematic transfer, though issues of scalability, quality, and usability remain [69].

LLMs also support KG construction and maintenance by extracting entities and relations from documents, modeling procedural knowledge, and updating ontologies. KGs, in turn, improve LLM performance in RAG applications by providing domain-specific grounding. Human expertise remains essential to ensure AI supports rather than replaces human roles. In practice, pipelines for named entity recognition and relation extraction often combine pretrained models with domain expert input [70]. Human-in-the-loop (HIL) approaches allow experts to intervene in each cycle, for instance, in interactive Q&A dialogs [70]. This reduces manual effort [71], ensures validity, and supports continuous curation as knowledge bases evolve [72, 73]. When these requirements are met, KGs provide a foundation for structuring explicit and procedural knowledge and can support transfer, for example, through KG-enhanced RAG chatbots that deliver context-specific recommendations and explanations.

Taken together, these technological approaches, combined with theoretical foundations and human-centered design, constitute the current state of the art in knowledge management. They show how conceptual models, socio-technical frameworks, and emerging technologies interact to capture, structure, and transfer knowledge in industrial contexts. The convergence of these elements underlines both the complexity of knowledge as a resource and the potential of integrated, interdisciplinary approaches to ensure its effective and sustainable use in production.

3. Concepts

This section builds on the fundamentals and state-of-the-art by presenting three complementary concepts for systematically recording, structuring, and digitally utilizing experiential knowledge in industrial assembly. These concepts were developed for settings with a high degree of manual work, where procedural knowledge has largely remained implicit. The first concept focuses on visual recording and standardized motion modeling; the second on language- and cognition-oriented methodologies. Together, they capture both the physical execution of actions and the decision-making logic behind them. A third concept integrates these perspectives into AI-supported modeling linked to enterprise systems. **Figure 2** illustrates how the three concepts complement and converge, combining methods for capturing experiential knowledge with approaches for its modeling and digital representation. This integration enables systematic utilization in assembly contexts.

3.1 AI-supported motion modeling for capturing implicit assembly knowledge

The first concept addresses the challenge of capturing procedural knowledge from manual assembly and transforming it into a measurable, digitally usable form. It combines AI-based object recognition with standardized motion modeling to

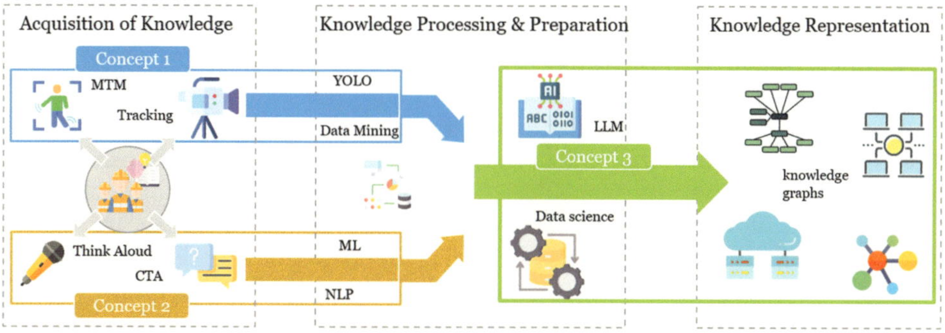

Figure 2.
AI-supported concepts for acquisition, processing, and representation of experiential knowledge in assembly – own illustration.

create structured representations of action sequences. Applications include process evaluation, derivation of best practice sequences, context-sensitive assistance, and model-based training.

In a prototype environment, interactions between operators, tools, and parts are captured via camera-based object recognition and converted into a formal structure. High-resolution RGB video sequences serve as input for a YOLO-based object recognition model, fine-tuned to the assembly domain using annotated datasets of typical objects (e.g., tools, parts, hand positions). Data augmentation techniques, such as perspective distortion, contrast variation, or background noise, improve robustness under real workshop conditions. The trained model detects relevant objects in real time, outputs confidence values per frame, and enables downstream motion analysis.

For structured analysis, the methods-time measurement (MTM) system is applied. Recent studies demonstrate how MTM can be digitized and automated, for example, through VR-based transcription using digital twins [74] or the integration of Industry 4.0 technologies into MTM-1 analysis for automotive applications [75]. Converting visually recognized actions into MTM elements produces a structured movement plan that allows systematic recording, comparison with best practices, and alignment with specifications.

To connect visual recognition with MTM analysis, mappings between object states and movement patterns must be defined. A rule base interprets sequences such as "hand approaches tool → hand covers tool → hand moves away with tool" as a grasping event. This logic can be refined with additional sensors or classifiers. The resulting profiles can then be stored in a structured form and compared with target sequences, process descriptions, or experience profiles.

These data can later inform training requirements, error analysis, or digital assistance services. Examples include context-sensitive assistance based on actual best practices or simulation-based learning tailored to individual weaknesses. Aggregated data can be integrated into semantically structured knowledge bases, supporting standardization and broad availability of experiential knowledge. Crucially, the concept follows a human-centered paradigm: Technologies are not used to monitor or evaluate workers but to document and make explicit skills previously embedded in the practice of experienced specialists. The combination of AI-based vision, ergonomic structuring, and semantic analysis thus opens new potential for digitalizing implicit assembly knowledge.

3.2 AI-supported speech processing and semantic modeling for capturing implicit assembly knowledge

The second concept targets the capture and structuring of procedural and strategic knowledge in complex manual assembly, for example, fixed-position assembly in special-purpose machinery. In such contexts, process knowledge, adaptation to unforeseen events, and optimal sequencing of work steps are highly dependent on individual expertise. The goal is to record this knowledge during real work, analyze and formalize it, and make it digitally accessible.

Methodologically, the approach builds on the think-aloud method and cognitive task analysis (CTA). In think-aloud, specialists verbalize thoughts, decisions, and perceptions while working, directly documenting cognitive processes. CTA complements this through structured post-hoc interviews, particularly using the critical decision method, which captures critical events, decision alternatives, and reference points, and makes action variants explicit. Primary data sources include audio recordings from think-aloud sessions and CTA interviews, optionally supported by video for complex or safety-critical cases. Audio is transcribed with speech-to-text systems enriched with domain-specific terminology to ensure recognition accuracy.

Transcripts are processed with NLP and LLMs, which filter irrelevant content, segment text into knowledge units, recognize entities (tools, components, test criteria), and map them to process steps. They also extract causal relations and decision logic, linking reasons to actions. LLMs further enrich vague passages with known process details, ensuring completeness.

The enriched content is linked with contextual parameters such as environment, resources, time constraints, or disruptions. This produces process-related knowledge objects that combine activity with decision logic. Ontologies or KGs then provide semantic representation and interlinking. This allows context-based searches, automated connection derivation, pattern recognition, and recommendations.

The approach aims to produce a consistent, machine-readable knowledge base supporting digital assistance, targeted training, and process analysis. Knowledge remains anchored in its original context of action and thus retains situational relevance. By combining human-centered data collection with AI-supported processing and semantic modeling, implicit knowledge is sustainably externalized and made available across the organization.

3.3 AI-supported knowledge modeling with human validation for reliability

The third concept extends these perspectives toward scalable integration into enterprise systems (e.g., ERP). Its objective is to structure experiential knowledge so it becomes machine-processable, context-sensitive, and expandable. Input sources include visually recorded actions, linguistically captured decision-making logic, and process data. These heterogeneous sources are integrated into a uniform semantic model, enabling adaptive, AI-assisted application.

Modeling begins with NLP and LLMs for entity recognition, relation extraction, and context assignment. Semantic normalization ensures consistency, while rule- and confidence-based checks, combined with expert validation, guarantee sufficient reliability. Human experts play a central role in validating outputs, contributing situational knowledge, and providing feedback [76, 77]. Versioning mechanisms ensure traceability and currency.

Structured content is represented in KG, enabling scalable analysis of relations and dependencies. Combined with RAG, this allows LLMs to deliver context-specific answers, recommendations, and process descriptions [70, 78]. Outputs follow a human-centered paradigm, tailored to user roles, expertise, and situational needs.

Integration potentials include ERP, MES, and PLM systems, creating continuous information chains across enterprise levels. This links operational data, experiential knowledge, and digital infrastructures. For example, motion data and decision rules from assembly can be connected with MES process records and ERP production orders, ensuring that experiential knowledge directly informs scheduling, resource planning, and quality documentation. Embedded validation, feedback, and continuous updating ensure that the approach combines AI efficiency with human expertise, producing reliable, up-to-date, and expandable knowledge models for sustainable use in production.

Together, the three concepts complement one another. The first provides objective, visually based representations of actions; the second captures cognitive processes and decision logic; and the third integrates both into semantically structured knowledge bases connected to enterprise systems. This results in a holistic digital knowledge base covering both execution and decision-making, forming a foundation for analysis, optimization, assistance, and training in assembly environments.

To operationalize these concepts in practice, **Figure 3** illustrates the core pipeline that summarizes the steps from elicitation to continuous improvement.

All data collection (video, audio, sensor) was conducted with the informed consent of participants. Recordings were securely stored, anonymized where possible, and retained only for the duration required for analysis. Access to raw data was strictly role-based and limited to authorized researchers. HIL validation ensured that data were used exclusively for knowledge elicitation and transfer, preventing surveillance drift or misuse for performance monitoring.

4. Best practices

Beyond conceptual approaches, current research projects already show how AI methods can be applied in practice. The following best-practice cases illustrate different ways of using such technologies in knowledge-intensive processes, ranging from structured learning and competence development to operational decision-making.

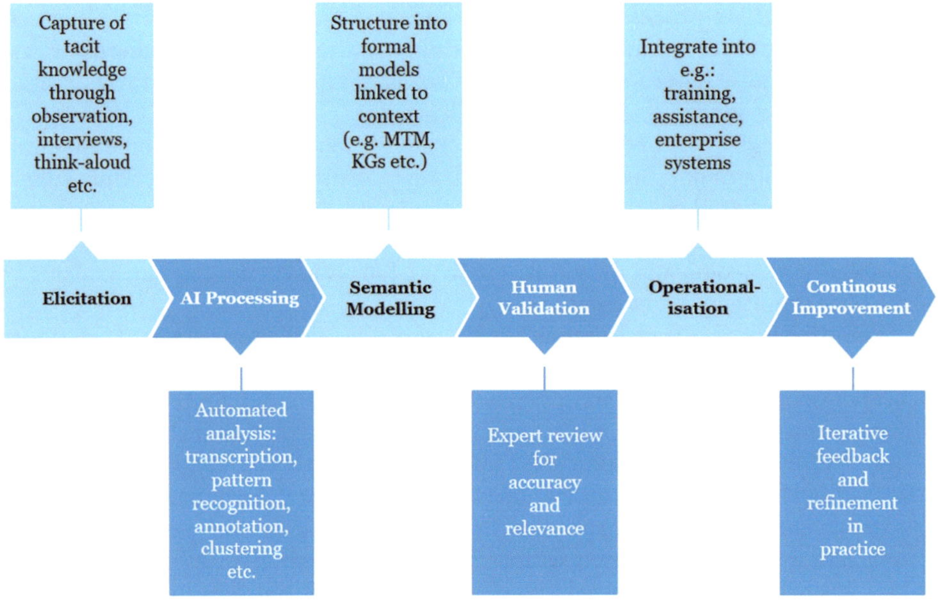

Figure 3.
Summarized pipeline for knowledge elicitation and operationalization – own illustration.

4.1 Customizable digital learning environments for AI-enhanced knowledge management

Within the KOMATRA project, the focus is on supporting sustainable and value-oriented transformation in SMEs. One central theme is learning and competence development as a prerequisite for long-term organizational change. In this context, digital learning platforms were explored as a means of structuring knowledge transfer and skill acquisition. The open-source learning management system (LMS) Moodle was evaluated as a best-practice example for integrating AI-based methods into corporate learning.

Moodle can be adapted to specific organizational needs, including the integration of locally hosted AI solutions. While the current version supports the OpenAI and Azure APIs, on-premises deployment of an LLM allows companies to maintain full control over data and infrastructure. Such integration enables AI-assisted content creation, image generation where models are available, and summarization tools that help learners condense and understand materials more effectively. Beyond AI functions, Moodle can be customized through user interface adjustments, additional plugins, and workflow-specific modifications. Structured guidelines further support the creation of interactive content, such as exercises and drag-and-drop activities, enhancing learner engagement.

The open-source Adapt authoring tool complements Moodle by creating SCORM-compliant learning packages for any LMS. With JavaScript expertise, the Adapt tool can be customized and extended with additional functions, enabling content creation beyond the standard framework. Possible customizations include UI modifications and integration of locally hosted LLMs. Generated outputs can be transferred to the web interface or stored directly in the database via API. Workflows may include direct prompt-based generation, guided content creation from predefined topics, or RAG integration with external context. Relevant corporate documents can be indexed to increase contextual awareness, with configurations defined at individual, departmental, or organizational levels. API keys provide a means of access control, ensuring that only authorized roles can interact with the system. However, secure handling of sensitive data additionally requires encryption, secret management, and strict access policies.

AI integration also extends to automated assessment. LLMs can generate quizzes based on selected course content, with parameters such as question type, number, and focus configurable. Communication between the LMS and AI can be set up for the automatic transfer of grading results. To comply with data minimization, any RAG-generated content is deleted from the LLM after use, preventing unnecessary accumulation. **Figure 4** illustrates how these elements converge in a customizable digital learning framework: Moodle as the core LMS, complemented by the Adapt tool for SCORM-compliant content, with AI functionalities such as LLM and RAG integrated through modular connections.

From a knowledge management perspective, the combined use of Moodle and Adapt with LLM and RAG enhances the capture, organization, and delivery of knowledge. Automated content creation, context-aware enrichment, and adaptive assessment improve efficiency while systematically preserving and disseminating organizational knowledge. For SMEs, such solutions are valuable as accessible, customizable and secure tools for building long-term knowledge infrastructures aligned with transformation objectives. In comparison, annual costs of unmanaged loss of knowledge in SMEs can exceed €80,000 [8]. By

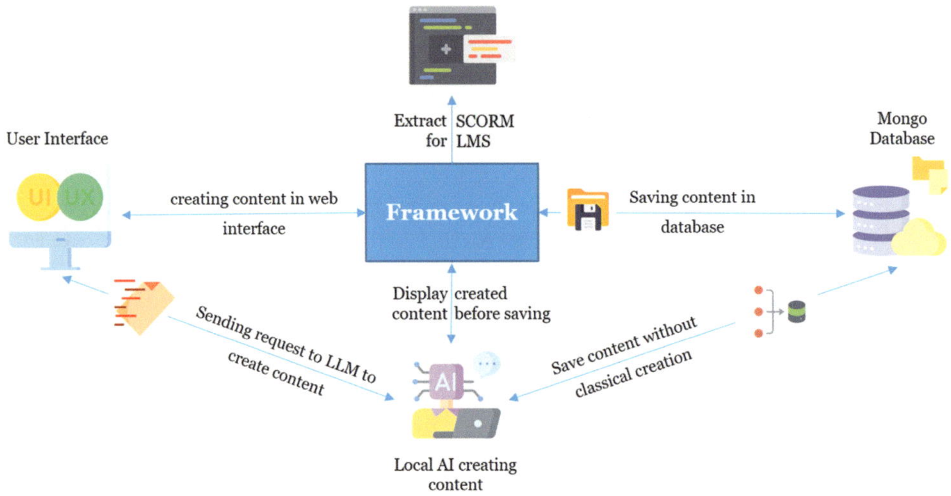

Figure 4.
Framework for customizable digital learning environments with integrated AI functionalities – own illustration.

contrast, implementing a self-hosted, AI-enhanced learning system involves one-time investments of around €12,500 and annual operating costs of approximately €8,000, making the system a clearly more cost-efficient option for sustainable knowledge management.

4.2 Human-centered AI for order allocation

As part of the ViSAAR research project, an application-oriented study explored the use of AI in knowledge-intensive logistics operations. The scenario focused on manual order picking, a process marked by short-term orders, high product variety, and fluctuating workloads. Under such conditions, allocation decisions typically rely on the implicit expertise of shift supervisors. Conventional systems and static algorithms often reach their limits, especially when rapid rescheduling or balanced workload distribution is required.

The project aimed to develop an adaptive, human-centered decision-support system for order allocation, as illustrated in **Figure 5**.

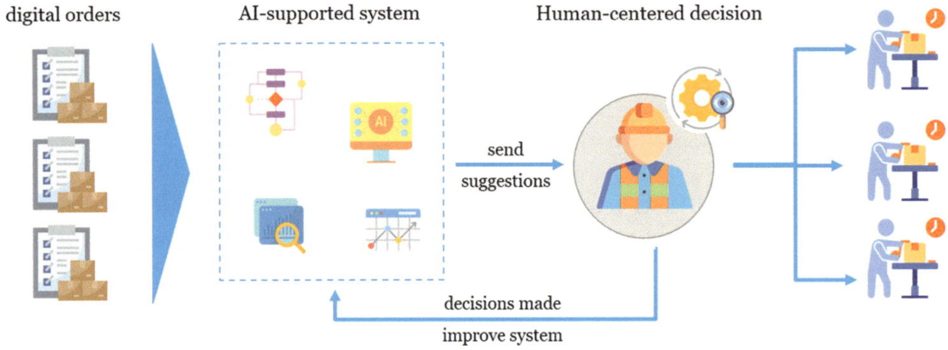

Figure 5.
Human-centered AI system for order allocation – own illustration.

A key challenge lies in formalizing decision-making processes previously driven by implicit knowledge. By combining human expertise with AI-based optimization, the system automated routine tasks while reducing cognitive load. As a result, process efficiency improved, idle time decreased, and workloads were distributed more evenly, enhancing both performance and employee satisfaction.

Reinforcement learning formed the methodological core. Here, an artificial agent developed decision strategies in a simulated environment. Training combined historical data with synthetic scenarios to establish a baseline, which was refined continuously using real-time inputs and structured human feedback. Crucially, final decision authority remained with operational staff, ensuring that domain knowledge and situational judgment stayed decisive.

To simulate realistic conditions, the environment modeled key logistics parameters such as order arrivals, workstation capacities, workload variability, and time-critical priorities. Continuous feedback in the form of reward signals guided the agent whenever its allocation strategies improved measurable performance. Through iterative learning, the system adapted progressively, improving allocation quality by integrating both operational outcomes and human input. For example, when urgent orders arrived on top of existing workloads, traditional allocation rules tended to overload one station. The RL-based system instead redistributed tasks dynamically across all workstations, balancing urgency and capacity. This reduced delays and idle time while improving on-time performance.

A distinctive element of the approach was the integration of workplace-specific experiential knowledge. Qualitative feedback and structured assessments from staff informed the design of the reward function, allowing the AI to internalize human expertise over time. This enabled systematic digitalization and targeted reuse of implicit knowledge within the optimization process.

The resulting system proved flexible, transparent, and adaptive. Decision-making capabilities were enhanced, while supervisors were relieved of routine tasks, enabling them to focus on complex cases where human judgment is essential. The study shows that AI can not only increase efficiency but also formalize, preserve, and apply experiential knowledge in a human-centered manner. Initial simulations indicated that the optimized allocation system could reduce overall working time by around 17% per shift, while subsequent pilot tests in real operations confirmed a sustained reduction of approximately 11% compared to the baseline. These findings demonstrate that measurable efficiency gains can be achieved both under controlled conditions and in practice. The system is currently undergoing further testing and refinement in operational environments to continuously improve robustness, accuracy, and integration into existing workflows, while keeping human expertise at the core [79, 80].

The two cases demonstrate complementary ways AI contributes to systematic knowledge management. The KOMATRA case highlights how LLM- and RAG-supported learning environments structure and disseminate knowledge, while the ViSAAR case shows how AI can capture and operationalize implicit expertise in dynamic logistics environments. Both follow a human-centered paradigm, ensuring AI supports rather than replaces expertise, and thus contribute to sustainable, adaptive knowledge infrastructures in SMEs.

5. Outlook and future developments

The best-practice examples show that AI-based technologies are no longer confined to theory but are already applied successfully in knowledge-intensive environments. They demonstrate how experiential knowledge can be externalized, structured, and utilized through digital systems while ensuring that human expertise remains central. Building on the presented concepts and implementations, several research directions and technological developments will shape the future of knowledge management in industrial contexts.

A major trend is the growing integration of multimodal data sources. Current solutions mostly combine text, structured data, and video, but future developments are expected to focus on vision language models capable of processing visual and linguistic information jointly. These models could support a more holistic representation of experiential knowledge by linking recorded actions with verbal explanations or contextual annotations in real time. In assembly, video data could be enriched with reasoning or procedural descriptions, creating semantically structured multimodal knowledge objects. For SMEs, this approach may simplify entry into knowledge management by reducing reliance on complex documentation or written records.

Scalability and interoperability of knowledge models form another key perspective. Ontologies and KGs already provide a strong foundation, yet seamless integration with enterprise systems such as MES, ERP, or PLM remains challenging. Advances in standardization and semantic web technologies are likely to be crucial in enabling cross-organizational knowledge exchange, especially for SMEs that require flexible but reliable infrastructures.

Human-centered design principles must continue to guide these developments. The examples from KOMATRA and ViSAAR have shown that acceptance, transparency, and employee participation are essential for successful implementation. Future systems must, therefore, focus not only on technical performance but also on governance structures, validation mechanisms, and participatory processes that safeguard trust and ensure responsible use of AI in operations.

The convergence of AI subfields – ranging from NLP and LLMs to reinforcement learning, computer vision, and multimodal modeling – will likely produce increasingly adaptive and autonomous systems. Rather than replacing human expertise, these systems will extend it, providing targeted support for decision-making, training, and process optimization. The long-term vision is the creation of sustainable knowledge ecosystems in which experiential, procedural, and contextual knowledge is continuously captured, structured, and made available across organizational boundaries.

In summary, the state of the art, together with the presented concepts and applications, underlines the transformative potential of AI for knowledge management in industrial production. Future research must refine methods for multimodal knowledge capture, strengthen human-centered integration, and advance scalable, interoperable infrastructures. These efforts will pave the way toward resilient knowledge systems that preserve and expand experiential expertise, ensuring its long-term availability for innovation, efficiency, and sustainable value creation.

Acknowledgments

This research was carried out within the framework of the ViSAAR project, "Virtual Work Design and Technologies for Innovation in Structural Change" (funding number 02L20B088), funded by the German Federal Ministry of Education and Research (BMBF), and the KOMATRA project, "Competence Center for Work Research: Shaping Transformation in the Saar-Palatinate Corridor – Value-Oriented Work and Learning in the Circular Economy" (funding number 02L22C005), supported by the German Federal Ministry for Research, Technology, and Space (BMFTR). The authors gratefully acknowledge this support.

The authors also thank Felix Weber and Abbas Nouri for their support and contributions to the implementation of this book chapter.

Author details

Jannis Pfister[1*], Daniela Schmidt[1], Matthias Jenner[1], Till Günther[1] and Rainer Müller[2]

1 ZeMA – Centre for Mechatronics and Automation Technology gGmbH, Saarbrücken, Germany

2 University of Saarland – Chair of Assembly Systems, Saarbrücken, Germany

*Address all correspondence to: jannis.pfister@zema.de

References

[1] Statistisches Bundesamt (Destatis). Dauer der Beschäftigung beim aktuellen Arbeitgeber. 2025. https://www.destatis.de/DE/Themen/Arbeit/Arbeitsmarkt/Qualitaet-Arbeit/Dimension-4/dauer-beschaeftigung-aktuell-Arbeitgeber.html [Accessed: 2025-August-26]

[2] Bureau of Labor Statistics. Employee Tenure in 2024

[3] International Labour Organization. World Employment and Social Outlook. Geneva: ILO; 2023

[4] OECD/OCDE. OECD Employment Outlook 2025. OECD Publishing; 2025 DOI: 10.1787/194a947b-en

[5] Gittleman M. The "Great Resignation" in perspective. https://www.bls.gov/opub/mlr/2022/article/the-great-resignation-in-perspective.htm [Accessed: 2025-August-28]

[6] Hauptmann A, Sirries S, Stepanok I. Arbeitskräftefluktuation im Verarbeitenden Gewerbe: In exportierenden Betrieben ist die Beschäftigung stabiler

[7] Zentrum für Europäische Wirtschaftsforschung GmbH. Mitarbeiterfluktuation Ist Schädlich Für Die Produktivität Junger Unternehmen. ZEW news: Zentrum für Europäische Wirtschaftsforschung GmbH. 2017

[8] KNOWRON. Die Kosten des Wissensverlusts: Mit KI-gestütztem Lernen gegen Wissensverlust und steigende Kosten. 2025

[9] Moon K. Employee turnover costs more than you think: A new study from wharton's ken moon reveals the hidden cost of employee turnover. 2022. https://knowledge.wharton.upenn.edu/article/why-employee-turnover-costs-more-than-you-think/#:~:text=%2A%20Each%20percentage,hundreds%20of%20millions%20of%20dollars. [Accessed: 2025-August-21]

[10] Massingham P. Measuring the impact of knowledge loss: More than ripples on a pond? Management Learning. 2008;**39**:541–560. DOI: 10.1177/1350507608096040

[11] Orth R, Finke I, Voigt S. Wissen greifbar machen: Den Umgang mit Wissen beschreiben und bewerten. 2008

[12] Lindner D, Leyh C. Digitalisierung von KMU – Fragestellungen, Handlungsempfehlungen sowie Implikationen für IT-Organisation und IT-Servicemanagement. HMD Praxis der Wirtschaftsinformatik. 2019;**56**:402–418. DOI: 10.1365/s40702-019-00502-z

[13] Böhl J. Wissensmanagement in Klein- Und Mittelständigen Unternehmen der Einzel- Und Kleinserienfertigung [Dissertation]. TU München; 2000

[14] Giwoleit D. Bridging The Gap: Der Weg vom impliziten zum expliziten Wissen durch semantische Spracherkennung auf Basis eines ontologischen Wissensmanagementsystems [Abschlussarbeit Master of Arts]: Fachbereich Wirtschaft der Fachhochschule Kiel; 13. Februar 2015

[15] Müller O. Einführung kollaborativer Roboter in kleinen und

mittleren Unternehmen: Leitfaden und Praxishilfe für die partizipative Gestaltung. April 2022. https://www.best-saarland.de/downloads/publikationen/leitfaden-einfuehrung-kollaborativer-roboter-in-kleinen-und-mittleren-unternehmen. [Accessed: 2025-July-28]

[16] Konrad K. Verfahren Zum Semantisch Unterstützten Anlagenanlauf von Montagesystemen. Stuttgart: Fraunhofer Verl.; 2012

[17] Barbara DEML, Manuel ZAREMSKI. Analyse von Augen- und Blickbewegungen zur Beschreibung von Handlungswissen in der manuellen Demontage

[18] Lin T-C, Chang CL, Tsai W-C. The influences of knowledge loss and knowledge retention mechanisms on the absorptive capacity and performance of a MIS department. Management Decision. 2016;**54**: 1757–1787. DOI: 10.1108/MD-02-2016-0117

[19] Offergelt F, Hofreiter S, Steiner T. Wissensmanagement in Modernen Organisationen. Berlin, Heidelberg: Springer Berlin Heidelberg; 2024

[20] Dragusanu G. Wissensmanagement: Sicherung und Weitergabe des Wissens beim Stellenwechsel

[21] Kohl H, Mertins K, Seidel H. Wissensmanagement Im Mittelstand. Berlin, Heidelberg: Springer Berlin Heidelberg; 2016

[22] Mader I. Wissensmanagement Erfolgreich Umsetzen. Berlin, Heidelberg: Springer Berlin Heidelberg; 2023

[23] North K, Maier R. Wissen 4.0 – Wissensmanagement im digitalen Wandel. HMD Praxis der Wirtschaftsinformatik. 2018;**55**:665–681. DOI: 10.1365/s40702-018-0426-6

[24] Probst GJB, Raub S, Romhardt K. Wissen Managen: Wie Unternehmen Ihre Wertvollste Ressource Optimal Nutzen. 6th ed. Wiesbaden: Gabler; 2010

[25] DIN Deutsches Institut für Normung e. V. DIN-ISO-30401-2021-02-1: Wissensmanagementsysteme - Anforderungen. DOI: 10.31030/3215028

[26] DIN Deutsches Institut für Normung e. V. DIN SPEC 91281:2012-04: Einführung von prozessorientiertem WM in KMU

[27] Gronau N. Modellieren des Umgangs mit Wissen für Industrie 4.0. Industrie 4.0 Management. 2021;**2021**:6–10. DOI: 10.30844/I40M_21-3_S6-10

[28] Polanyi M. The I Dimension. 1966

[29] Reinmann G, Mandl H. Wissensmanagement und Weiterbildung. In Tippelt R, von HA, editors. Handbuch Erwachsenenbildung/Weiterbildung. Wiesbaden: VS Verlag für Sozialwissenschaften; 2011. p. 1049–1066. DOI: 10.1007/978-3-531-94165-3_66

[30] Sandkuhl K, Smirnov A. Context-oriented knowledge management in production networks. Applied Computer Systems. 2018;**23**:81–89. DOI: 10.2478/acss-2018-0011

[31] DIN Deutsches Institut für Normung e. V. Systematisches Wissensmanagement für KMU DIN

[32] North K. Wissensorientierte Unternehmensführung: Wissensmanagement Im Digitalen Wandel. 7th ed. Wiesbaden, Heidelberg: Springer Gabler; 2021

[33] Nonaka I, Takeuchi H. Die Organisation Des Wissens: Wie Japanische Unternehmen Eine Brachliegende Ressource Nutzbar Machen. 2nd ed. Frankfurt am Main: Campus; 2012

[34] Bürgel HD, editor. Wissensmanagement: Schritte Zum Intelligenten Unternehmen. Berlin, Heidelberg: Springer; 1998

[35] Schmidt D, Kuhn D, Becker D, Müller R, Bierwirth K, Hielscher V. Menschzentrierte Digitalisierung im Sondermaschinenbau. Zeitschrift Für Wirtschaftlichen Fabrikbetrieb. 2023;**118**:717–722. DOI: 10.1515/zwf-2023-1128

[36] Institut DGB-Index Gute Arbeit. 2022. https://www.dgb.de/gute-arbeit/ [Accessed: 2025-September-01]

[37] Ding Y, Ma J, Luo X. Applications of natural language processing in construction. Automation in Construction. 2022;**136**:104169. DOI: 10.1016/j.autcon.2022.104169

[38] Assainar Hafnan PP, Mohan A. Summary-Based Document Classification. In Sa PK, Bakshi S, Hatzilygeroudis IK, Sahoo MN, editors. Recent Findings in Intelligent Computing Techniques. Singapore: Springer Singapore; 2018. p. 153–160. DOI: 10.1007/978-981-10-8633-5_16

[39] Varagantham C, Reddy J, Yelleni U, Kotha M, Rao P. Text summarization using NLP. International Journal Of Trendy Research In Engineering And Technology. 2022;**06**:26–30. DOI: 10.54473/ijtret.2022.6405

[40] Hirschberg J, Manning CD. Advances in natural language processing. Science. 2015;**349**:261–266. DOI: 10.1126/science.aaa8685

[41] John Basha M, Vijayakumar S, Jayashankari J, Alawadi AH, Durdona P. Advancements in Natural Language Processing for text understanding. E3S Web of Conferences. 2023;**399**:4031. DOI: 10.1051/e3sconf/202339904031

[42] Yu W, Zhu C, Li Z, Hu Z, Wang Q, Ji H, Jiang M. A survey of knowledge-enhanced text generation. ACM Computing Surveys. 2022;**54**: 1–38. DOI: 10.1145/3512467

[43] Kelbert P, Siebert J, and Jöckel L. Was sind Large Language Models? Und was ist bei der Nutzung von KI-Sprachmodellen zu beachten? 2023. https://www.iese.fraunhofer.de/blog/large-language-models-ki-sprachmodelle/ [Accessed: 2025-August-27]

[44] Bünnagel W. Wissenskommunikation, Maschinelles Lernen Und Sprachmodelle. Berlin, Heidelberg: Springer Berlin Heidelberg; 2024

[45] Kalweit Maria KG. Warum wir neu lernen müssen, mit Maschinen zu sprechen – Eine Momentaufnahme der generativen KI im Januar. Ordnung der Wissenschaft. 2024;**2024**:125–138

[46] de LF, Monti F, Mecella M. Engineering Information Systems with Large Language Models. Cham: Springer Nature Switzerland; 2025

[47] SAP. Was ist Was ist ein großes Sprachmodell. 2024. https://www.sap.com/germany/resources/what-is-large-language-model [Accessed: 2025-September-01]

[48] KI Lexikon. Was sind Large Language Models(LLM)? https://www.dogado.de/ki-lexikon/large-language-models [Accessed: 2025-September-01]

[49] IBM. Was sind große Sprachmodelle (Large language Models, LLMs)? https://www.ibm.com/de-de/think/topics/large-language-models [Accessed: 2025-September-02]

[50] Baron S. Abschlusbericht LLMs für Unternehmen. 2024. https://redlink.at/docs/Redlink-2024-LLMs-fuer-Unternehmen.pdf. [Accessed: 2024-October-22]

[51] Klesel M, Wittmann HF. Retrieval-Augmented Generation (RAG). Business & Information Systems Engineering. 2025;**67**:551–561. DOI: 10.1007/s12599-025-00945-3

[52] Kothapalli M. The practical applications of retrieval-augmented generation in AI. Journal of Computer Science and Software Development. 2024;**3**. DOI: 10.17303/jcssd.2024.3.205

[53] Honroth T, Siebert J, Kelbert P. Open Source Large Language Models selbst betreiben. 2024. https://www.iese.fraunhofer.de/blog/open-source-large-language-models-selbst-betreiben/ [Accessed: 2025-August-29]

[54] Balaguer A, et al. RAG vs fine-tuning: Pipelines, tradeoffs, and a case study. *arXiv preprint*. 2024. DOI: 10.48550/arXiv.2401.08406

[55] Baranowska V. Warum die Verwendung lokaler LLMs möglicherweise besser ist als ChatGPT, Claude oder Gemini: Eine einfache Kostenaufschlüsselung. https://scand.de/unternehmen/blog/local-llms-vs-chatgpt-cost-comparison/ [Accessed: 2025-August-22]

[56] codecamp. LLM im Selbstbetrieb: Lohnt sich das? https://www.codecamp-n.com/blog/llm-im-selbstbetrieb-lohnt-sich-das [Accessed: 2025-August-28]

[57] Lee K, et al. Enhancing security and applicability of local LLM-based systems: An offline retrieval-augmented QA approach. Electronics. 2025;**14** (17):3407. DOI: 10.3390/electronics14173407

[58] Κουκάρας Π, Koukaras P Interdisciplinary data science methods using machine learning for enhanced knowledge acquisition; 2021

[59] Varshney M, Garg S, Rajpoot JAK. A study on issues, challenges and application in data science. International Journal of Trend in Scientific Research and Development. 2017;**1**:526–533. DOI: 10.31142/IJTSRD2340

[60] Galan N. Knowledge loss induced by organizational member turnover: A review of empirical literature, synthesis and future research directions (Part I). The Learning Organization. 2023;**30**:117–136. DOI: 10.1108/TLO-09-2022-0107

[61] Dai S, Meng F. Addressing modern and practical challenges in machine learning: A survey of online federated and transfer learning;2022

[62] Hartmann T, Moawad A, Fouquet F, Nain G, Klein J, Le Traon Y, Jezequel J-M. Model-driven analytics: Connecting data. Domain Knowledge,

and Learning. 2017. DOI: 10.48550/arXiv.1704.01320

[63] Ördek B, Borgianni Y, Coatanea E. Machine learning-supported manufacturing: A review and directions for future research. Production & Manufacturing Research. 2024;**12**. DOI: 10.1080/21693277.2024.2326526

[64] Gao K, Mei G, Piccialli F, Cuomo S, Tu J, Huo Z. Julia language in machine learning: Algorithms, applications, and open issues. Computer Science Review. 2020;**37**:100254. DOI: 10.1016/j.cosrev.2020.100254

[65] Gendy W, Patel D. Advancements in computer vision: A comprehensive survey of image processing and interdisciplinary applications. Academic Journal of Science and Technology. 2024;**13**:28–34. DOI: 10.54097/5e1cqw59

[66] Tian Y, Ye Q, Doermann D. YOLOv12: attention-centric real-time object Detectors: arXiv; 2025

[67] Andreas G. Entwicklung einer Ontologie zur funktionsorientierten Lösungssuche in der Produktentwicklung. 2010

[68] Mansfield M, Tamma V, Goddard P, Coenen F. Capturing expert knowledge for building enterprise SME knowledge graphs. In: Gentile AL, Gonçalves R, editors. K-CAP '21: Knowledge Capture Conference; 02 12 2021 03 12 2021; Virtual Event USA. New York, NY, USA: ACM; 2021. p. 129–136. DOI:10.1145/3460210.3493569

[69] Hogan A, Blomqvist E, Cochez M, D'amato C, de MG, Gutierrez C, et al. Knowledge graphs. ACM Computing Surveys. 2022;**54**:1–37

[70] Celino I, Carriero VA, Azzini A, Baroni I, Scrocca M. Procedural knowledge management in industry 5.0: Challenges and opportunities for knowledge graphs. Journal of Web Semantics. 2025;**84**:100850. DOI: 10.1016/j.websem.2024.100850

[71] Tsaneva S, Käsznar K, Sabou M. Human-centric ontology evaluation: Process and tool support. In Corcho O, Hollink L, Kutz O, Troquard N, Ekaputra FJ, editors. Knowledge Engineering and Knowledge Management. Cham: Springer International Publishing; 2022. p. 182–197. DOI: 10.1007/978-3-031-17105-5_14

[72] Tsaneva S, Sabou M. Enhancing human-in-the-loop ontology curation results through task design. Journal of Data and Information Quality. 2024;**16**:1–25. DOI: 10.1145/3626960

[73] Cappelli MA, Di Marzo Serugendo G. Methodological exploration of ontology generation with a dedicated large language model. Electronics. 2025;**14**:2863. DOI: 10.3390/electronics14142863

[74] Gorobets V, Billeter R, Adelsberger R, Kunz A. Automatic MTM-Transcription in Virtual Reality Using the Digital Twin of a Workplace. *Proceedings of the International Conference on Human Interaction and Emerging Technologies (IHIET-AI 2024)*. 2024. DOI: 10.1007/978-3-031-64724-6_42

[75] Markov D, Vitliemov E. Industry 4.0 technologies for MTM-1 analysis improvement for the automotive industry. Engineering Proceedings. 2025;**4**(1):34. DOI: 10.3390/engproc2025004034

[76] McKay MH, Shneiderman B, Guimbretière F, et al. Realizing the Promise of AI Governance Involving Humans-in-the-Loop. In: Helmut, Degen, and Stavroula, Ntoa. Human-Computer Interaction – HCII 2024. Lecture Notes in Computer Science. Cham: Springer; 2024. p. 123–135. DOI: 10.1007/978-3-031-76827-9_9

[77] Wilchek L, Zur M, Avital A. Human-in-the-loop for computer vision assurance: A survey. Engineering Applications of Artificial Intelligence. 2023;**123**:106270. DOI: 10.1016/j.engappai.2023.106270

[78] Buchgeher G, Gabauer D, Martinez-Gil J, Ehrlinger L. Knowledge graphs in manufacturing and production: A systematic literature review. IEEE Access. 2021;**9**: 123456–123478. DOI: 10.1109/ACCESS.2021.3072345

[79] Kuhn D, Pfister J, Jenner M, Eichenwald M, Müller R. KI in der Kommissionierung. Zeitschrift Für Wirtschaftlichen Fabrikbetrieb. 2025;**120**(s1):274–279

[80] Kuhn D, Schmidt D, Pfister J, Bashir A, Müller R. Aufbau eines menschzentrierten Unterstützungstools zur optimierten Zuweisung von Aufträgen zu Arbeitsplätzen in der Kommissionierung mittels Linear Integer Programming. VDI Mechatroniktagung 2024

[81] Amtsblatt der Europäischen Union. Commission Recommendation of 6 May 2003 concerning the definition of micro, small and medium-sized enterprises

[82] Europäischen Union. SME definition. https://single-market-economy.ec.europa.eu/smes/sme-fundamentals/sme-definition_en?prefLang=de [Accessed: 2025-September-05]

[83] Luxemburg: Amt für Veröffentlichungen der Europäischen Union, 2020. Benutzerleitfaden zur Definition von KMU. https://www.ihk.de/blueprint/servlet/resource/blob/3279018/3fda3b99a60558d115d3758af5388283/kmu-definition-in-deutsch-data.pdf [Accessed: 2025-August-26]

Chapter 6

Filling Data Gaps with Operational and Experiential Knowledge from Employees – Opportunities and Challenges for AI-Based Solutions

Günter Müller-Czygan and Viktoriya Tarasyuk

Abstract

Artificial intelligence (AI) solutions are heavily dependent on sufficiently qualified data. In technical processes, this data is usually available in analog or digital form and can be used to train AI. In many existing complex technical processes, especially those that have remained unchanged for decades, there is not enough data available that can be used to train AI models. In these processes, most of the knowledge lies with the employees who plan, implement, and operate them. This often so-called "implicit knowledge" of the employees is a real data advantage if it can be made available for the training of AI models. Using the example of complex wastewater management processes, it is shown in this work how process controls based on AI systems, such as case-based reasoning, were developed with the help of operational and experiential knowledge of employees, including the question of how the existing employee knowledge was made available. The biggest challenge here is to document the tacit knowledge of particularly experienced employees who know a lot about special and emergency cases of the processes, but which are rarely documented, even though they represent a great value for process operations. In addition, the other positive effects of generating employee experience and the role of generative AI in these processes are presented.

Keywords: generative AI, tacit/implicit knowledge, water management, case-based reasoning, human-in-the-loop, human–AI interaction

1. Introduction

In the modern knowledge economy, the success of organizations increasingly depends on a crucial but often overlooked resource: the so-called "tacit knowledge." This knowledge is deeply rooted in the personal experience, intuition, and cognitive abilities of individual subject matter experts [1]. It is described as a kind of "black box" of an organization, as it is not easy to document or transfer but represents a fundamental value for sustainability and competitiveness [2]. However, this valuable knowledge often exists in isolated form as so-called "tribal knowledge" within individual teams or among specific employees. The departure of experienced specialists, whether through retirement or a change of job, poses a significant risk that this unique expertise will be irretrievably lost [3]. At the same time, the rapid development in the field of artificial intelligence (AI), especially in the field of generative AI, is facing a new fundamental challenge: the so-called training data crisis. The development of large language models (LLMs) has largely exhausted the publicly accessible internet as a data source, which means that "fresh data is becoming scarce and expensive," and the supply of unused texts and information is running out [4]. This lack of high-quality, novel training data makes the exploitation of previously hidden sources a strategic requirement for the further development of AI. In this context, the systematic recording, digitalization, and utilization of hidden, implicit knowledge and its effective management and use with AI are coming into focus in technical infrastructures. Particularly in complex municipal infrastructures, such as water management, there is often a lack of strategic knowledge management systems or concepts to counter the acute risk of knowledge loss due to the departure of people with experience. At the same time, the legal, social, and organizational challenges in these sectors are increasing, particularly due to the increasing effects of climate change [5].

For example, initial approaches to the use of generative AI as part of knowledge management in maintenance processes are described by [6] using the methodology of design science research. Based on several cycles of design and evaluation, a solution approach for the selected problem of maintenance processes was developed by achieving theoretical knowledge and applicable design options through the personnel and researchers involved. For problem identification and motivation [6], conducted the workshops with selected electricity distribution network operators from Germany and identified the "Pains and Gains" of maintenance work. Generative AI and, in particular, LLMs were selected as technologies for the definition of associated solution objectives. A similar approach was taken by [7] to capture implicit knowledge information for the digitization of preventive maintenance work on street drains to absorb rainwater during heavy rainfall. In order to prevent the overflowing of street drains during heavy rainfall, the knowledge of maintenance personnel was recorded in special workshops, transferred into a structured form, and transferred as a matrix into an automated process analysis. Based on various algorithms, by integrating digitally available precipitation forecasts, a prediction was created by linking with the developed experience matrix to determine which street drains are in danger of overflowing depending on rainfall. At that time, current possibilities of generative AI were not available for that application. Nevertheless, this example is well-suited to analyze the prerequisites for capturing implicit knowledge and is presented in detail below. In addition to other applications presented in this article, these examples served as the basis for further considerations and developments by the authors in various projects, which are summarized below.

2. Implicit knowledge as input data – How correct and comprehensive recording can succeed?

In order to adequately incorporate the inherent complexity of tacit knowledge when capturing and using it, it is essential to differentiate precisely between the various forms of knowledge.

Category	Category description
Explicit knowledge	This form of knowledge is documented, structured, and easy to communicate and share. It is available in formal sources such as manuals, instructions, reports, and databases. In terms of digitization, it can be processed by software programs such as decision support systems or automatic text analysis systems [8].
Implicit knowledge	Implicit knowledge is often understood as the application of explicit knowledge. It is not directly documented but is usually conveyed through informal communication, social interactions, and the application of skills [9]. It represents a kind of subconscious understanding or assumption that can potentially be made explicit with effort or reflection.
Tacit knowledge	This is the deepest and most personal form of knowledge. It is the knowledge that is most difficult to articulate or put into a tangible form, as it stems from personal experience, intuition, judgment, and cultural circumstances [10]. This knowledge is context-dependent and subjective, which makes it difficult to standardize and generalize [11].

Table 1.
Categories of knowledge management (own presentation).

According to **Table 1**, knowledge management research primarily distinguishes between three categories. Classic IT-based knowledge management focuses primarily on category 1, explicit knowledge, where knowledge is documented and structured as information that only has to be transformed from its analog into digital form. This is ideally suited for integration into AI-based systems. The implicit knowledge (category 2) that builds on this can also be integrated into AI-based systems in a comparatively manageable way due to its structured basis. Although the effort involved may be considerable, conversion into machine-readable form is achievable. This article focuses on Category 3, the tacit knowledge, as the deepest and most personal, but also the most valuable form of experiential knowledge in relation to complex framework conditions.

A knowledge building and sharing process is multi-layered, dynamic, and circular. To understand this process, it is important to classify the role of tacit knowledge in organizations and make it usable as data input for digital systems. Nonaka and Takeuchi have dealt with the structure and process of knowledge formation and have developed the so-called SECI model (Socialization, Externalization, Combination, Internalization), a model of how knowledge circulates and multiplies between its tacit and explicit forms [12]. According to [13], Nonaka and Takeuchi describe knowledge creation and sharing as four phases: socialization (implicit to implicit via shared experience), externalization (tacit to explicit through articulation), combination (linking explicit knowledge to create new knowledge), and internalization (explicit to tacit through learning by doing). The example of the SECI model shows that technological solutions alone are not sufficient to provide relevant knowledge without taking into account the process of knowledge processing between and within people. An exclusive focus on IT infrastructure to manage knowledge as an object neglects the crucial human–social phases of knowledge conversion. An effective knowledge management strategy must therefore consider technological as well as procedural, cultural, and human aspects. How this can work and the challenges that arise will be illustrated using examples from the water industry that focus on capturing tacit knowledge as an input variable in digitally based processes. In the first example, implicit knowledge serves as a static input variable for a digital alerting system, and in the second example, as input and training data for an AI-based evaluation and decision support system.

2.1 Example 1: Digital-based preventive control and maintenance of street drains without sensor data

The city of Siegen in North Rhine–Westphalia (Germany) is a city with around 100,000 inhabitants which, due to its mountainous topography, regularly experiences flooding events in some parts of the city during heavy rainfall. As a result of climate change, heavy rainfall events are increasing, and with them, the risk of roads and properties being flooded [14]. Of particular importance here is the functionality of street drains, where rainwater runs off from road surfaces and flows into the municipal sewer system. If these drains are blocked or damaged, the water remains on the road surface, accumulates at low points, and leads to flooding. There are around 14,000 street drains in the city of Siegen. An analysis of heavy rainfall events for Siegen has shown that flooding frequently occurs due to blocked street drains [7]. Although preventive maintenance and inspections are carried out regularly, blocked street drains and thus unexpected flooding occur time and again. The available data did not allow a reliable prediction based on past data as to which street drain could be particularly affected and no longer be able to drain the water. As a result, maintenance and inspections were carried out based on spatially and temporally defined route planning, regardless of possible urgent maintenance requirements with a view to expected extreme events. As part of the KOMMUNAL 4.0 research project, funded between 2016 and 2019 by the German Federal Ministry for Economic Affairs and Climate Protection [15], a sensor-based risk forecast was to be created and tested for the city of Siegen, and a forecast was to be calculated using digital data from a web-based precipitation forecast as to whether or not a street drain would be at risk of flooding during the next expected heavy rainfall event. This would make it possible to preventively check and, if necessary, maintain the street drain at risk to ensure full functional safety when a heavy rainfall event starts. However, the available sensor technology at that time was too expensive on the one hand, and on the other, its informative value was inadequate for the task at hand. Individual sensor data could not capture the many factors influencing the functional reliability of a street drain facility (e.g., debris discharge from nearby fields, forest paths, or other unpaved areas; runoff obstacles such as construction site material; leaves on the road; or debris from adjacent field or forest paths). To realize the pilot project goals nevertheless, the operating personnel responsible for checking and maintaining the street drains were asked which criteria can influence the functionality of street drains and how these are influenced by heavy rainfall. At the beginning, more than 20 different criteria were named by the staff. In a workshop, these criteria were then discussed together with scientists and the management of the sewer system to select the 10 most important criteria.

The evaluation of the selected criteria was based on two aspects. On the one hand, it was determined which criterion generally represented a high risk (so-called risk class), and on the other hand, the (impact) effect that the selected criteria could have at the specific location (so-called location effect) was assessed using examples of locations. The arithmetic mean then resulted in the relevant so-called risk value of the respective criterion.

Criterion	Risk class	Location effect	Risk value
Foliage	4–5	4	18
Prolonged rain	2–3	2	4–6
Heavy rain	4–5	4	18
Root ingrowth	2	1	2
Influence of dirty roads	4–5	4	18
High hydraulic load	4–5	5	22,5
Surface gradient	2	2	4
Damage	4	3	12
Snow	4	3	12
Reduced flow rate	4	3	12
	Total risk factor		11,9

Table 2.
Exemplary criteria evaluation for a road inlet by the operating personnel (1 = not relevant; 5 = highly relevant) (own presentation; data adapted from [16]).

Table 2 shows an example of an assessment for a selected street drain. The assessment was carried out directly in the asset management system of the city of Siegen for the respective digitally recorded street drain structure, where the further calculation for preventive inspection and maintenance was also performed. Subjective assessments were thus quantified and converted into a form that could be used for further automated processing in the asset management system. As shown in **Figure 1**, this visualization illustrates the translation of tacit operational knowledge into structured, machine-readable inputs.

In the next step, an algorithm created a forecast probability of whether or not a road would overflow in the event of an expected heavy rainfall event. For this purpose, all street drains with a total risk factor > 15 were first selected and divided into two priority levels (level 1 > 20, level 2 > 15). After that, selected street drains were linked with forecast data from the precipitation data portal NiRA.web. The operating staff determined that all expected heavy rainfall of > 20 mm within 60 minutes constituted relevant events (threshold value) and should be included in the calculation. The NiRA.web portal, which is connected to the asset management system, sent current precipitation forecasts to the algorithm at 5-minute intervals. The algorithm in the asset management system checks whether the specified threshold value could be expected within the next 24 hours. If this was the case, the asset management system automatically created a control/maintenance order (see **Figure 2**) for all high-risk street drains and created the corresponding control route based on the priorities, which had to be completed by the operating staff within 12 hours. This automation demonstrates how digitized staff knowledge, combined with real-time weather data, directly initiates preventive action in municipal operations. Although this solution appeared simple at first glance in terms of technical implementation and was positively received at conferences, many municipalities approached by the KOMMUNAL 4.0 project management team decided against implementing a digital solution based on employee knowledge. At the time of the project, there

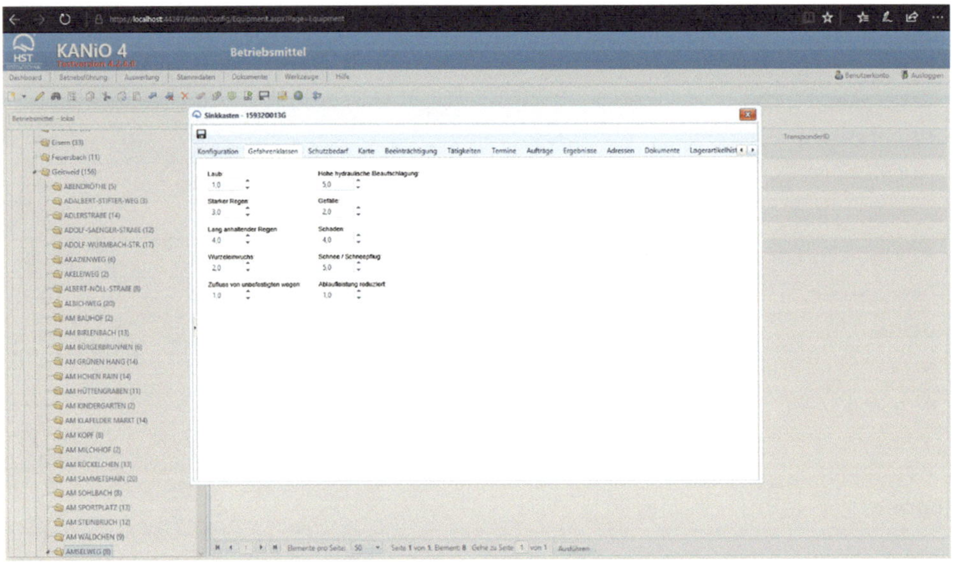

Figure 1.
Evaluation of the road drains in the asset management system [16].

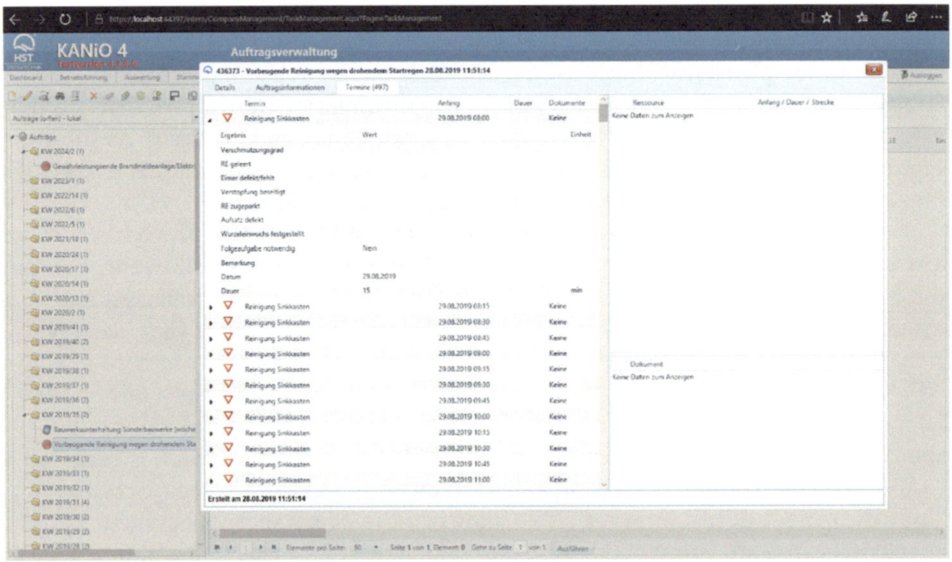

Figure 2.
Automatically generated maintenance order in the asset management system [16].

was greater trust in sensor-based data than in the knowledge of the municipalities' own employees [16].

2.2 Example 2: Implicit knowledge as input data for simulation models

Another example of the effective use of the expertise of experienced knowledge carriers in digital-based water management applications is the integration of employee knowledge into simulation models. In water management, as in many other technical fields of application, simulation models are increasingly being used to develop and test specialized processes and transfer them to reality based on the simulation results [17, 18, 19]. As a rule, operating data or other data measured or recorded by sensors are used as input information for building such models and performing simulations. For some water management applications, the data input is often insufficient. These include, for example, the planning and optimization of sewer network systems or the rehabilitation of small and medium-sized wastewater treatment plants. In these cases, only a small amount of valid measurement data is available. On the one hand, there is a lack of sufficient sensor density; on the other hand, some of the necessary measurement campaigns were only carried out as examples, were not representative enough, or were carried out at long intervals, so the data validity for a reliable simulation is questionable for this type of project. In two research projects, the authors have investigated the inclusion of expertise from experienced knowledge carriers in simulation models. In both cases, the AI method case-based reasoning (CBR) was used. This involves comparing historical use cases with current operating situations to use optimal operating settings from the past for current operations without having to carry out time-consuming individual analyses and comparisons. As there was a lack of sufficient historical operating data for the AI model, the first step was to record the required operating situations with the help of the employees' experience and then simulate them. The data generated by the simulation could then be incorporated into the AI model of CBR.

CBR is a knowledge-based problem-solving method. It is based on the idea that new situations can be better understood and overcome by referring to previously solved cases [20]. Instead of relying exclusively on abstract models or large amounts of data, the CBR methodology allows the use of experience, which is systematically documented in the form of cases and stored in a case database. A typical case comprises the description of a specific problem situation, the approach chosen to solve it, and the results achieved. By comparing current tasks with these previous experiences, proven strategies can be identified and adapted to new framework conditions.

The method offers the opportunity to make the difficult documentation of implicit knowledge of experienced experts accessible in a structured form and to use it specifically for the creation of digital scenarios. The case base can be based on both discrete and continuous parameters that serve as identifiers for individual cases. A case is not only represented as a textual description but also through the combination of parameters such as load and flow variables, pollutant concentrations, energy consumption data, or characteristic operating states. In this way, descriptive experiences of employees can be combined with existing records (e.g., laboratory and sensor data as well as SCADA archives) and defined, checked, and validated by experts before final documentation. This ensures that a case base to be documented is based on a high-quality foundation and captures both successful solution strategies and failed approaches that can serve as an indication of errors to be avoided. Once the desired cases have been defined, described, and documented in the desired form (e.g., by simulation) with the help of the employees, these cases can be used as a database for the CBR process.

The methodological core of CBR is a cyclical process that is divided into four phases (see **Figure 3**). Phase 1 involves case identification (retrieval), in which new problems are compared with the stored cases based on relevant parameters, and similar situations are identified. This is followed by reuse in phase 2: The solution of the most similar case is adopted and, if necessary, adapted to the current situation. The third phase is revision, in which the proposed solution is validated by experts or tested and verified in a further simulation. Fourthly and finally, retention involves the inclusion of the new or adapted case in the case base so that the system continuously learns and dynamically expands its empirical knowledge. **Figure 3** illustrates this iterative cycle and highlights why CBR is particularly well-suited to codifying tacit knowledge: it allows experiential cases to be systematically documented, reused, and refined, creating a dynamic bridge between human expertise and AI-based decision support.

The following examples focus on the process of determining the necessary case descriptions based on the implicit knowledge of the responsible operating personnel using simulation. For further descriptions of the use of the generated case descriptions in the respective CBR models, please refer to the corresponding publications (e.g., [5, 21]).

2.2.1 Case 1: Wastewater treatment plant control

Case 1 focused on the process control of a small wastewater treatment plant in Heringhausen (North Hesse, Germany) at the Diemelsee swimming lake [22]. In summer, during the main tourist season, the wastewater load is up to 5,000 PE; in winter, it is only 400 PE. The wastewater treatment plant, which was more than 40 years old, had to be replaced by a modern treatment technology that can be

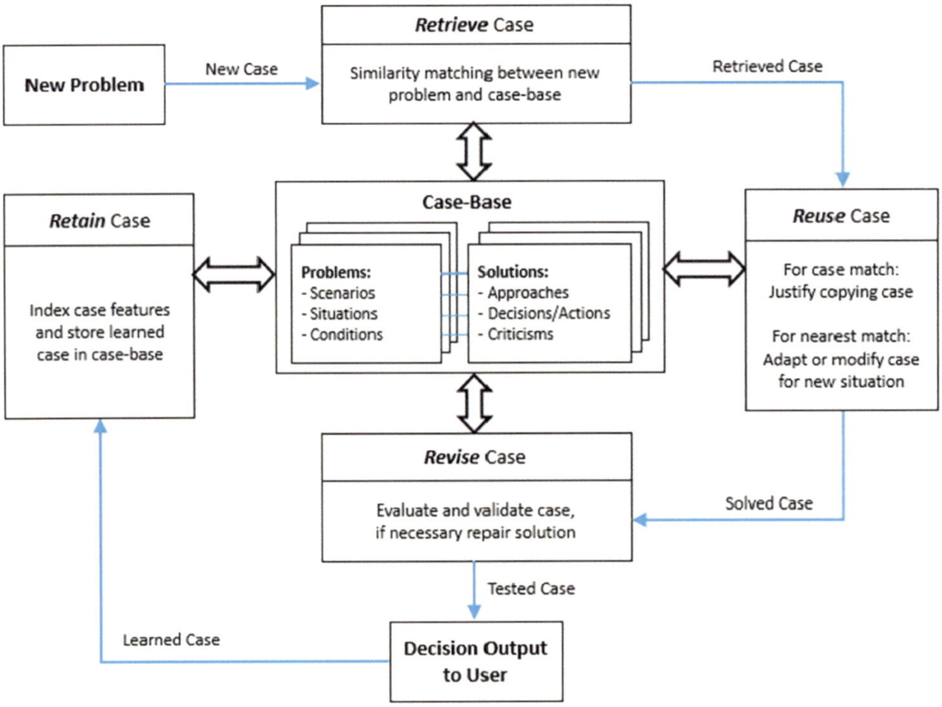

Figure 3.
General model of CBR (own illustration, adapted from [20]).

automatically adapted to highly fluctuating wastewater inflows. The old plant had no sensor technology to speak of, and only a small amount of manually recorded operating data was available. However, in order to develop effective plant technology and modern process control for the extreme inflow fluctuations, a comprehensive simulation of the various operating situations was required, which was carried out using the SIMBA# software simulation system. **Figure 4** illustrates the resulting basic model of the sequencing batch reactor (SBR) wastewater treatment plant, showing how tacit operational knowledge was embedded into the simulation framework to enable robust process control design.

As the data available was insufficient for a professional simulation, the operational knowledge of the employees responsible had to be recorded. Similar to the aforementioned example of street drains, the employees were asked about their operating experience, and the relevant input data for the further simulation was documented. Here too, the quality of the input data depended on the willingness of the operating staff to cooperate and the methods used to record knowledge. In contrast to the street drains, the data required was far more complex, so a simple survey was not sufficient, which is why the operating personnel became an important part of the development team and were directly involved in model creation and validation over the entire project period. With the help of the operational staff, all the necessary information for the operational situations in question was determined iteratively, simulated in the model, the results checked, and any necessary adjustments made. A total of 20 different operating situations were identified, described, simulated in the model, and the most effective process parameter settings documented based on the experience of the operating personnel. They formed the basis for the CBR model's case database.

The identified and finalized evaluation parameters included the incoming pollutant load (with and without rainwater), the incoming wastewater volume (with and without rainwater), the wastewater temperature, and the calendar assignment (weekday, weekend, peak or off-season, no season). The optimal use of simulation information is achieved when the load case data is permanently compared with the real situation, and another system evaluates this data (in this case, a CBR module)

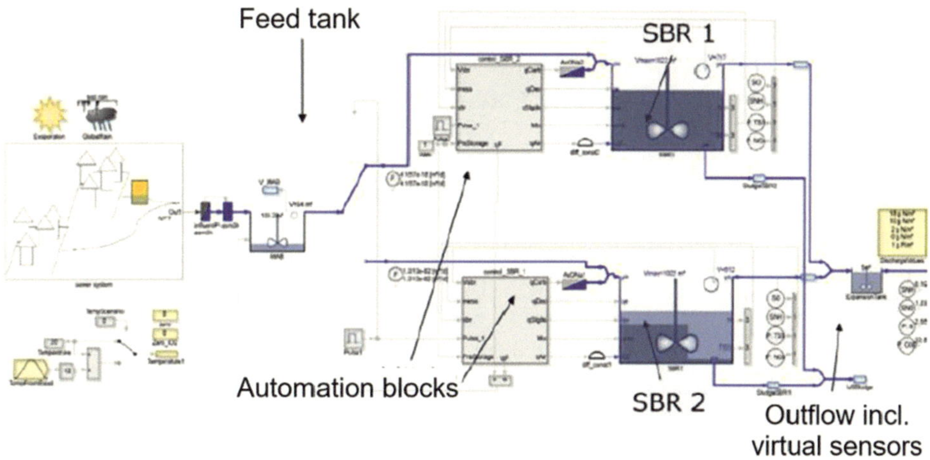

Figure 4.
Basic model of the SBR wastewater treatment plant in SIMBA# [22].

and shows the operating personnel which load cases can be expected in the near future in order to adapt complex wastewater systems to new load cases in good time. Previous attempts to react to changing inflow situations with neural networks or fuzzy logic have been tested in wastewater technology and implemented for individual processes such as air control but were only suitable to a limited extent for dynamic adaptation to strongly fluctuating operating conditions [23]. Moreover, their application is not desirable for complex overall systems, as these solutions do not offer sufficient transparency in the determination of results [5]. The intensive cooperation between SIMBA# model developers and operating personnel has shown that, in the pilot project presented, a high degree of variability of the entire wastewater treatment process was necessary for the subsequent adaptation of the system control to real situations, as a large number of different inlet conditions (load cases) had to be taken into account, which would not have been recognizable and thus definable without the targeted query of implicit knowledge of the experience carriers from the staff side. **Table 3** shows the effects of the presented AI system, based on staff knowledge, from a one-year monitoring conducted by Hof University of Applied Sciences in 2023/2024. It shows that the annual energy consumption of the CBR solution is only around one-third of the annual energy consumption of a conventional control system and approximately half that of a modern wastewater treatment plant control system.

2.2.2 Case 2: Sewer network control

The city of Jena faced a situation similar to the Heringhausen example regarding the data situation when it sought to better adapt its sewer network to extreme weather events. Extreme weather events are increasingly causing problems, such as the discharge of untreated wastewater from the sewer into the River Saale during heavy rainfall or the occurrence of foul gas odors from the sewer due to the increasing build-up of deposits during prolonged dry periods. Technical measures, such as additional storage basins to retain rainwater, regular removal of deposits by manually flushing the sewer, or the addition of odor-reducing chemicals, are possible but very costly. The city of Jena wanted a more effective and, above all, preventative management of the wastewater situation in the sewer system during extreme weather conditions. This was to be achieved with the help of modern machine and control technology that automatically makes better use of the storage volume available in the sewer during heavy rainfall and automatically removes deposits by flushing during dry periods [21]. The selection of suitable machine technology was not a problem, and control of the processes in conjunction with a modern process analysis and SCADA system was also available. The control system was also to be

Kind of control system	Energy consumption (kW per year)
Classical control system without AI	171.832
Modern control system without AI	129.346
Modern control system with AI (CBR)	52.756

Table 3.
Annual energy consumption of the AI solution with CBR compared to solutions without AI (Source: HOF University of Applied Sciences).

supported by a new CBR model in the comparative analysis and evaluation of historical load cases. As with the Heringhausen example, the historical data situation in Jena was also inadequate, meaning that the knowledge of the operating personnel also had to be evaluated. There was already a comprehensive digital image of the Jena sewer network that could be used for a simulation. However, this model had last been updated more than 10 years ago and had to be revised to reflect the current state of the sewer system. As the Jena sewer network only had a small number of sensors, the existing database was insufficient for simulating various operating situations. Accordingly, as in Heringhausen, the missing data had to be supplemented with the experience of the operating staff.

By combining existing data (sensor data and manually recorded data) with the operational knowledge of the employees, it was possible to check the behavior of a sewer network, for example, in relation to extraneous water situations, flushing surge effects, rainwater run-on or inflow volumes to the connected wastewater treatment plant, as well as solids and organic pollutant concentrations and loads. Existing storage structures can be better evaluated in terms of retention time and discharge behavior, so that even a static optimization of these structures derived from this (e.g., by increasing the water level or fixed throttle discharges in individual sewer sections) results in a significantly better use of the existing sewer network volume.

Further and more detailed simulation studies identified additional unused storage volume or showed suitable locations for new basins and their potential for effective sewer network management. A total of 13 different exemplary use cases were selected from an analysis of 180 individual case studies, some of which were very close to each other, which serve as the basis for the comparative analysis with the actual status quo. **Table 4** shows the classification of identified scenarios into three main groups (dry weather, rainy weather, extreme weather). Each scenario contains specific control parameters such as flap/weir position, throttle values, or flushing intervals. The table shows the logical link between the forecast weather situation and the operating measures derived from it.

The scenario development was not only based on more than 180 simulations but also on intensive knowledge management with the specialists at the JenaWasser organization. Interviews, online surveys, and discussion rounds were used to systematically collect empirical values and transfer them to the case database. This represents a digital, reproducible form of operational knowledge and creates

Scenario group	No.	Brief description	Objective
Dry weather and flushing	1–3	Normal operation, medium, and heavy flushing.	Deposit prevention, odor, and corrosion protection.
Rainy weather management	4–11	Graduated strategies depending on rain intensity and season.	Utilization of storage space, reduction of runoff, and protection of the wastewater treatment plant.
Extreme weather scenarios	12–13	Protection of plant technology, preventive flushing before heavy rainfall.	Damage limitation, ensuring technical integrity.

Table 4.
Overview of the 13 CBR scenarios in Jena (presented according to [21]).

a transparent and comprehensible decision-making logic through integration into the CBR system. **Table 5** shows the detailed information for each scenario, based on the experience of operational staff and simulation results.

In the application-related part of the project, the developed, controllable flap/ weir system was tested in a real section of the sewer in the city of Jena. The Storm Water Management Model simulation software was used to simulate possible operating situations that had already occurred, select the cases relevant for the future, and use these results to find the optimum location for machine technology. In the test phase, the simulation results were then compared with the real operating results and, if necessary, the result parameters of the individual scenarios were adapted to reality. In the next step, the determined and verified operating states and information on the behavior of the sewer network can be used for suitable management strategies. The correlations were complex in some cases, as the operating situations in the sewer network were not only very dynamic depending on the rainfall situation, but even the wastewater from commercial and residential sources without rain showed strong daily fluctuations. In the simulation studies, rain volumes were determined that could be managed well and dealt with by storage in the sewer network and delayed onward transportation to the wastewater treatment plant. Other identified operating situations with critical rainfall amounts were less easy to manage and led to a sewer overflowing. In these cases, the rapid onward

Scenario	Objective	Criterion precipitation	Category
Scenario 1	Dry weather – preventive flushing	1 mm total rainfall (forecast 1).	
Scenario 2	Dry weather – preventive flushing – medium flushing	Less than 1 mm total rainfall (forecast 1) and days without average flushing.	
Scenario 3	Dry weather – management – heavy flushing	<1 mm rainfall (forecast 1) and days without heavy flushing.	
Scenario 4	Rainy weather – management	1–5 mm total rainfall (forecast 1).	4 (summer)
Scenario 5	Rainy weather – cultivation	1–5 mm total rainfall (forecast 1).	3 (winter)
Scenario 6	Rainy weather – cultivation	5–10 mm total rainfall (forecast 1).	4 (summer)
Scenario 7	Rainy weather – cultivation	5–10 mm total rainfall (forecast 1).	3 (winter)
Scenario 8	Rainy weather – relief avoidance	10–15 mm total rainfall (forecast 1).	2 (long-term)
Scenario 9	Rainy weather – relief avoidance	10–15 mm total rainfall (forecast 1).	1 (short-term)
Scenario 10	Rainy weather – relief avoidance	15–20 mm total rainfall (forecast 1).	2 (long-term)
Scenario 11	Rainy weather – relief avoidance	15–20 mm total rainfall (forecast 1).	1 (short-term)
Scenario 12	Extreme weather – relief avoidance	Greater than 20 mm total rainfall (forecast 1).	
Scenario 13	Extreme weather – preventive flushing	Greater than 20 mm total rainfall (forecast 1).	

Table 5.
Structural overview of the 13 CBR scenarios for the Jena sewer network [21].

transport of the wastewater volumes to the treatment plant had to be prioritized in order to avoid or reduce the overflow into the river Saale and thus water pollution. Extreme weather situations with prolonged drought led to low volumes of wastewater in the sewer, resulting in heavy deposits and thus unpleasant odors. Two representative application scenarios were identified for this purpose, which form the basis for a control module for predictive sewer flushing during longer dry phases. The implementation was carried out on two special sewer structures, which were equipped with modified flushing and backwater gates. In combination with the forecast data from the web-based precipitation data portal NiRA.web and the real-time values of the sewer from the SCADA system, previously unusable storage volumes can be used in a targeted manner during heavy rainfall events and flushing waves during dry periods.

Initial evaluations based on sensor data and implicit staff knowledge show that this reduces combined sewer overflows into the River Saale by up to 50%. Preventive sewer flushing during dry periods proved to be particularly successful, minimizing deposits and reducing odor pollution.

The constant involvement of staff throughout the entire course of the project, right through testing, shows that CBR control is not just based on a one-off extraction of knowledge, but that employees were continuously involved in the validation of new scenarios as well as the test cases with real machines, sensors, and control units. This corresponds to a dynamic approach to knowledge generation. The Jena and Heringhausen case studies are examples of the possibilities offered by linking an AI-based system design and a methodical recording of implicit employee knowledge. By transferring the results of knowledge generation into a CBR control system, for example, results from detailed simulations can be combined with the experience of the operating personnel and transferred into a reproducible decision-making logic.

The approaches presented to date for the use of empirical knowledge in complex infrastructures, such as water management, can also be transferred to other urban sectors, such as energy supply or urban planning. From today's perspective, these approaches can be significantly improved. As already noted, a case database based on implicit knowledge can quickly become outdated without continuous feedback. The rapid development of generative AI enables the further development of the previous use of tacit knowledge presented above, as well as new and more effective ways of tapping into missing data for AI models based on empirical knowledge. Captured tacit knowledge not only serves as data input for AI systems, but generative AI systems, in particular, can be used directly to capture and process tacit knowledge. However, this also requires effective human–AI interaction as an important component of further developments for the use of tacit knowledge, which is examined in more detail in the following.

Beyond the domain of water management, the same pattern of leveraging tacit operational knowledge can be observed in other sectors. In manufacturing, operator heuristics and experiential rules inform machine set-up, fine-tuning of process parameters, and early detection of quality deviations [24]. In logistics, dispatchers and drivers rely on implicit knowledge for exception handling, such as rerouting during disruptions or optimizing load distribution under dynamic conditions [25]. In the energy sector, maintenance personnel use experiential heuristics for preventive outage management, for example, recognizing weak signals of equipment fatigue or weather-induced risks [26, 27]. Each of these domains requires adaptation of the underlying taxonomy of tacit knowledge, alignment with sector-specific constraints (e.g., throughput and cost in manufacturing, delivery time in logistics, reliability and regulation in energy), and explicit consideration of safety requirements to ensure that experiential rules do not compromise operational integrity. These analogues underline that the structured capture and integration of tacit knowledge is a cross-sector challenge and opportunity, not limited to municipal water infrastructures. In addition, the degree of formalizability of tacit knowledge varies: manufacturing heuristics are often more standardizable than the situational judgment of dispatchers or the safety-critical expertise of energy operators. Sectoral organizational cultures and incentive structures also shape whether employees are willing to externalize their knowledge and how quickly it can be integrated into AI models.

Finally, the issue of scalability differs: while heuristics in manufacturing can be replicated across identical lines, logistics and energy often demand highly context-specific adaptations.

3. The role of human–AI interaction

AI systems differ fundamentally in the way they integrate employees over time. While traditional systems are often based on one-off knowledge extraction, the human-in-the-loop (HITL) approach aims to continuously involve users via various forms of feedback loops in order to develop further together with the users. As shown in the previous examples, in one-off knowledge extractions, empirical knowledge is collected in workshops or interviews and then transferred to a mostly static knowledge base; knowledge is effectively "frozen." For employees, this means that their role ends with the initial transfer of knowledge. In dynamic environments, this quickly leads to models becoming obsolete, as new situations, special cases, or changed framework conditions are not recorded [28].

On the other hand, there are continuous HITL systems. Here, employees remain a permanent part of the process: they provide feedback on forecasts, correct misclassifications, add new cases, and thus actively shape the knowledge base [29]. It is explicitly recommended to design systems in such a way that users can correct errors and make desired settings [30]. Studies show that such feedback loops significantly improve model quality. Such a system is therefore not just a one-off transfer medium but an adaptive tool that continues to evolve in interaction with employees. This makes it clear that the time dimension is crucial: Only where employees remain continuously involved can the system develop dynamically and generate the necessary results for complex issues and tasks. As already described in the SECI model by [12], the knowledge spiral only remains alive if there is continuous interaction.

As illustrated in **Figure 5**, the dynamics of knowledge sharing differ fundamentally between one-off extraction and continuous human–AI interaction. In the one-off model, knowledge transfer occurs only once, effectively interrupting the organizational knowledge spiral and leading to rapid obsolescence, as new cases or changing conditions are not captured. By contrast, in HITL systems, employee feedback and updates sustain the spiral, ensuring that tacit knowledge remains dynamic and continuously enriches the knowledge base. **Figure 5** thus highlights why continuous interaction is essential: it maintains adaptability, fosters trust, and prevents AI models from becoming static or outdated.

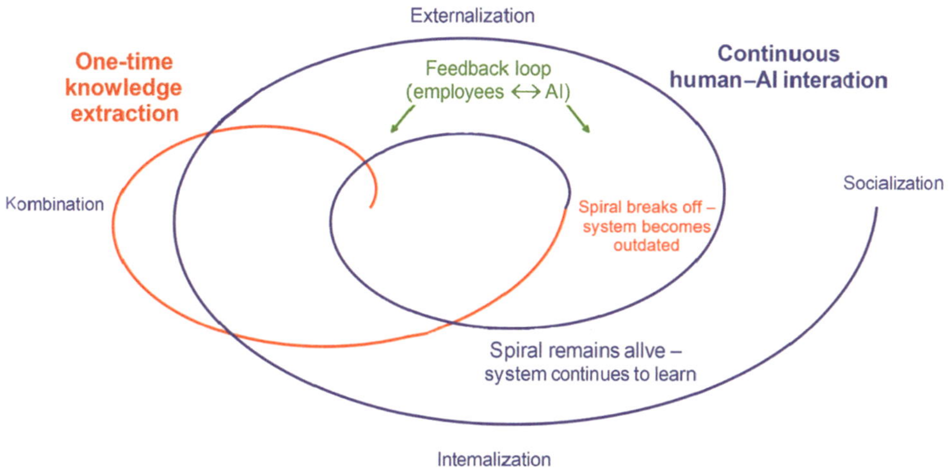

Figure 5.
Knowledge spiral with one-off vs. continuous human–AI interaction (own illustration).

The temporal dimension is also of central importance in organizational and work research. Systems that only involve employees once are often perceived by them as a "black box" and are therefore met with mistrust [31]. Conversely, studies show that employees who are regularly involved in training and feedback processes are significantly more willing to share their knowledge and adopt new technologies [5, 32]. These studies also underline that both the leadership culture and the basic technophile attitude within an organization have a significant influence on whether AI adoption promotes or blocks knowledge sharing. The consequence of this perspective is that the HITL approach fundamentally changes the role of employees over time. They are not reduced to the function of one-time knowledge providers but develop into active co-authors of the system design [33]. This is demonstrated using the example of predictive maintenance, where interactive systems transform employees from passive recipients of results into active co-creators of solutions and confirm the results of the project presented on the predictive control of road inlets [34]. In this context, complementarity is emphasized: the strengths of humans and machines do not compete but complement each other, so that functional system services are optimally linked with human expertise. This makes it clear that continuous human–AI interaction not only increases technical robustness but also has a lasting impact on the social dimension of acceptance and trust. Such a HITL approach therefore offers various advantages. On a technical level, robustness is increased because the system is continuously adapted and improved through ongoing feedback. At the same time, acceptance among employees increases as they see their expertise taken into account in the system and develop trust in the results [5].

In case studies from the water industry, such as Siegen, Heringhausen, and Jena, systems were primarily designed as one-off knowledge repositories. The introduction of generative AI opens the possibility of permanently integrating employees according to the HITL approach. Systems are therefore not developed as static instruments for knowledge transfer but as dynamic tools that learn together with employees. Whether an AI system evolves over time or not depends largely on the role of the employees.

4. Methodology of knowledge documentation and structuring

As shown, the recording and structuring of experience and implicit knowledge is a critical factor for the future process management of complex infrastructures. It is also a key methodological task in order to make this knowledge usable for AI-based systems. Previous challenges and opportunities were demonstrated using several examples from the water industry. The decisive question here is not only how knowledge is collected, but also how it is standardized, structured, and ultimately converted into machine-readable formats, and how it is suitable for the training and operation of AI systems. Implicit knowledge, as described by [12], is deeply anchored in the routines, intuitions, and situational patterns of action of professionals and is therefore difficult to articulate explicitly. Nevertheless, it is precisely this knowledge that is indispensable for safe and efficient decisions in knowledge-intensive environments and situations. Against this background, it is necessary to combine different methodological approaches to bridge the gap between subjective experiential knowledge and standardized, machine-readable data formats [35, 36]. **Table 6** shows possible methods of knowledge documentation that can also be used when employing AI-based models to generate implicit knowledge. For further information, please refer to the relevant literature.

A comparison of the approaches in **Table 6** shows that traditional methods, such as databases and CBR, primarily offer transparency and direct reusability, while modern formats, such as knowledge graphs and LLM-based systems, open new horizons of flexibility, automation, and semantic networking. The decisive factor, however, is not so much the choice of a single method as their combination; only the combination of standardization, dynamics, and semantic depth creates a representation system that not only preserves implicit empirical knowledge but also makes it usable as a living resource for adaptive AI systems. The result shows

Method	Aim	Objective advantages	Limitations
Interviews	Tapping into individual experience	Rich in context, narrative	Subjectivity, reconstruction possible
Workshops	Collective prioritization and consensus building	Consensus and balancing perspectives	Dominant voices possible
Online surveys	Broad, standardized distribution of knowledge	Representative, structured data	Less depth, possible response bias
Scoring systems	Quantification of qualitative assessments	Comparable, traceable	Loss of contextual details
Simulation	Testing/refining model-specific knowledge	Realistic validation	Effort, resource-intensive
CBR	Store and reuse experience cases	Dynamic use of knowledge	Quality dependent on case base
Knowledge graphs	Map semantic knowledge structures	Flexible, inference-capable	Complex modeling
LLM assistance systems	Integrated into everyday life for knowledge utilization	Intuitive, adaptive	Black box, bias

Table 6.
Methods of knowledge documentation and structuring (own presentation).

that knowledge documentation is neither a one-off nor a linear task, but rather a cyclical process in which different methods build on and complement each other [12]. To illustrate this integration, **Figure 6** summarizes the capture-to–AI pipeline, mapping the sequence from elicitation and structuring to validation, enrichment, operationalization, and continuous learning. The latter stage also encompasses essential elements of data governance and ethics, ensuring consent and attribution, managing disagreements through adjudication and versioning, and auditing updates in HITL systems. For the case studies presented (Siegen, Heringhausen, and Jena), this means that the methods described are not to be understood in isolation but as elements of an integrated process. The next section presents what future solutions for complex environments, such as water management, could look like. The framework shown in **Figure 6** illustrates the methodological sequence by which tacit knowledge is elicited, structured, validated, enriched, and operationalized via CBR with HITL, and continuously updated under governance safeguards to remain adaptive, accountable, and trustworthy.

In addition to methodological techniques, the governance and ethical dimension of codifying tacit knowledge requires systematic attention. Across water management, manufacturing, logistics, and energy, several issues recur. First, the principle of informed consent and attribution must be ensured: employees contributing their heuristics and situational judgments should be aware of how this knowledge will be stored and reused, and they should receive appropriate recognition for their intellectual input [37]. Second, disagreements and divergent practices are unavoidable. Operators may follow different rules-of-thumb for plant set-up, dispatchers may prioritize routes differently, or maintenance engineers may disagree on preventive interventions.

Structured procedures for adjudication and versioning are, therefore, required to document and preserve such diversity rather than prematurely enforcing homogenization [38]. Third, in HITL settings, every modification – whether the correction of a forecast, the insertion of a new case, or the override of a recommended action – should be audited and traceable. This auditability guarantees accountability and supports long-term trust in dynamic knowledge bases [39]. Finally, codification processes must remain aligned with sector-specific constraints and safety requirements: in manufacturing, throughput and quality

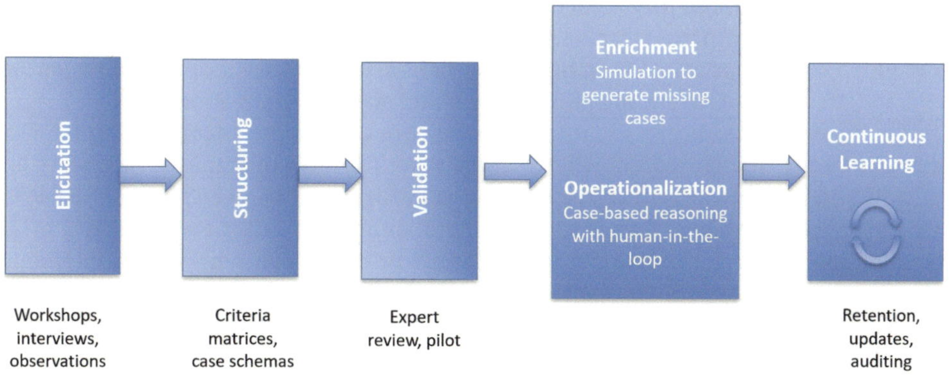

Figure 6.
Capture-to-AI pipeline (own illustration).

standards; in logistics, time-critical deliveries; in energy, strict regulatory and safety obligations; and in water management, environmental compliance, and public safety. Governance mechanisms that combine transparency, traceability, and ethical safeguards ensure that the codification of tacit knowledge is not only technically effective but also socially legitimate and sustainable [40].

5. Generative AI – possible uses and optimization of existing applications for the integration of implicit knowledge

Generative AI is of particular interest against the background of capturing tacit knowledge because its model architectures – from generative adversarial networks (GANs) to variational autoencoders (VAEs) and LLMs – are based on the ability to recognize patterns, make hidden correlations visible, and generate new, plausible knowledge. This property distinguishes generative approaches from purely analytical methods, which are primarily based on the identification and evaluation of existing data [3, 41]. In contrast, generative models are not only able to draw conclusions from existing data but also to generate synthetic scenarios and explanatory representations. This opens up the possibility of transferring empirical knowledge, which is reflected in the operating protocols, conversation notes, or situational decisions, into machine-readable forms and making it usable for knowledge-based systems [42]. LLMs, in particular, have the potential to integrate heterogeneous data sources, from text documents to audio recordings and sensor data, and to tap into previously unstructured knowledge. In combination with methods such as retrieval-augmented generation (RAG) or knowledge graphs, both explicit specialist information and implicit empirical values can be integrated into decision-making processes [43].

Against this background, the use of generative models in water technology processes appears to be promising, but at the same time remains a highly experimental approach. Fully developed applications are hardly available to date. However, ongoing research and pilot projects indicate that generative AI has the potential to optimize existing processes, tap into new sources of knowledge, and systematically integrate implicit expert knowledge into decision-making processes [44, 45]. While classical NLP approaches mainly extract entities and relations [3], LLMs can also make implicit correlations visible that are not explicitly noted in any individual document. This makes decision-making routines, which were previously only accessible through direct exchanges with experts, machine-readable and reproducible. The process begins with the pre-processing of multimodal data. Audio and video recordings can be reliably transcribed using automatic speech recognition and assigned to a specific time and person using speaker diarization [46].

On this basis, LLMs can identify thematic cores, condense them, and transfer them into semi-structured knowledge units. A well-known example of this is the so-called knowledge cards, which document the event, context, decision, and result in a standardized form and thus systematically map implicit experience routines [47]. Ontologizing, that is, the transfer of the results into a formal representation system for the respective task, then plays a decisive role. This involves transferring the knowledge units generated by Generative AI into formal, domain-specific ontologies that enable permanent storage and semantic searchability.

One example of such an architecture is the context-driven feedback framework presented by [47]. This concept illustrates how generative AI can be used reliably in knowledge-intensive domains by combining document management, context-driven query enrichment, security layers, and continuous expert feedback. This also shows that generative AI can create a reliable knowledge base in combination with contextualized knowledge structures and human expertise. The result is an end-to-end process that ranges from the extraction of implicit knowledge from multimodal sources and pre-structuring by generative AI to ontologizing and storage in knowledge graphs. These knowledge graphs not only enable permanent storage but also a wide range of connection options, for example, for semantic search systems, decision support, the transfer of empirical knowledge in the course of succession planning for departing employees, or integration into digital twins of, for example, water management systems. The use of retrieval-augmented generation (RAG) [48] and more recent developments such as GraphRAG [49], which ensure that generative models always link answers to source citations and make global patterns visible across entire corpora (extensive collections of texts or spoken language), is particularly important. In addition, modern generative systems are increasingly multimodal. They not only process text but also speech, images, and videos. This is highly relevant for, for example, water management, as knowledge gained from experience will, in the future, be communicated in the form of spoken comments during plant operation or in gestural demonstrations. By automatically analyzing such multimodal data, non-verbal elements can also be converted into structured knowledge representations.

The previous explanations show that generative AI unfolds its strengths where implicit patterns and knowledge, which are difficult to access, play a central role. Traditional data-driven methods often reach their limits; for example, in water management, sensors and operating data are fragmentary, empirical knowledge is rarely documented systematically, and complex processes exhibit numerous non-linear dependencies [3]. Against this backdrop, generative AI opens the possibility of closing existing information gaps and uncovering hidden correlations that can hardly be tapped using traditional methods.

6. Key findings

The paper demonstrates that tacit knowledge from employees is essential to closing data gaps in complex infrastructures. Generative AI offers new ways to systematically capture and leverage this knowledge – provided that human experts remain actively involved in a continuous feedback loop.

The main key findings are:

- Tacit knowledge is a critical but often untapped data source in technical infrastructures. Retirement or staff turnover poses a high risk of knowledge loss.

- Parallel AI models require large volumes of high-quality data, but freely available sources are largely exhausted. To compensate for this lack, employees' experiential knowledge can be used as a strategically valuable alternative source.

- Tools for knowledge documentation include interviews, workshops, surveys, scoring systems, simulations, CBR, knowledge graphs, and LLM-based assistance systems. No single method is sufficient – a combination of standardization, dynamics, and semantic depth is required.

- AI as an application for knowledge documentation can reveal implicit patterns, process multimodal data (text, audio, video), and transform it into structured units (e.g., knowledge cards, ontologies, knowledge graphs). It is useful for simulating rare events, filling data gaps, and reconstructing decision logic previously held only by experts.

- Using AI, one-off knowledge extraction leads to outdated knowledge bases. HITL solutions are needed. Employees evolve from mere knowledge providers to co-creators of AI systems.

7. Conclusions

Using three research projects as examples, previous approaches to the use of implicit knowledge from complex environments as part of digital solutions were demonstrated using the example of water management. In accordance with the respective state of AI solutions for the effective use of empirical knowledge, mainly static or partially dynamic knowledge processes were installed. Effective solutions were developed that were suitable for closing missing data gaps and developing effective overall solutions for the digitalization of water management processes. The emergence of generative AI solutions, especially LLM systems, opens new possibilities. On the one hand, existing solution concepts can be improved. On the other hand, the risk of aging knowledge bases for intelligent systems can be reduced through a continuous and automatable feedback loop between humans and AI-based knowledge generation. Current limits and risks, as well as some challenges of generative AI that still need to be solved, such as hallucinations, data sovereignty, data protection, or liability issues, must be kept in mind. In addition, generative AI models and systems are constantly evolving. It is, therefore, very important to capture the respective contextual situations in relation to the task as completely as possible.

This also includes the integration of possible AI models for the generation and subsequent application of implicit empirical knowledge in the day-to-day work of the selected applications. It is already becoming apparent today that there will be similar non-technical obstacles to implementation as in the general digitalization of complex infrastructures. The focus is on the factors for the successful implementation of digitalization solutions in the water industry, recorded by [5], such as added value, acceptance and integration into everyday working life, transparency, or the overall orientation of the project. In contrast to previous digitalization, developments in generative AI also mean that employees are more affected by these solutions in terms of the safety of their workplace [50], while at the same time having to integrate AI tools into their own work tasks [51]. In addition to a basic understanding of the functionalities, limits, and possibilities of generative AI, the competence of so-called prompting, that is, the type of information input into a generative AI, is increasingly decisive for the result. According to [52], there is a considerable need for learning here, especially in the area of complex tasks such as water management. For the water industry, this means that the next few years will be characterized by the implementation of targeted pilot projects. Only through close cooperation between research institutions, technology providers, and technical experts can generative systems be developed or adapted to the respective task and tested, validated, and further developed in real operation.

Such projects are not only used for technical testing but also for the development of practical validation standards, which are essential for acceptance and regulatory safety. At the same time, the people involved in the project must be adequately trained (see [52]). Only when it is possible to combine technical performance and human expertise in an integrative process can generative AI mature into a reliable tool in the operational operation of water management systems and effectively leverage and use the data treasure trove of implicit knowledge.

The examples presented from research and application show that, although generative models in water management are still predominantly in the development and pilot stage, they already offer methodological added value today. For example, they can fill gaps in measurement series, simulate rare events, or reconstruct decision-making logic that was previously anchored exclusively in the intuition of experienced specialists. This opens up the possibility of using generative processes to transfer existing but difficult-to-access empirical knowledge into standardized and transferable knowledge units. The future of the use of generative AI in water management will, therefore, largely depend on pilot projects in which technological possibilities are brought together with real operational requirements. In summary, it can be said that generative AI has considerable potential for integrating implicit knowledge into technical infrastructure processes. Its strength lies in its ability to link heterogeneous data sources, including text, speech, images, and videos, to make implicit patterns visible and to make them permanently usable through processes such as ontologizing and knowledge graphs. This not only enables a new form of knowledge documentation but also creates a basis for systematically integrating empirical knowledge into process control systems, for example.

Conflict of interest

The authors declare no conflicts of interest.

Author details

Günter Müller-Czygan* and Viktoriya Tarasyuk
Hof University of Applied Sciences, City of Hof, Germany

*Address all correspondence to: guenter-mueller-czygan@hof-university.de

References

[1] Chadha A. Capturing and Converting Tacit Knowledge for. [Online] 23, October 2023. https://www.phpkb.com/kb/pdf-343.html. [Accessed: 2025-August-23]

[2] Jamshidi E, Nedjat S, Nedjat S, Nikooee S, Rostamigooran N, Majdzadeh R, et al. How to utilize tacit knowledge in health organizations: An Iranian perspective. Medical Journal of the Islamic Republic of Iran. 2018;**32**:678–685

[3] Zaoui Seghroucheni O, Lazaar M, Al Achhab M. Using AI and NLP for Tacit Knowledge Conversion in Knowledge Management Systems: A Comparative Analysis. Technologies. 2025;**13**:87

[4] Villalobos P, et al. Will we run out of data? Limits of LLM scaling based on human-generated data. arXiv 2022

[5] Müller-Czygan G, et al. Die deutschsprachige Wasserwirtschaft im Jahr 2020/21 – Metastudie „WaterExe4.0" zeigt Erfolgsfaktoren und Erwartungen für die digitale Zukunft auf. Österreichische Wasser- Und Abfallwirtschaft. 2022;**74**:241–250

[6] Zur Heiden P, Kaltenpoth S. Wissensmanagement für Wartung und Instandhaltung im Verteilnetz– Konzeption eines Assistenzsystems basierend auf einem Large Language Model. HMD Praxis der Wirtschaftsinformatik. 2024;**61**:911–926

[7] Müller-Czygan G. Von vernetzten Kläranlagen bis zur Vorhersage von Gullyüberflutungen. 2019 wwt .Modernisierungsreport.2019/2020, pp. 31–34 [Accessed: 2020-May]

[8] Göllner J, Mak K, Woitsch R. Intellectual Capital Management using Knowledge Scorecards: The Austrian National Defence Academy Showcase. International Conference on Electronic Government and the Information Systems Perspective. Berlin, Heidelberg: Springer; 2010. pp. 249–256

[9] Summerscales J. Harvesting tacit knowledge for composites workforce development. Composites Part A: Applied Science and Manufacturing. 2024;**185**:108357

[10] Egger S. Stummes Wissen: Die Bedeutung Impliziter Vermittlung Im Gestaltungsprozess. Basel: Birkhäuser; 2022

[11] Gertler MS. Tacit knowledge and the economic geography of context, or The undefinable tacitness of being (there). Journal of Economic Geography. 2003;**3**:75–99

[12] Nonaka I, Takeuchi H, and Umemoto K. A Theory of Organizational Knowledge Creation. International Journal of Technology Management. 1996;**11**:833–845

[13] Bernie C. An Investigation into the Tacit Knowledge Transfer Process in an Open Plan Office Environment. Level. 2015;**3**:1–39

[14] Kropp K. Überlastung der Kanalisation, Überflutung durch Rückstau. Hlh. 2022;**73**:32–35

[15] Müller-Czygan G. Industrie 4.0 für die deutsche Wasserwirtschaft. In: Innovation Race - Wegweisende Prinzipien Für Das Management von

FuE-Projekten. Dortans P, (Hrsg.) Hamburg: Murmann Publishers GmbH; 2021. p. 160–177

[16] Müller-Czygan G. KOMMUNAL 4.0 - Customized Services Für Die Wasserwirtschaft (Schlussbericht Zum Fördervorhaben Im Rahmen desTechnologieprogramms "Smart Service Welt I". Meschede: HST Systemtechnik GmbH & Co. KG; 2020

[17] Ruiz LM, Pérez JI, Gómez MA. Practical review of modelling and simulation applications at full-scale wastewater treatment plants. Journal of Water Process Engineering. 2023;**56**:104477

[18] Rosen R, et al. Die Rolle der Simulation im Kontext des Digitalen Zwillings. Atp Magazin. 2021;**63**:82–89

[19] Klose A, et al. Virtuelle Inbetriebnahme modularer Prozessanlagen. Atp Magazin. 2021;**63**:74–81

[20] Aamodt A, Plaza E. Case-based reasoning: Foundational issues, methodological variations, and system approaches. AI Communications. 1994;7 (1):39–59

[21] Oeltze H, Simancas Suárez JA, and Wiese J. Effizienter Kanalbetrieb mit Einsatz von KI. Wasser Und Abfall. 2025;**27**:44–48

[22] Müller-Czygan G, et al. Energieeinsparung auf Kläranlagen mit fallbasierter Steuerung (Cased Based Reasoning). [book auth.]. In Krupka D, and Federrath H, editors (Bonn: Gesellschaft für Informatik). Daniel Demmler. INFORMATIK 2022 – INFORMATIK in den Naturwissenschaften. 2022;1447–1471

[23] Bongards M. Improving the efficiency of a wastewater treatment plant by fuzzy control and neural network. Water Science and Technology. 1999;**62**:8719–8944

[24] Michels AS, Costa AM, et al. Model and heuristics for the multi-manned assembly line worker integration and balancing problem. International Journal of Production Research. 2024;**62**:8719–8744

[25] Gao S, and Xu D An intelligent agentassisted logistics exception management decision support system: A design science approach». In Hart, DN, and Gregor, SD, editors, (Salisbury South: Griffin Press) Information Systems Foundations: the Role of Design Science. 2010:189–199

[26] Refaiy M, Labib A. The effect of applying tacit knowledge on maintenance performance: An empirical study of the energy sector in the UK and Arab countries. Knowledge Management Research & Practice. 2009;7(3):277–288

[27] Phaladi MP. Framework for Tacit Knowledge Loss Mitigation in South African Public Sector Enterprises. Administrative Sciences. 2025;**15**(6):218

[28] Mosqueira-Rey E, et al. Human-in-the-loop machine learning: A state of the art. Artificial Intelligence Review. 2023;**56**:3005–3054

[29] Amershi S, et al. Guidelines for Human-AI Interaction. In Proceedings of the Conference on Human Factors in Computing Systems 2019:1–13

[30] Wu X, et al. A Survey of Human-in-the-loop for Machine Learning. Future Generation Computer Systems **135**. 2022;364–381

[31] Mirbabaie M, et al. The rise of artificial intelligence – Understanding the AI identity threat at the workplace. Electronic Markets. 2022;**32**:73–99

[32] Hu X, et al. How and when artificial intelligence adoption promotes employee knowledge sharing? The role of paradoxical leadership and technophilia. Frontiers in Psychology. 2025;**16**:1573587

[33] Amaliah NR, Tjahjono B, Palade V. Human-in-the-loop XAI for Predictive Maintenance: A Systematic Review of Interactive Systems and Their Effectiveness in Maintenance Decision-Making. Electronics. 2025;**14**:3384

[34] Huchler N. Die Mensch-Maschine-Interaktion bei künstlicher Intelligenz im Sinne der Beschäftigten gestalten: Das HAI-MMI-Konzept und die Idee der Komplementarität. Digitale Welt. 2020;**4**:30–33

[35] Davenport TH, Prusak L. Working Knowledge: How Organizations Manage What They Know. Boston: Harvard Business School Press; 1998

[36] Yan A, and Cheng Z. A review of the development and future challenges of case-based reasoning. Applied Sciences. 2024;**14**(16):7130

[37] Papagiannidis E, et al. Responsible artificial intelligence governance: A review and research framework. The Journal of Strategic Information Systems. 2025;**34**:101885

[38] Batool A, Zowghi D, Bano M. AI governance: A systematic literature review. AI Ethics. 2025;**5**:3265–3279

[39] Salloch S, Eriksen A. What are humans doing in the loop? Co-Reasoning and practical judgment when using machine learning-driven decision aids. The American Journal of Bioethics. 2024;**24**(9):67–78

[40] Chen X. Ethical Governance of AI: An Integrated Approach via Human-in-the-Loop Machine Learning. Computer Science and Mathematics Forum. 2023;**8**:29

[41] Kaczorowska-Spychalska D, et al. Generative AI as source of change of knowledge management paradigm. Human Technology. 2024;**20**:131–154

[42] Xu Y, and Omitaomu O. Generative AI and digital twins for urban infrastructure: A scoping review. Sustainable Cities and Society. 2024;**3** (1):29

[43] Rothfarb S, Keller M, Hansen J. Multi-agent large language model frameworks: Unlocking new possibilities for optimizing wastewater treatment operation. Environmental Research. 2025;**275**:121401

[44] Allen R, and Cordiner J. Knowledge-enhanced frameworks for energy-efficient wastewater treatment. Computers and Chemical Engineering. 2025;**194**:1–16

[45] Xu Y, Zhang L, and Patel R. Benchmarking large language models for water research. Frontiers in Water. 2024;**6**:122–135

[46] Radford A, et al. Robust speech recognition via large-scale weak supervision. arXiv. 2022

[47] Pulsipher DW. Capturing Tacit Knowledge through Generative AI: A Context-driven Feedback Architecture for Reliable Output. Santa Clara: Intel; 2024

[48] Lewis P, et al. Retrieval-augmented generation for knowledge-intensive NLP tasks. Advances in Neural Information Processing Systems. 2020;**33**:9459–9474

[49] Microsoft. GraphRAG: Retrieval-augmented generation on knowledge graphs. [Online] September 06, 2025. https://www.microsoft.com/en-us/research/publication/graph-rag. [Accessed: 2025-August-26]

[50] Brynjolfsson E, Chandar B, Chen R. Canaries in the Coal Mine? Six Facts about the Recent Employment Effects of Artificial Intelligence. Stanford, CA: Stanford University; 2025

[51] Weilbacher JC. Eine neue Arbeitswelt. Changement. 2024;**8**(3):17–21

[52] Müller-Czygan G, Frank J, and Tarasyuk V. Can GenAI promote complexity skills of scientists? – A hypothetical observation. Annals of Social Science and Management Studies. 2025;**11**(2):1–13

Section 2

Digitalisation of SMEs, Education and Skills

Chapter 7

Perspective Chapter: Factors Influencing Digitalisation in Small and Medium-Sized Enterprises

Lilia Dvořáková and Marek Segeč

Abstract

The aim of the chapter is to analyse and evaluate the factors that lead to the effective implementation of digital business processes and models in small and medium-sized enterprises (SMEs) and to propose recommendations that will contribute to timely and effective digitalisation. Digitalisation in SMEs is linked to two inseparable topics: Industry 4.0 and enterprise information systems (IS). The concept of research questions, questionnaire surveys, and statistical processing of the obtained data is aimed at analysing and assessing the factors influencing successful digitalisation in SMEs. The questionnaire survey in SMEs is mainly focused on the segment – the current state of digitalisation activities and IS—and on the human resources segment. The number of IS types used was identified as a significant factor for building a digitalisation environment in SMEs, as the company gains higher automation, integration, and visibility across business processes by increasing useful IS. Another finding was the identification of a significant trend in the number of IS vendors – as the number of IS vendors increases, the chance of adopting a digitalisation concept decreases. The ratio of the number of experienced senior employees to the total number of employees, their involvement in digitalisation activities, including the preservation and transfer of their knowledge base, proved to be another significant factor for the readiness and implementation of digitalisation activities. The research results and the proposal of recommendations for SMEs in the field of digitalisation are limited by the absence of the inclusion of the category of microenterprises.

Keywords: digitalisation, industry 4.0, information systems (IS), human resources (HR), small and medium-sized enterprises (SMEs)

1. Introduction

Industry 4.0 and the digitalisation of business processes, which is one of the components of the fourth industrial revolution, have the same goals: they aim for higher efficiency, flexibility and competitiveness. We see Industry 4.0 as a vision and concept, while digitalisation in small and medium-sized enterprises (SMEs) is a practical step towards its fulfilment. SMEs that digitalise their processes are approaching the principles of Industry 4.0 and thereby increasing their resilience and ability to innovate, as Industry 4.0 serves as a framework for strategic innovation in SMEs. Digitalisation in SMEs involves the introduction of digital technologies into all areas of business – from production, through logistics, to customer service and internal corporate processes. Digital technologies are technologies that process, store, or transmit information in digital (numerical, binary) form. They use computing, software, electronic devices, and internet platforms to automate, communicate, analyse and control various processes. As selected demonstrative examples of digital technologies, we can mention ERP and CRM systems for company management, the transition from paper to electronic and digital documentation, the use of cloud services, online marketing and e-commerce, sensors in production, chatbots, predictive analysis, image recognition, industrial robots, automated lines, firewalls, antiviruses, encryption, training simulations, 3D modeling, video conferencing, chat applications, and many others. The relationship between Industry 4.0 and digitalisation in SMEs is very close and important, especially for maintaining the competitiveness of these companies, for increasing their productivity, for faster response to changes in the goods and services market and to changes in the labour market, and for improving access to the collection, analysis and evaluation of data and information in the digital economy.

In theory and practice, there is still no clear consensus on which technologies are part of the Industry 4.0 concept. The opinion on whether a technology is innovative and has the potential to be applied to be considered a part or pillar of Industry 4.0 varies among different authors. Some authors emphasise the importance of cyber-physical systems, the internet of things [1], and cloud computing [2], while other authors consider augmented reality [3], digital twins [4], collaborative robots [5], or blockchain [6] to be essential. Of these studies, the concept of the so-called 9 pillars of Industry 4.0 (big data, autonomous robots, simulation, system integration, internet of things, cybersecurity and cyber-physical systems, cloud technologies, additive manufacturing, augmented reality) defined by Rüssman et al. has received the greatest acceptance in 2015 [7], and was subsequently taken over or expanded by other pillars of Industry 4.0 (artificial intelligence, digitalisation of production processes, Automatic Identification and Data Capture) by a number of other authors [8–13].

Currently, there is no uniform definition of the content for the concepts of Industry 4.0 – Society 4.0. This is mainly due to the rapid development of new technologies and the different transnational and national perspectives, documents, and programs addressing this issue. They all share similar visions, but each of them also focuses on different specific goals, set with regard to the historical development of the industry, the current position in the market, and the level of education. The authors of this chapter, based on a search of literary sources and the results of research conducted, are inclined to define the concepts of Industry 4.0 and Society 4.0 with the following content: **These are concepts for automation and robotisation, electronisation, digitalisation of current production, non-production, and creative human activities, for the transformation of the labor market, and for the transformation of education. These concepts support the increase in the speed and efficiency of the creation of higher-quality, more reliable, and cheaper products and services, more efficient use of natural and material resources, and the environmental friendliness of production and**

non-production processes in all sectors of the national economy, in households, and in professional and private human life, with the aim of improving the quality of life [14].

The path to fulfilling the concept of Industry 4.0 – Society 4.0 represents a large-scale societal change. The advent of new technologies and their use changes entire value chains, creates conditions for new production and business models, but also exerts pressure on flexibility, personalisation, automation, and digitalisation of production and services, as well as increased demands on cybersecurity and interdisciplinarity of approach in all sectors of the economy. We see the key to the successful practical implementation of Industry 4.0 – Society 4.0 primarily in interdisciplinary and innovative content and approaches to education, with an emphasis on flexibility and creativity at all levels of the educational system, in lifelong education of adults of all generations, with an emphasis on general and professional digital skills, and the related change in the way of thinking of the entire society – the so-called Thinking 4.0.

The motivation for our research was the limited attention in the field of theory and practice paid to selected factors of digitalisation in SMEs related to information systems and human resources. SMEs, which often face limited financial and human resources and a lack of know-how, nevertheless play an irreplaceable role in a market economy in many areas: economic, social, environmental, as well as regional and societal development. In the case of economic and other crises, it is precisely SMEs that have the potential to slow down economic stagnation or decline. Small and medium-sized enterprises in the EU make up 99% of companies and employ approximately 100 million people, making them one of the basic sources of entrepreneurial spirit and innovation, which are of fundamental importance for competitiveness.

The aim of the study is to identify, analyse and evaluate selected factors that influence the degree and method of digitalisation in SMEs, to support effective digitalisation in SMEs, and to eliminate barriers resulting from a lack of understanding of the need for digital transformation of business processes. The main objective and its decomposition into three research questions are focused on three segments, with the aim of identifying and evaluating digitalisation factors related to enterprise information systems (IS), human resources (HR), and the future perception of digitalisation in SMEs.

Research question 1 (RQ 1): What factors significantly influence the likelihood of SMEs being successful in implementing digitalisation and information systems?

Research question 2 (RQ 2): What factors influence the effective readiness of human resources in SMEs to manage digitalisation and technological processes?

Research question 3 (RQ 3): What factors influence the likelihood that a company will be more successful in digitalisation in the future compared to its current state?

In logical connection with the main objective and research questions, the second subchapter elaborates on the theoretical basis for the issue of digitalisation in SMEs. The third subchapter documents the methodological approach and procedure for implementing qualitative and quantitative research. The fourth subchapter evaluates and discusses the results of a questionnaire survey in SMEs, including recommendations for digitalisation in the areas of IS and HR.

2. Theoretical background

The theoretical foundations in this subchapter are aimed at the issue of digitalisation in SMEs within the framework of the Industry 4.0 concept, with a specific focus on corporate information systems and human resources, based on a search of peer-reviewed sources published in the period 2015–2025, especially in the WoS and Scopus databases.

2.1 Digitalisation in SMEs

Digitalisation is the introduction of digital technologies into business processes to increase the flexibility, efficiency, decision-making, competitiveness, and sustainability of a company in the digital economy. In the context of Industry 4.0, digitalisation is understood as a comprehensive transformation involving the integration of information and communication technologies into all areas of business, the comprehensive transformation of organisational processes, and the development of digital competencies among employees. Industry 4.0 is referred to as a catalyst for digital transformation in SMEs [15]. Digitalisation plays a specific role in SMEs, as these enterprises often have limited technological, human, financial, and information resources compared to large enterprises. At the same time, digitalisation offers opportunities and possibilities to overcome the structural disadvantages of SMEs and addresses concerns about the disruption of established processes in connection with digital transformation. SMEs often focus on partial steps in digitalisation, such as the digitalisation of accounting processes, the introduction of cloud services, the digital archiving of company documents, or the use of AI to optimise storage and retrieval systems, which can be a gradual process leading to the overall digital transformation of the company.

Current research studies document digitalisation in SMEs as a multi-level process in which technological innovations interact with organisational decisions, human resource management, and financial capabilities. The success of digital transformation in SMEs does not primarily depend on the mere adoption of technologies but on motivation, learning, and the gradual implementation and alignment of technologies with production and business strategies and processes. These works also emphasise that SMEs should take into account their initial conditions and invest in continuous education and development of employee competencies and skills to exploit the potential benefits of digitalisation [16].

In terms of theoretical frameworks, the digital transformation of SMEs is often interpreted within the strategic framework of the Resource-Based View (RBV) and Dynamic Capabilities. RBV provides a basic framework for identifying which resources are valuable for the firm, while Dynamic Capabilities demonstrate how the firm effectively uses, develops, and adapts these resources to ensure resilience and long-term success in a constantly changing environment. More recent approaches are complemented by socio-technical perspectives. These frameworks allow for an explanation of why the same technologies lead to different outcomes and benefits of digitalisation in different contexts within SMEs [17, 18].

2.2 Information systems

Information systems (IS) form the basis of digitalisation. They represent technologies and tools that ensure the collection, storage, processing, and distribution of data within the company. For SMEs, IS is particularly important in the following areas:

1. Decision support: Information systems provide management with relevant and up–to-date data that enable a more efficient decision-making process.

2. Efficiency and automation: Thanks to the implementation of Enterprise Resource Planning (ERP), Human Resources Information System (HRIS), or Document Management System (DMS) systems, companies can automate routine activities, eliminate duplicate activities, and reduce administrative burden.

3. Flexibility and scalability: Cloud and modular IS allow SMEs to introduce technologies gradually according to their needs, financial capabilities, and human resources.

4. Customer service and competitiveness: Customer Relationship Management (CRM) systems support the management of customer relationships and increase customer loyalty, which is crucial for SMEs to remain competitive in the market.

ERP, HRIS, DMS, and CRM systems are different types of business information systems that help with the management and automation of specific areas of a company. An ERP system integrates finance, accounting, production, purchasing, and sales; CRM focuses on the management and analysis of customer data, including their contacts, purchase history, and communication; DMS serves as a digital archive for the management, organisation, and search of electronic and scanned documents within the company; HRIS is a system designed to automate and manage processes related to human resources, such as employee registration, payroll management, recruitment, performance evaluation, or employee training.

Empirical and review studies on IS adoption in SMEs describe various motivations for adoption, such as process standardisation, data transparency, and compliance, as well as barriers to IS adoption, such as costs or lack of ICT competencies and skills of internal human resources. Success factors for IS adoption in SMEs include clear and measurable goals, change management, and post-implementation support. On the contrary, underestimation of business process reconfiguration and timely human resource training leads to an erosion of benefits [19, 20]. The latest research studies also show that the impacts of IS digitalisation change decision-making rights, coordination, and autonomy in the workplace, which can bring both greater efficiency and new forms of burden if the socio-technical side of implementation is not managed [21].

2.3 Human resources

The human factor plays a crucial role in the digitalisation process. Information systems are the technical core, but their effective use depends on the capabilities, skills, and digital readiness of employees. Digitalisation is changing the structure of jobs and qualification requirements – the demand for analytical, technical, and creative skills is growing, while manual and routine activities are increasingly replaced by automation.

The digitalisation of human resources takes place on two levels:

1. Digitalisation of HR processes – the implementation of HR information systems (HRIS).

2. Development of digital competencies across the enterprise – employees must possess digital literacy and the ability to work with modern tools, which represents an investment in training and systematic education.

Empirical studies from recent years demonstrate that digital HR practices support productivity and innovation, both directly and indirectly through employee engagement and education, and are effective where complementarity with strategy and leadership is ensured [22, 23]. HRIS adoption is explained by the Technology Acceptance Model (TAM) and the Unified Theory of Acceptance and Use of Technology (UTAUT), where the main determinants are perceived usefulness, ease of use, social influence, and organisational support. These determinants are more pronounced in SMEs due to the limitations of financial resources and the less specialised job roles in these enterprises. The TAM and UTAUT models are theoretical models in computer science that explain why users accept or reject a certain technology and how their acceptance can be improved. TAM focuses on perceived usefulness and perceived ease of use, while UTAUT is an extension of TAM that includes other factors such as social influences, environmental conditions, and psychological expectations, and is more comprehensive in predicting user behaviour.

The combination of IS and HR development creates a synergy that allows SMEs to respond effectively to market changes. While IS provides the technical background for data processing and process optimisation, human capital is key to the adaptation and innovative use of these tools. Successful digitalisation in SMEs is, therefore, not only a technical but also an organisational and cultural process. The combination of IS and HR approaches is described through the AMO (Ability, Motivation, Opportunity) framework, which shows that the benefits of IS only materialise if they are complemented by the redesign of business and work processes, motivation, and development of digital skills of human resources [16, 18, 24].

2.4 Implications for practice and research

The results of the theoretical research suggest that a gradual transformation is appropriate for SMEs. SMEs should start with low-cost and less complex solutions (ERP, cloud), which provide relatively quick benefits, and gradually move to more complex Industry 4.0 technologies (internet of things, big data, AI) when organisational capability and digital competencies are acquired. The digitalisation of SMEs must be understood as a socio-technical process that combines technology and human capital. The implementation of IS brings significant opportunities for improving efficiency and competitiveness, but without the systematic development of employees' digital skills and adequate change management, the expected benefits may not be realised. The current research agenda is shifting from the question of 'What technology to use' to the question of 'How can SMEs align technology with people, organisational structures, culture, and finances, and how to manage the post-implementation phase' [15–18, 20, 24, 25].

3. Methodology

The methodological approach to the study is based on a mixed research strategy and is divided into two stages with the intention of fulfilling the main objective: To identify, analyse, and evaluate selected internal factors that influence the degree and method of digitalisation in small and medium-sized enterprises (SMEs). Both stages of the research (qualitative and quantitative) interact with each other and are combined in the preparation of the research design, data collection, and in the analysis, evaluation, and interpretation of the results. An exploratory sequential design was chosen as the main design of the mixed research, during which qualitative data collection and discussion of its results created the basis for quantitative data collection and analysis, which resulted in the interpretation of the results and the formulation of recommendations in the field of digitalisation of small and medium-sized enterprises and the identification of the focus of future research.

I. Stage: Descriptive research – analysis of theoretical foundations

For the theoretical framework, a synthesis of the results of peer-reviewed articles indexed, in particular, in the Web of Science and Scopus databases during the period 2016–2025 was carried out. Keywords included the following terms: digitalisation, information systems, SMEs, HRM, Industry 4.0. The search included systematic reviews, empirical studies, and theoretical contributions with high citation rates. We focused, in particular, on sources related to the digitalisation of SMEs, the implementation of information systems (IS), and the role of human resources (HR) in the digital transformation process.

II. Stage: Empirical research

The main objective and its decomposition into three research questions focus on three segments, aiming to identify and assess the factors of digitalisation related to information systems, human resources, and the future perception of digitalisation in SMEs.

Main research questions (RQ 1–3)

RQ 1: What factors significantly influence the likelihood of SMEs being successful in implementing digitalisation and information systems?

RQ 2: What factors influence the effective readiness of human resources in SMEs to manage digitalisation and technological processes?

RQ 3: What factors influence the likelihood that a company will be more successful in digitalisation in the future compared to its current state?

3.1 Questionnaire survey, respondent selection, and data collection

The questionnaire survey among SME representatives was carried out in 2025, during the period of January–March. The questions in the questionnaire are divided into an introductory section and three professional segments (sections) aligned with the objectives of the three main research questions. The introductory section of the questionnaire includes questions designed to obtain basic information about the respondent (enterprise size category, business sector, location, region). **In professional Section 1 (Current IS equipment of the enterprise and the state of digitalisation)**, the main questions are aimed primarily at determining the current state of IS use, IS types, number of IS vendors, use of cloud technologies, integration of AI and machine learning technologies in IS, percentage of the financial budget allocated for IS maintenance and development and digitalisation, stages of digitalisation development, use of subsidies for IS acquisition and implementation of a digitalisation project, and application of the Industry 4.0 concept in building a digitalisation environment.

In the second segment of the questionnaire, **in the professional** Section 2 **(Development of digitalisation principles and IS technologies from a personnel perspective)**, the main questions focus on determining the number of employees working in IS, IT systems, and digitalisation projects; the percentage of the financial budget allocated for the salaries of these employees; the length of adaptation for a new employee to work in IS, IT systems, and digitalisation projects; the ratio of experienced senior employees to the total number of employees in IS and digitalisation projects; internal and external employee training and their focus; as well as the existence, content, and distribution of information about the digital strategy and its evaluation by employees.

In the third segment of the questionnaire, **in the professional Section 3 (Plans in the field of IS and digitalisation)**, the main questions are aimed at the planned number and types of IS in the next three years, the scheduled number of IS vendors, the planned percentage of the financial budget for IS maintenance and development and digitalisation, the planned expansion of cloud technologies, the planned expansion of the integration of AI and machine learning technologies in IS, the plan for the implementation of follow-up or new digitalisation projects and IS, the planned use of subsidies, the planned growth in sales, the number of employees, and the number of customers in connection with digitalisation. Each professional section (1–3) of the questionnaire contains 15–18 questions, which are divided into main and additional questions (closed, semi-closed, and open). The main questions of the questionnaire were designed to extract data from them to answer three research questions. Additional questions in the questionnaire serve to expand the main questions with accompanying information and for statistical processing.

In each professional section (1–3), the questions are evaluated with the help of one final question for the respective section. The final question in each section requires the respondent to decide, based on a combination of previous answers and their professional judgment, whether the company has been or will be successful in the given segment. This choice of the final question for each segment was made to clearly determine whether the respondent perceives the direction of the company in the given area as positive or negative. This eliminates ambiguity. In addition, because it is a combination of the respondent's judgment and the questionnaire questions, the option 'I don't know' can be omitted in this final question. The respondent for whom the questionnaire is intended should be able to communicate a negative or positive impression of this process in the context of a specific business environment. The issue of a certain degree of subjectivity on the part of the respondent is partly eliminated by the fact that the questionnaire was content-targeted and distributed in such a way that only a worker with a sufficiently high level of qualification could answer it, and therefore, with a higher probability of objectivity resulting from experience and comprehensive problem-solving. Missing answers were addressed by using the Multiple Imputation method functionality, which searches for links across data rows and columns, evaluates the composition of answers in the row where the answer was given in a valid way, and, on this basis, performs the most appropriate addition and evaluation for the missing value.

The Orbis database was used to select respondents. It contains a range of information on more than 400 million companies from all over the world. Among other things, it also includes contact emails of both a public and non-public nature (available only to paid database users), which were used to select respondents. **The location (Czech Republic) and the category of companies (small and medium-sized) were chosen as the criteria for selecting respondents.** For small companies, the relevant categorisation criteria were specified to exclude micro companies. To partially mitigate adverse effects, a sorting mechanism was added that prioritised companies for which data is actively and regularly updated in the Orbis database, thereby reducing the obsolescence of contact data. The involvement of micro companies was excluded due to the significant differentiation in the levels of digitalisation compared to small and medium-sized companies. While this is undoubtedly a very interesting area of digitalisation, it would be more appropriate to assess this category of companies separately, especially considering the larger gap in terms of technology, personnel, and financial resources. It is also worth noting that many studies in the field of digitalisation provide statistics specifically for small and medium-sized enterprises, often excluding or not mentioning micro enterprises. The Orbis database used does not directly include the micro-enterprise category and instead associates this category with small enterprises. However, a crucial factor in the exclusion of micro-enterprises is the high absence of contact data or the outdated nature of data for this category of enterprises. As part of the research, respondents were informed that this was a survey related to the use and level of implementation of their IS systems and digitalisation activities. They were also informed that the survey was conducted for educational and research purposes (non-business purposes) within the University of West Bohemia. Respondents were assured that their data would be completely anonymised. The questionnaire was created using Google Forms, and the only entity capable of identifying the respondents is Google, as respondents were required to log in with a Google account to complete the questionnaire. Respondents were also informed that by submitting the questionnaire, they consented to the processing of their data. Additionally, they were notified that the usable and processable form of the data would be retained by the co-author of this study for a maximum of six months after the conclusion of the questionnaire survey.

In the first and second rounds of the online questionnaire survey, a total of 1,000 small business respondents and 1,000 medium-sized business respondents from 14 regions in the Czech Republic were addressed across all sectors of the economy. **The final number of respondents who participated in the questionnaire survey was 182, with 144 valid answers for statistical processing being evaluated. The sector mix of the 144 respondents,** according to the Statistical Classification of Economic Activities EU (NACE Revision 2), is mainly from the sections: C-Manufacturing industry (17% of enterprises), F-Construction activities (13%), G-Wholesale and retail trade (8%), H-Transport and storage (8%), J-Information and communication activities (11%), M-Professional, scientific and technical activities (8%), and N-Administrative and support activities (8%). The other enterprises are from the sections A-Agriculture, forestry and fishing; E-Water supply, sewerage, waste management and remediation activities; I-Accommodation, catering and hospitality activities; L-Real estate activities; O-Public defense and administration, compulsory social security; P-Education; Q-Health and social care; and R-Cultural, entertainment and recreational activities, and their share in the sector mix is from 1% to 6%.

3.2 Data processing methods and tools used

The statistical method of logistic regression was used for data analysis and processing, which is employed to predict the probability of a binary outcome based on one or more independent variables. To answer research questions that assess factors affecting the success of digitisation, this method is suitable for data evaluation. For these purposes, a dichotomous variable with only two states was chosen, the result of which is influenced by a number of independent variables. The Statistica and Excel programs were used to process statistical tests.

Procedure for processing information and data from the questionnaire survey:

- Transformation (recoding) of variables used for performing logistic regression.
- Finding extreme values - Multivariate outliers report and Mahalanobis distance.
- Removing distorting extremes – The Box-Tidwell test.
- Evaluation of the dependence between independent variables – Spearman test.
- Performing logistic regression.

  For the main research of the three segments, odds ratios (OR) with 95% confidence intervals were calculated using logistic regression. In the first segment, the number of IS types used had the most significant effect (OR = 4.14), which means that each additional IS type increases the chance of a positive (successful) result of digitalisation by four times. On the contrary, a higher number of IS vendors reduced the chance of a positive perception of the digitalisation process (OR = 0.74). For another significant factor, which was the phase of digitalisation development, the odds ratio was (OR = 0.029) in the case of the initial stage of digitalisation, and the odds ratio was (OR = 2.61) in the case of advanced or full integration. In the second segment (human resources), the existence of a company's digitalisation strategy and a satisfactory number of experienced senior employees (OR = 9 and OR = 9.3) had a strong positive effect, while the absence of a strategy or an undersized state of senior employees significantly reduced the chance (OR = 0.108 and 0.11). In the third segment (future of digitalisation), the factor of planning multiple IS types (OR = 0.52) was significant, which in turn reduces the chances of success from the perspective of the future. The second significant variable, number of planned IS vendors, also showed a negative trend within the future state with an odds ratio (OR = 0.8). The models achieved a Nagelkerke R^2 coefficient in the range of 0.26–0.6 and correctly classified 75–84% of cases, which corresponds to medium to good predictive power and robustness of the presented estimates.

- Model quality testing – Global null hypothesis testing, Nagelkerke R^2 coefficient, case classification with ROC curve. Case classification with accompanying

ROC curve – it confirms the quality of the model in terms of its ability to correctly predict a randomly selected observation due to the relationships of independent variables with the dependent one. The classification matrix shows that the model is able to correctly predict the output for 117 out of 139 observations, which corresponds to a model quality of 84% (for RQ 1), for 109 out of 143 observations, which corresponds to a model quality of 76% (for RQ 2), and for 100 out of 134 observations, which corresponds to a model quality of 75% (for RQ 3).

4. Results and discussion

This subchapter summarises the findings obtained from a questionnaire survey conducted in SMEs in the Czech Republic across 14 regions. From a sectoral perspective, SMEs were represented across all sectors of the economy, with SMEs from the manufacturing industry and construction being more represented in percentage terms. The results of statistical analyses of the obtained data are interpreted and discussed to answer research questions and propose recommendations for SMEs in the field of digitalisation, specifically focusing on IS, human resources, and the future of digitalisation in SMEs. **To use the proposed recommendations in business practice, it is necessary to take into account the specific situation of SMEs in the field of digitalisation and the specifics of SMEs in individual sectors or industries.** An internal or external digital audit in SMEs is a tool for assessing the current state of digitalisation or the readiness of SMEs for digitalisation, which consists of a thorough mapping of the specific business environment, thereby also contributing to the elimination of possible erroneous steps and decisions when managing digitalisation activities.

RQ 1. What factors significantly influence the likelihood that SMEs are successful in implementing digitalisation and information systems?

Significant factors identified: number of IS types used, number of IS vendors, and current stage of digitalisation.

1. **The first significant factor** in the success of digitalisation initiatives in a company was **the number of implemented types of information systems.** The empirical findings indicate that the increasing diversity and number of used IS increase the probability of a positive decision to implement digitalisation changes, which can be explained by a higher level of technological readiness of the organisation. This relationship is in line with the concept of the digital maturity framework, according to which broader integration of IS contributes to achieving a higher level of process optimisation and organisational flexibility of the company [26, 27]. A prerequisite for the effective implementation of IS is its appropriate selection. SME management should choose systems that bring both immediate and long-term benefits to the company, while the decision-making process can be understood in terms of the Technology Acceptance Model (TAM) or Diffusion of Innovations Theory [28, 29].

   To verify the real usability of IS, pilot projects or test implementations can be used, which allow verification of IS benefits and provide measurable outputs for decision-making processes. The practical implications of this finding can be demonstrated, for example, in the implementation of Supply Chain Management Systems, which contribute to the optimisation of logistics and the streamlining of order fulfilment. Similarly, management information systems, when appropriately integrated with other business applications, enable advanced analytics and automated data processing, thereby supporting the identification of demand trends, the detection of bottlenecks in processes, or the evaluation of employee performance [30].

Recommendation: Before implementing IS, verify and evaluate the immediate and future usability of IS in advance. If this assumption is met, the finding can be confirmed that by increasing the number of useful IS, a company gains greater transparency, automation, and integration across business processes, which are determinants of digitalisation.

2. **The second significant finding** of the analysis is the identification of **the relationship between the number of IS vendors and the probability of adopting a digitalisation concept.** The results of the empirical investigation showed the opposite trend to what was originally expected – with an increasing number of IS vendors, the chance of effective implementation of a digitalisation strategy decreases. This finding corresponds to theoretical approaches that recommend that companies cooperate with a maximum of two to three IS vendors, with individual IS vendors specialising in specific IS segments and, at the same time, ensuring compatibility and connectivity with other solutions within the corporate environment [31, 32]. Another important aspect is ensuring substitutability with at least two IS vendors, which eliminates the key disadvantage of a low number of vendors, known as vendor lock-in [33, 34]. In a situation where a company uses only one vendor without the possibility of substitution, there is a high dependence on its solution, which poses a serious risk to the continuity, flexibility, and development of the digital infrastructure. From the analysed data of our research, it is found that currently, small businesses have an average of three IS vendors and medium-sized businesses an average of 5 IS vendors, which also corresponds to the results documented in the European Commission report [35].

   Recommendation: Conduct internal audits focused on implemented IS solutions. These IS audits should identify areas where it is possible to replace the existing IS with alternative solutions while maintaining the quality and functionality of the currently running IS solutions. This recommendation is particularly relevant for medium-sized businesses, since, due to their longer stay on the market and the influence of experience, they should be more effective in this regard than small businesses, if only from the principle of flexibility, when their planning and management of digitalisation must be more far-sighted. Appropriate changes in medium-sized businesses cannot be made as dynamically as in small businesses, where IS solutions of full-fledged or complex proportions are often not implemented. The results of IS audits should at least lead to the initiation of discussions with the current IS vendors on the possibilities of optimising cooperation, or possibly reducing their number. Reducing redundant IS vendors contributes to achieving higher efficiency, simplifying the digitalisation architecture and improving IS governance. It is necessary to emphasise that a higher number of vendors not only complicates IS management and integration but also limits one of the key benefits of – interoperability [36]. This problem may not be obvious in the short term, but in the context of dynamic technological development and increasing diversification of solutions from individual vendors, it can become a significant limitation to the future growth and competitiveness of the company. If this recommendation is ignored, companies risk economic losses, since switching to a new IS vendor without sufficient opportunity to test and adapt its solution to business needs is associated with high costs and risks [37].

3. **The third significant finding** of the analysis conducted in the first segment of the research is the influence of **the currently achieved stage of digitalisation** on the success of the implementation of digitalisation concepts. A positive trend was observed for companies that had reached an advanced stage of digitalisation, while companies with digitalisation concepts in the initial stages showed a negative trend. This phenomenon can be explained mainly by the accumulation of experience from previous stages of digitalisation and the gradual increase in the rate of implementation of digitalisation processes. Companies in the initial stages of digitalisation often have limited knowledge and experience among the organisers of the digitalisation process, which leads to a perception of the complexity and scope of preparatory activities without sufficient understanding of their strategic significance. This lack of insight can lead to neglecting important steps, with subsequent negative consequences often interpreted as a failure of digitalisation as a whole. On the contrary, companies that have passed the initial stages and reached the advanced stage of digitalisation have the opportunity to reflect on previous steps, fine-tune the system, and monitor the specific benefits of implementation. These benefits are manifested, for example, in the form of adding new IS, transitioning to digitalised business and management models, or redesigning the functioning of the enterprise, which allows for more effective use of digitalisation investments and the achievement of strategic goals [26, 38–40].

   Recommendation: From a practical point of view, the recommendation resulting from these findings is to thoroughly plan the initial steps of digitalisation, especially with regard to their logical sequence and level of detail. A carefully structured plan allows for a smooth transition to the subsequent stages of digitalisation and paves the way for the successful completion of the digitalisation process.

RO 2: What factors influence the level of readiness of human resources in SMEs to manage digitalisation and technological changes?

Significant factors identified: Number of senior employees and the existence of a digitalisation strategy.

1. **The first significant factor in the area of personnel management was identified as the proportion of experienced senior employees in the total number of employees.** A low number of experienced senior employees negatively affects the readiness of the entire agenda and reduces the effectiveness of the implementation of digitalisation activities, while a sufficient number of experienced senior employees shows the opposite effect. This finding is in line with research studies in the field of HR digitalisation [41, 42].

This phenomenon can be explained by the fact that the absence of key competencies impairs the ability to map processes and the detailed specification of requirements for IS and digitalisation changes. The result is a limited level of integration, inaccurate planning and reduced usability of the digitalisation potential.

Recommendation: Systematically support experienced senior employees and their knowledge base. Practical measures can include, for example, the creation of a knowledge repository, regular documentation of procedures, reports, and records from project meetings. Such an approach ensures the continuity of the process, even in the case of the temporary or permanent absence of an individual employee, and enables the preservation and transfer of knowledge and experience across the company. The company should emphasise respectful treatment of its key employees, behind whom there is a demonstrable and measurable result of their efforts toward the development of the corporate digitalisation environment.

2. **The second important factor** in the area of the personnel agenda is **the existence and active communication of the digitalisation strategy**, which ensures employee awareness of the ongoing digitalisation processes. Insufficient distribution of information, which is still a common problem in many companies, leads to a negative perception of digitalisation activities, distrust and even sabotage of the implementation.

   Recommendation: Regularly update the digitalisation strategy and distribute it across the company. The rate and scope of information sharing should be clearly defined in the corporate strategic framework to ensure transparency and support employee involvement in digitalisation processes. Another important aspect is the implementation of a digitalisation audit, which provides an overview of the company's readiness for digitalisation and helps to choose appropriate strategies and goals [38, 39, 43, 44].

RQ 3: What factors influence the likelihood that a company will be more successful in digitalisation in the future compared to its current state? Significant factors identified: Planned number of IS vendors, planned number of implemented IS, and artificial intelligence.

1. To answer the third research question, a conclusion can be stated that strongly correlates with and confirms the factors identified in the first segment of the research. **The first factor** for the future success of digitalisation is **the reduction in the number of IS vendors in the corporate environment.** Here, a negative trend was identified from the present to the future in the studied SMEs. The trend for medium-sized enterprises, from the perspective of the future, shows improvement and has shifted from the average use of five IS vendors to four vendors. For small enterprises, on the contrary, we observe a deterioration and a shift from three to four IS vendors on average.

   Recommendation: The identified trend signals that the involvement of a larger number of IS vendors does not provide a company with a higher chance of successfully achieving a fully digitalised environment, which is also consistent with the recommendations in published studies on the optimisation of vendor management in SMEs [45, 46]. This means that applying a multi-vendor strategy will not give the company a higher chance of achieving a positive effect in attaining and expanding a fully digitised environment.

2. **The second finding** is the **planned number of implemented IS**; here, the trend has changed from the present to the future and shows a higher probability of positive acceptance and expansion of digitisation using a smaller number of IS connections. However, this may be an effect when companies currently perceive the insufficient usability of some systems and plan to make their environments more efficient.

   Recommendation: In terms of recommendations for SMEs, this means not using those systems and concepts whose use does not bring added value, which can subsequently lead to conclusions about the need for digitisation. It should also be noted that testing the third segment demonstrated a lower quality of the tested data model. For a more accurate prediction and formulation of recommendations for digitisation with a view to 3 years ahead, it would be necessary to analyse a larger research sample and conduct follow-up research in this segment. The results of our empirical findings are consistent with the conclusions of current studies, which document that less IS involvement supports positive adoption and expansion of digitalisation. This effect is related to the need to streamline the business environment, remove redundant systems, and focus only on those IS and concepts that truly bring added value [44].

3. Based on the analysis comparing the present and the future, a positive trend was revealed **for the incorporation and expansion of AI** into digitalisation concepts in small businesses. Furthermore, a trend was demonstrated in medium-sized businesses in abandoning the use of subsidies to finance their digitalisation concepts, which may be due to the complexity factor that large-scale digitalisation projects require in terms of administration and strict adherence to implementation deadlines for drawing subsidies, which can be limiting, especially for large digitalisation projects [47–50].

4.1 Graphical processing of selected outputs from the questionnaire survey

This subchapter graphically presents selected outputs (significant factors) from the questionnaire survey, which show the numerical structure of enterprises separately for the category of small enterprises and separately for the category of medium-sized enterprises. **Figure 1** documents the current number of IS vendors in SMEs (segment 1, current status). This questionnaire question (independent variable) shows a significant influence on the decision to positively accept the concept of digitalisation for the current development of this concept (the average number of vendors found for small businesses is 3, and for medium-sized businesses is 5).

Figure 2 presents the planned distribution of the number of IS vendors represented in the surveyed SMEs (segment 3, future status – outlook for 3 years ahead). It is this question (independent variable) that showed a significant influence on the decision to positively adopt the concept of digitalisation in the future development and expansion of this concept (the planned average number of vendors in small businesses is 4, and in medium-sized businesses, it is also 4).

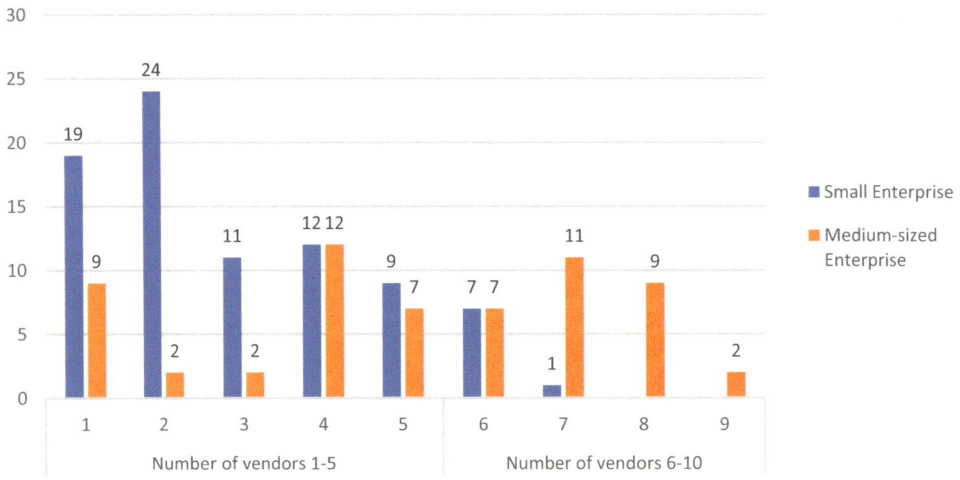

Figure 1.
Current number of IS vendors in SMEs according to the contingency table. Source: own processing, 2025.

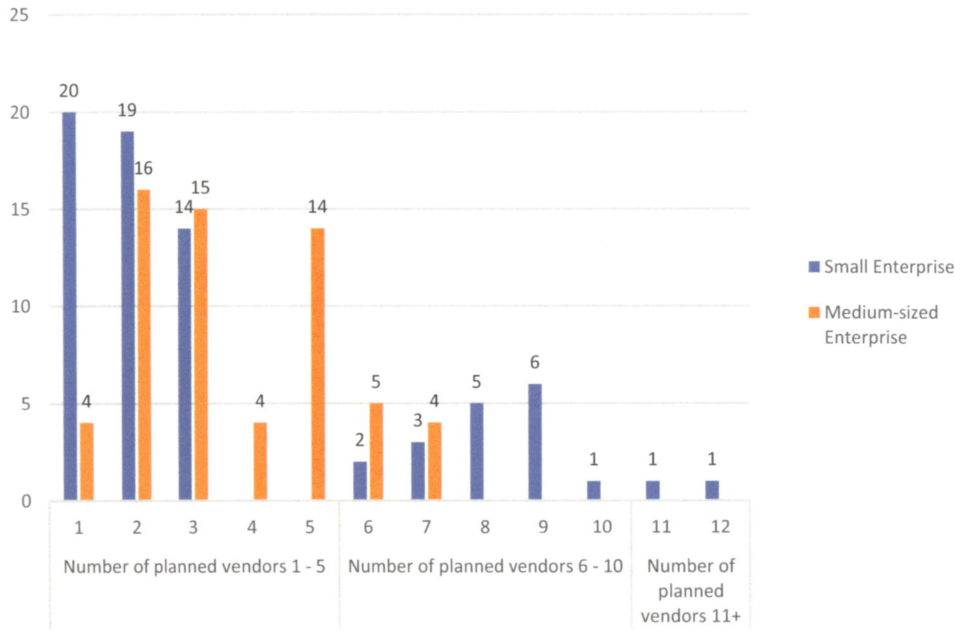

Figure 2.
Number of planned IS vendors in SMEs according to the contingency table. Source: own processing, 2025.

Figure 3 presents an overview of the individual stages of the digitalisation concept across the surveyed companies (segment 1, current status). This question (independent variable) also showed a significant influence on the decision to positively accept the digitalisation concept in the current building and expansion of this concept. More than half of the surveyed SMEs already have their digitalisation concepts in full or advanced integration, which indicates that they have been actively

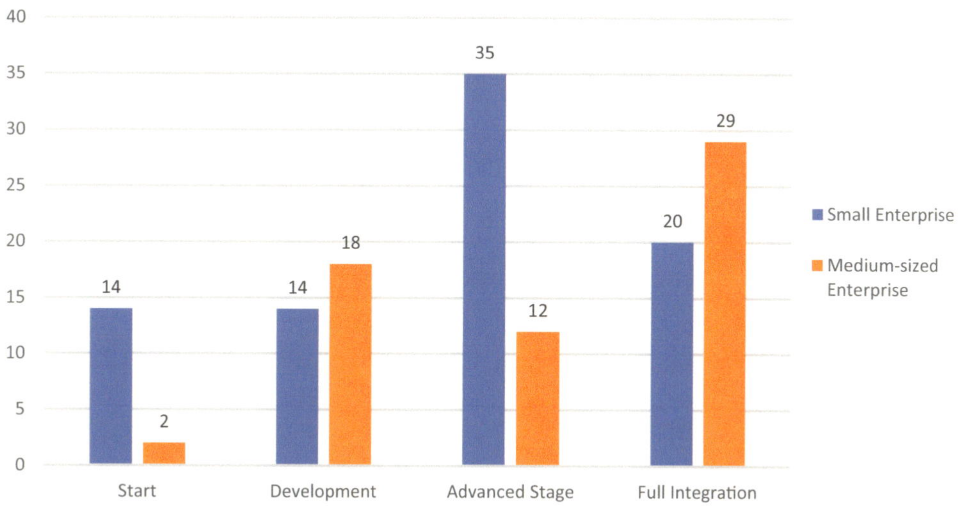

Figure 3.
Phases of in SMEs according to the contingency table. Source: own processing, 2025.

interacting with the digitalisation concept for several years and have experience in this area.

4.2 Summary of findings and recommendations for smes within Industry 4.0

The results of the research show that the success of digitalisation in SMEs is conditioned not only by technological factors but also by organisational readiness and high-quality human resource management. **Based on the results of the data analysis, several significant factors were identified, the connection of which with Industry 4.0 and with the proposed recommendations for SMEs is detailed in Table 1.**

On the technological level, it has been confirmed that a wider diversity of IS used increases the organisation's readiness for digitalisation. At the same time, it has been shown that it is not the quantity that is decisive, but the purposeful use of IS and their interconnection. On the contrary, a large number of IS vendors becomes a barrier to interoperability and effective management of digitalisation. This confirms the importance of a rational approach to vendor management, which allows for reducing complexity and increasing transparency of processes. The advanced stage of digitalisation brings SMEs not only experience but also measurable added value. On the contrary, SMEs that are just starting out are exposed to a higher risk of failure due to underestimating planning and a lack of knowledge. This confirms that digitalisation is a path of gradual steps, where a quality plan and a logical sequence of implementation are of fundamental importance. The human factor remains a key prerequisite for the success of digitalisation in SMEs – a sufficient number of experienced senior employees ensures continuity and a precise definition of digitalisation requirements. Equally important is clear communication of the digitalisation strategy within the entire SME so that employees understand the meaning of the changes and actively participate in them. Predictive factors for the future point to the growing importance of artificial

Significant factor	Relevance to Industry 4.0 technological groups	Recommendation for SMEs
Number of implemented IS	Process automation, ERP/MIS systems, cloud platforms	Conduct an audit of the current IS. Launch a pilot project with a new IS and assess its real contribution before scaling up
Number of IS vendors	Platform integration, interoperability, and interface standardisation	Audit current IS vendors and negotiate to reduce their number while maintaining flexibility
Stage of digitalisation	Digital maturity frameworks, IoT integration, and data-driven management	Develop a digitalisation roadmap with milestones and responsibilities
Number of experienced senior employees	Knowledge management, e-learning platforms, HR technologies	Create a digital knowledge repository and enforce project procedure documentation
Existence of a digitalisation strategy	Strategic management, corporate information portals	Update the digitalisation strategy and communicate it through an internal workshop
Planned number of IS vendors	Cloud services, SaaS integration, and cybersecurity standards	Set a maximum number of IS vendors (e.g., 3) and select based on interoperability
Planned number of implemented IS	Lean digitalisation, process automation	Identify and eliminate low-value systems, focusing on core IS
Artificial intelligence	Predictive analytics, intelligent sensors, decision-making automation	Test a simple AI tool (e.g., demand prediction or chatbot)

Table 1.
Linking factors, Industry 4.0, and recommendations for SMEs. Source: Own processing, 2025.

intelligence as a technology that allows SMEs to increase their competitiveness without the need for massive investments in infrastructure. At the same time, there is a noticeable tendency for companies to streamline their IS environment and eliminate redundant applications. **This shift corresponds to the principles of Industry 4.0, which is based on smart integration of technologies, not their quantity.** Overall, it can be stated that SMEs have the potential to use digitalisation as a tool to increase flexibility, transparency, and innovation. However, success depends on the ability to balance technological readiness, rational management of IS and IS vendors, and support of human resources through strategic communication and knowledge management.

5. Conclusion and future research lines

This chapter aims to identify, analyse and evaluate significant internal factors that influence the positive acceptance and success of digitalisation projects in SMEs. Today's turbulent environment places increased demands on the speed, reliability, protection, and distribution of input and output business data, which are key to the smooth growth of SMEs in the context of digitalisation as a part of the Industry 4.0 concept. SMEs are forced to adapt increasingly quickly due to the dynamics of the times, and thus, there is pressure for timely and, above all, effective implementation and use of the benefits of Industry 4.0 technologies. The main objective of the study and its decomposition into three research questions are focused on three segments with the aim of identifying and evaluating digitalisation factors related to enterprise information systems (IS), human resources (HR), and the future perception of digitalisation in SMEs.

The theoretical background of the study presents and discusses the digitalisation process in SMEs, with a specific focus on IS and HR, based on sources from the period 2015–2025. The results of the theoretical research indicate that a gradual digital transformation is suitable for SMEs. SMEs should start with low-cost and less complex solutions (ERP, cloud), which provide relatively quick advantages and benefits, and gradually move to more complex Industry 4.0 technologies (internet of things, big data, AI) when organisational capability and digital competencies are acquired. The digitalisation of SMEs must be understood as a socio-technical process that combines technology and human capital. The implementation of IS brings significant opportunities for improving efficiency and competitiveness, but without the systematic development of employees' digital skills and adequate change management, the expected benefits may not be realised.

The subsequent practical part of the study seeks answers to the objectives and research questions of the study based on the statistical processing of data obtained through the questionnaire survey method in SMEs. Based on the analysis and evaluation, the research results are presented and discussed in the context of significant factors of digitalisation in the areas of IS and HR, and recommendations for SMEs are proposed. These recommendations can support digitalisation activities and the perception of digitalisation as a tool for improving corporate governance, efficiency, flexibility and competitiveness.

The number of IS types used was identified as a significant factor for building a digitalisation environment in SMEs, since by increasing useful IS, a company gains higher automation, integration, and visibility across business processes. Another finding was the identification of a significant trend in the number of IS vendors – as the number of IS vendors increases, the chance of adopting a digitalisation concept decreases. The ratio of the number of experienced senior employees to the total number of employees, their involvement in digitalisation activities, including the preservation and transfer of their knowledge base, also proved to be another significant factor for the readiness and implementation of digitalisation activities. The presence of other factors, such as the existence of a current digitalisation strategy and the continuous distribution of information across the company, has a positive effect on the success of digitalisation.

We are aware that we have not covered all aspects of the broad and heterogeneous issue of SME digitalisation in the context of Industry 4.0. In the follow-up research, we aim to confront the outputs of the addressed issues more deeply with the reality of business practice and supplement the outputs and recommendations with the opinions of management and experts on digitalisation from SMEs, using the method of controlled expert interviews. We also intend to respond to other current needs of practice, focusing on specific digitalisation solutions in micro, small, and medium-sized enterprises across individual sectors in the manufacturing and service industries, both in a national and international context.

Acknowledgments

This paper was created within the project SGS-2023-007, 'Current Challenges and Problems of Modern Society from the Perspective of Finances and Accounting', at the University of West Bohemia, Faculty of Economics.

Conflict of Interest

The authors declare no conflict of interest.

Author details

Lilia Dvořáková[1*] and Marek Segeč[2]

1 University of West Bohemia, Pilsen, Czech Republic

2 University Hospital, Pilsen, Czech Republic

*Address all correspondence to: lila.dvorakova@seznam.cz

References

[1] Trappey AJC, Trappey CV, Hareesh Govindarajan UH, Chuang AC, Sun JJ. A review of essential standards and patent landscapes for the Internet of Things: A key enabler for Industry 4.0. Advanced Engineering Informatics. 2017;**33**:208–229. DOI: 10.1016/j.aei.2016.11.007

[2] Zhong RY, Xu X, Klotz E, Newman ST. Intelligent manufacturing in the context of industry 4.0: A review. Engineering. 2017;**3**(5):616–630. DOI: 10.1016/j.eng.2017.05.015

[3] Egger J, Masood T. Augmented reality in support of intelligent manufacturing – A systematic literature review. Computers & Industrial Engineering. 2020;**140**: 106195. DOI: 10.1016/j.cie.2019.106195

[4] Novák P, Vyskočil J, Wally B. The digital twin as a core component for industry 4.0 smart production planning. IFAC-PapersOnLine. 2020;**53**(2):10803–10809. DOI: 10.1016/j.ifacol.2020.12.2865

[5] Schmidbauer C, Komenda T, Schlund S. Teaching cobots in learning factories – user and usability-driven implications. Procedia Manufacturing. 2020;**45**:398–404. DOI: 10.1016/j.promfg.2020.04.043

[6] Esmaeilian B, Sarkis J, Lewis K, Behdad S. Blockchain for the future of sustainable supply chain management in Industry 4.0. Resources, Conservation and Recycling. 2020;**163**: 105064. DOI: 10.1016/j.resconrec.2020.105064

[7] Rüßman M, Lorenz M, Gerbert P, Waldner M, Engel P, Harnisch M, Justus J. Industry 4.0: The Future of Productivity and Growth In Manufacturing Industries. Boston Consulting Group; 2015. Available from: https://www.bcg.com/publications/2015/engineered_products_project_business_industry_4_future_productivity_growth_manufacturing_industries [Accessed: 2025-September-27]

[8] Erboz G. How To Define Industry 4.0: Main Pillars Of Industry 4.0. In: Košičiarová I, Kádeková Z, editors. Managerial Trends in the Development of Enterprises in Globalization Era. Nitra: Slovak University of Agriculture; 2017; p. 761–767. Available from: https://www.researchgate.net/publication/326557388_How_To_Define_Industry_40_Main_Pillars_Of_Industry_40 [Accessed: 2025-September-27]

[9] Vaidya S, Ambad P, Bhosle S. Industry 4.0 – A glimpse. Procedia Manufacturing. 2018;**20**:233–238. DOI: 10.1016/j.promfg.2018.02.034

[10] Alcacér V, Cruz-Machado V. Scanning the industry 4.0: A literature review on technologies for manufacturing systems. Engineering Science and Technology, an International Journal. 2019;**22**(3):899–919. DOI: 10.1016/j.jestch.2019.01.006

[11] da Silva ER, Shinohara AC, Nielsen CP, de Lima EP, Angelis J. Operating digital manufacturing in industry 4.0: The role of advanced manufacturing technologies. Procedia CIRP. 2020;**93**:174–179. DOI: 10.1016/j.procir.2020.04.063

[12] Rehman SU, Ejaz S. An Implementation of 9 pillars of Industry 4.0 in conventional footwear industry model. International Journal of

Engineering Applied Sciences and Technology. 2020;**4**(12):283–286. DOI: 10.33564/IJEAST.2020.v04i12.047

[13] Polívka M, Dvořáková L. The importance of Industry 4.0 technologies when selecting an ERP system – An empirical study. E&M Economics and Management. 2023;**26**(3):51–69. DOI: 10.15240/tul/001/2023-3-004

[14] Dvořáková L, Horák J, Caha Z, Machová V, Hašková S, Rowland Z, Krulický T. Adaptation of small and medium-sized enterprises in the service sector to the conditions of industry 4.0: Evidence from the Czech Republic. Econ Ann-XXI. 2021;**191**(7–8):67–87. DOI: 10.21003/ea.V191-06

[15] González-Varona J, López-Paredes A, Poza D, Acebes F. Building and development of an organizational competence for digital transformation in SMEs. Journal of Industrial Engineering and Management. 2021;**14**(1):15–24. DOI: 10.3926/jiem.3279

[16] GH S, AI S, Sutopo W, Suryadi K. Toward SMEs digital transformation success: A systematic literature review. Information Systems and e-Business Management. 2024;**22**(4):667–719. DOI: 10.1007/s10257-024-00682-2

[17] Zamani SZ, Gerlitz L, Hack A, Høgevold NM, Theriou NG. SMEs facing an evolving technological landscape: A systematic review. European Journal of Innovation Management. 2022;**25**(7):263–286. DOI: 10.1108/EJIM-07-2021-0360

[18] Iden J, Bygstad B. Sociotechnical micro-foundations for digital transformation. European Journal of Information Systems. 2024. DOI: 10.1080/0960085X.2024.2347950

[19] Gessa A, Plaisent M, Bernard P. Exploring ERP systems adoption in challenging times. Insights of SMEs stories. Technological Forecasting and Social Change. 2023; **195**. 122795. DOI: 10.1016/j.techfore.2023.122795

[20] Butarbutar ZT, Handayani PW, Suryono RR, Wibowo WS. Systematic literature review of critical success factors on enterprise resource planning post implementation. Cogent Business & Management. 2023;**10**(3):1–29. DOI: 10.1080/23311975.2023.2264001

[21] Colombari R, Nudurupati SS, Bititci U. Disentangling the socio-technical impacts of digitalization: What changes for shop-floor decision-makers? International Journal of Production Economics. 2024; **264**:109377. DOI: 10.1016/j.ijpe.2024.109377

[22] Zhang Y, Sun Y, Wei J. Digitizing success: Leveraging digital human resource practices for transformative productivity in Chinese SMEs. Heliyon. 2024;**10**(17):e36853. DOI:10.1016/j.heliyon.2024.e36853.

[23] Wakabi R, Matovu M. Adoption of human resource information system across East African SMEs: A systematic review. International Journal of Multidisciplinary Studies in Management and Business Economics. 2024;7(2):37–54. DOI: 10.21608/ijmsbe.2025.302900.1015

[24] Mick MMAP, Kovaleski JL, Chiroli D. Developing a sustainable digital transformation roadmap for SMEs: Integrating digital maturity and strategic alignment. Sustainability. 2024;**16**(20):8745. DOI: 10.3390/su16208745

[25] Masood T, Sonntag P. Industry 4.0: Adoption challenges and benefits for SMEs. Computers in Industry. 2020; **121**. 103261. DOI: 10.1016/j.compind.2020.103261

[26] Vial G. Understanding digital transformation: A review and a research agenda. The Journal of Strategic Information Systems. 2019;**28**(2): 118–144. DOI: 10.1016/j.jsis.2019.01.003

[27] Westerman G, Bonnet D, McAfee A. Leading Digital: Turning Technology into Business Transformation. Boston: Harvard Business Review Press; 2014. p. 304

[28] Davis FD. Perceived usefulness, perceived ease of use, and user acceptance of information technology. MIS Quarterly. 1989;**13**(3):319–340. DOI: 10.2307/249008

[29] Rogers EM. Diffusion of Innovations. 5th ed. New York: Free Press; 2003. p. 576

[30] Laudon KC, Laudon JP. Management Information Systems: Managing the Digital Firm. 16th ed. Harlow: Pearson; 2020. p. 656

[31] Sambamurthy V, Zmud RW. Research Commentary: The Organizing Logic for an Enterprise's IT Activities in the Digital Era—A Prognosis of Practice and a Call for Research. Information Systems Research. 2000;**11**(2):105–114. DOI: 10.1287/isre.11.2.105.11780

[32] Chen DQ, Mocker M, Preston DS, Teubner A. Information systems strategy: Reconceptualization, measurement, and implications. MIS Quarterly. 2010;**34**(2):233–259. DOI: 10.2307/20721426

[33] Gefen D, Karahanna E, Straub DW. Trust and TAM in online shopping: An integrated model. MIS Quarterly. 2003;**27**(1):51–90. DOI: 10.2307/30036519

[34] Whitten D, Wakefield RL. Measuring switching costs in IT outsourcing services. Journal of Strategic Information Systems. 2006;**15**(3):219–248. DOI: 10.1016/j.jsis.2006.05.001

[35] European Commission. Directorate-General for Internal Market, Industry, Entrepreneurship and SMEs. Annual report on European SMEs 2021/2022: SMEs and environmental sustainability. Luxembourg: Publications Office of the European Union; 2022. DOI: 10.2826/451263

[36] Lehne M, Sass J, Essenwanger A, Schepers J, Thun S. Why digital medicine depends on interoperability. NPJ Digit Med. 2019;**2**:79. DOI: 10.1038/s41746-019-0158-1

[37] Lacity MC, Khan SA, Willcocks LP. A review of the IT outsourcing literature: Insights for practice. Journal of Strategic Information Systems. 2009;**18**(3):130–146. DOI: 10.1016/j.jsis.2009.06.002

[38] Kane GC, Palmer D, Phillips A, Kiron D, Buckley N. Strategy, not technology, drives digital transformation. MIT Sloan Management Review. 2015;**14**(1):1–25. Available from: https://sloanreview.mit.edu/projects/strategy-drives-digital-transformation [Accessed: 2025-September-27]

[39] Bharadwaj A, Sawy OA, Pavlou P, Venkatraman N. Digital business strategy: Toward a next generation of insights. MIS Quarterly. 2013;**37**(2):471–482. DOI: 10.25300/MISQ/2013/37:2.3

[40] Matt C, Hess T, Benlian A. Digital transformation strategies. Business & Information Systems Engineering. 2015;**57**(5):339–343. DOI: 10.1007/s12599-015-0401-5

[41] Ayoko OB. SMEs, innovation and human resource management. Journal of Management & Organization. 2021;**27**(1):1–5. DOI: 10.1017/jmo.2021.8

[42] Alhloul A, Kiss E. Industry 4.0 as a challenge for the skills and competencies of the labor force: A bibliometric review and a survey. Sci. 2022;**4**(3):34. DOI: 10.3390/sci4030034

[43] Ross J, Beath C, Sebastian I. How to develop a great digital strategy. MIT Sloan Management Review. 2017;**58**(2):7–9. Available from: https://sloanreview.mit.edu/article/how-to-develop-a-great-digital-strategy [Accessed: 2025-September-27]

[44] Zhang X, Xu Y, Ma L. Research on successful factors and influencing mechanism of the digital transformation in SMEs. Sustainability. 2022;**14**(5):2549. DOI: 10.3390/su14052549

[45] Chatterjee S, Chaudhuri R, Kumar A, Aránega AY, Biswas B. Development of an integrative model for electronic vendor relationship management for improving technological innovation, social change and sustainability performance. Technological Forecasting and Social Change. 2023; **186**. 122213. DOI: 10.1016/j.techfore.2022.122213.

[46] Gaaje NN, Ogundaini OO, Mlitwa NBW. Optimising data management systems for SME scalability in South Africa: A vendor perspective. South African Journal of Business Management. 2025;**56**(1):a4924. DOI: 10.4102/sajbm.v56i1.4924

[47] Oldemeyer L, Jede A, Teuteberg F. Investigation of artificial intelligence in SMEs: A systematic review of the state of the art and the main implementation challenges. Management Review Quarterly. 2024. DOI: 10.1007/s11301-024-00405-4

[48] Bettoni A, Matteri D, Montini E, Głądysz B, Carpanzano E. An AI adoption model for SMEs: A conceptual framework. IFAC-PapersOnLine. 2021;**54**(1):702–708. DOI: 10.1016/j.ifacol.2021.08.082

[49] Hao X, Chen X, Wang F. How Government Subsidies facilitate the digital transformation of suppliers. Sustainability. 2024;**16**(19):8652. DOI: 10.3390/su16198652

[50] Le Dinh T, Vu M-C, Tran GTC. Artificial intelligence in SMEs: Enhancing business functions through technologies and applications. Information. 2025;**16**(5):415. DOI: 10.3390/info16050415

Chapter 8

Adaptation of Bachelor's and Master's Students in Engineering to a Changing Environment

Dzhamilia Nugmanova, Inna Kozlova and Roman Kupriyanov

Abstract

This chapter examines how undergraduate and master's engineering students adapted to university life before and during the COVID-19 pandemic. The sample comprised 626 students majoring in physics and chemical engineering from two Russian universities (KFU and KNITU), with data collected in 2018 and 2021. Adaptation was assessed across four domains: sociocultural, physiological, sociopsychological, and academic. Before the pandemic, no significant differences were observed between student groups. However, during the pandemic, master's students demonstrated higher adaptation levels, particularly in the sociocultural and physiological domains. The strengthened correlation between sociocultural and physiological adaptation during the pandemic (from 0.27 to 0.49 for master's students and from 0.34 to 0.48 for bachelor's students) suggests that students' health and well-being became more closely linked to their sense of social and cultural integration under crisis conditions. Enhancing students' adaptation thus requires a holistic strategy that integrates cultural belonging, peer networks, well-being, and study skills. In an era of rapid technological and social transformation, these dimensions are central – not peripheral – to preparing engineers whose adaptability underpins the workforce readiness demanded by Industry 4.0.

Keywords: adaptation, engineering education, bachelor's students, master's students, COVID-19, student well-being, sociocultural integration

1. Introduction

Industry 4.0 – marked by digitalization, automation, and global integration – has transformed the role of engineers. Technical knowledge alone is no longer sufficient; graduates must also demonstrate resilience, flexibility, and the ability to work in multicultural teams [1]. With technologies evolving almost as quickly as the length of a degree program, universities must prepare students for the constant reorientation of skills and knowledge [2–4].

In Russia, engineering education has traditionally emphasized strong theoretical training in physics, chemistry, and mechanical engineering. While this provides a solid scientific base, it often leaves students less prepared for the fast-changing realities of modern industry. Recent studies confirm that engineering students must adapt not only to academic demands but also to broader sociocultural environments [5].

For today's engineering students, adaptation means more than meeting academic requirements. It includes mastering digital tools, keeping pace with technological innovation, and functioning effectively in collaborative environments. Flexible curricula and hands-on learning are crucial supports [6]. Employers increasingly rank adaptability alongside technical expertise: The Future of Jobs Report lists analytical thinking, resilience, flexibility, agility, leadership, and social influence among the top skills for the future workforce [7].

Research offers multiple perspectives on how students adapt to university life, underscoring the complexity of this process. Early models emphasized integration into academic and social contexts. Tinto argued that students' persistence depends on the degree of their academic and social integration [8]. Building on this idea, Baker and Siryk developed the Student Adaptation to College Questionnaire (SACQ), which distinguishes three domains of adaptation: academic, social, and personal-emotional [9].

Subsequent approaches broadened the concept. Kim described adaptation as a dynamic interaction between individuals and unfamiliar environments, highlighting adaptation as an ongoing process rather than a single transition [10]. Vlasova interpreted adaptation as part of socialization and personality development, while Morozov emphasized behavioral patterns shaped by psychophysiological and sociopsychological factors [11, 12].

Further research pointed to the multicomponent character of adaptation. Rean identified behavioral, cognitive, and subjective-personal components, while Friedlander demonstrated the importance of social support, self-esteem, and stress management for first-year students [13, 14]. Martin and Ertzberger later emphasized resilience as the capacity to regulate thoughts, emotions, and behavior under conditions of novelty and uncertainty [15].

Synthesizing these perspectives, Nugmanova et al. proposed a framework that views student adaptation as a multidimensional process encompassing physiological, sociocultural, sociopsychological, and academic domains. This model captures both individual characteristics and environmental demands, as well as the dynamic interaction between them [16].

1.1 The COVID-19 pandemic as a catalyst for change

The COVID-19 pandemic introduced unprecedented challenges that tested students' adaptive capacities. The sudden shift to online instruction disrupted established pedagogical practices and intensified problems related to social integration, mental health, and access to learning resources [17, 18]. In Russia, the pandemic became a driver of rapid digitalization in higher education. Universities redesigned teaching and assessment strategies, expanded the use of platforms such as Zoom and Microsoft Teams, and required students to adjust to new learning formats almost overnight [19]. While distance learning was initially seen as a temporary measure, by the autumn of 2020, it had become a new educational reality demanding sustained adaptive effort [20].

The crisis also revealed deep inequalities. Limited access to technology and insufficient digital competencies hindered many students' learning, with those from lower socioeconomic backgrounds facing the greatest difficulties [21, 22]. Thus, the pandemic not only accelerated digital transformation but also underscored the diverse conditions under which students experience adaptation.

1.2 Adaptation challenges for bachelor's students

For bachelor's students, adaptation is closely tied to the transition from school to university, a period marked by significant developmental and social changes. Entering higher education requires mastering new academic expectations, integrating into unfamiliar peer groups, and adjusting to new cultural and institutional norms.

Research on first-year students confirms the complexity of this stage. Friedlander showed that adaptation depends strongly on social support, self-esteem, and the ability to cope with stress [14]. These factors are particularly relevant for undergraduates, who often face challenges in establishing new support networks and managing academic demands simultaneously.

Age-related characteristics intensify these difficulties. Nebylitsyn described late adolescence as a period of heightened reactivity to external stimuli, while Hagquist noted the increased vulnerability of older adolescents to psychological problems [23, 24]. This makes undergraduates more sensitive to stressors arising from both academic transitions and changes in their social environment.

Thus, for bachelor's students, adaptation encompasses not only academic integration but also emotional regulation and social belonging. Theories that highlight the role of stress, social support, and personal-emotional adjustment [14, 25] are particularly applicable to understanding this group's challenges.

1.3 Adaptation challenges for master's students

Master's students face a different set of adaptation challenges than undergraduates. Having already completed a first degree, they usually possess experience in navigating academic demands and institutional structures. Their adaptation, therefore, centers less on transition into university life and more on professionalization, research activity, and social responsibility.

Theoretical perspectives on socialization and personal development help explain this stage. Ananyev identified the years up to age 25 as a crucial period of personality formation and entry into professional maturity [26]. Similarly, Vlasova framed adaptation as part of socialization, while Morozov emphasized the behavioral and sociopsychological patterns that shape students' adjustment [11, 12]. These views underscore that, for master's students, adaptation is tied not only to academic tasks but also to the consolidation of professional identity and social roles.

However, adaptation at this level is not without obstacles. Deniz and Yilmaz listed several factors that can hinder adaptation: regional traditions and languages, as well as practical issues such as housing and peer relationships [27]. We added to this the differences in national education systems and institutional expectations [16]. Such issues may complicate the adaptation process even for students with prior university experience, particularly in international or interdisciplinary programs.

In sum, adaptation among master's students is best understood through theories that highlight socialization, professional development, and the integration of academic and personal roles. While their challenges differ from those of undergraduates, they remain multifaceted and demand sustained flexibility.

1.4 Research focus of this chapter

Building on these distinctions, this chapter examines how physics and chemical engineering students – at both the undergraduate and master's levels – adapt to university life. The analysis draws on case studies from Kazan Federal University (KFU) and Kazan National Research Technological University (KNITU), covering the period before and during the COVID-19 pandemic.

Adaptation is considered across four interrelated domains – academic, sociocultural, sociopsychological, and physiological – following the multidimensional framework proposed by Nugmanova et al. [28]. Within this framework, bachelor's and master's students confront different challenges: undergraduates focus on transition, social integration, and emotional regulation, while master's students emphasize professional skills, professional development, and social responsibility. These contrasts are summarized in **Figure 1**, which presents the main adaptation challenges for each group.

This leads to the following research questions:

Q1. What differences exist in the adaptation components of bachelor's and master's engineering students?

Q2. How did the COVID-19 pandemic affect the adaptation of bachelor's and master's engineering students?

Q3. Did the relationships among adaptation components change differently for bachelor's and master's engineering students during the pandemic?

The answers to these questions aim to provide new insights for engineering educators and university administrators, with the goal of designing learning environments that not only transmit knowledge but also strengthen students' resilience and capacity to adapt in the era of Industry 4.0.

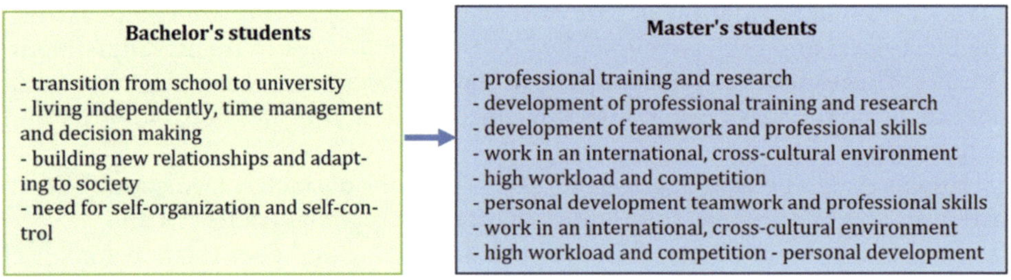

Figure 1.
Adaptation challenges: bachelor's and master's students.

2. Method

2.1 Study design

The study employed a non-experimental, comparative design with natural grouping variables [29–31]. Data were collected in two waves: before the COVID-19 pandemic and during the pandemic, which allowed the context of the pandemic to be treated as an additional grouping variable.

The study included two grouping variables: level of higher education (bachelor's or master's) and period (before vs. during the COVID-19 pandemic). The measured variables were the four components of adaptation: academic, sociocultural, sociopsychological, and physiological. The analysis further examined the relationships between these components.

Participants completed the *Assessment of Students' Adaptation to University* questionnaire [32]. The study was conducted in accordance with the Declaration of Helsinki and was approved by the Ethics Committee of Kazan Federal University (Russia). The research was carried out in three stages:

- Stage 1 (2018–2019): Baseline study at Kazan Federal University (KFU) and Kazan National Research Technological University (KNITU), with students completing paper questionnaires.
- Stage 2 (2020–2021): Follow-up study at the same universities during the COVID-19 pandemic, using online questionnaires distributed via Google Forms.
- Stage 3: Statistical analysis of survey data using *STATISTICA*.

Based on the theoretical framework and study design, the following hypotheses were formulated:

H1. Bachelor's and master's students differ in their adaptation to university.

H2. The COVID-19 pandemic influenced the adaptation of bachelor's and master's students in engineering in distinct ways.

H3. The COVID-19 pandemic altered the relationships between adaptation components for bachelor's and master's students.

To address these hypotheses, the study was guided by the following research objectives:

(1) To compare the adaptation components of bachelor's and master's engineering students.

(2) To examine how the COVID-19 pandemic affected these adaptation components in both groups.

(3) To analyze how the relationships between adaptation components changed for bachelor's and master's students during the pandemic.

2.2 Participants

The study included 626 first-year engineering students in physics and chemical engineering programs at Kazan Federal University (KFU) and Kazan National Research Technological University (KNITU). Data were collected before the COVID-19 pandemic (2018) and during the pandemic (2021). The average age of bachelor's students was 18.6 years, while master's students averaged 22.1 years. Participation was voluntary, with students informed of the study's aims and providing consent before completing the questionnaire.

2.3 Instruments

The study utilized the *Assessment of Students' Adaptation to University questionnaire* [32], which includes 25 items rated on a 7-point Likert scale and covers four domains: sociocultural, physiological, sociopsychological, and academic adaptation. Validation studies reported good internal reliability, with Cronbach's α values ranging from 0.78 to 0.84 [32, 33].

Sociocultural adaptation reflects students' engagement with university culture, campus life, and the broader social environment [34–39]. A sample item is: "How well are you informed about the university's social and cultural activities?"

Physiological adaptation captures perceptions of health, rest, nutrition, and the physical comfort of learning conditions [40]. A sample item is: "Assess your general physical condition since beginning your studies at the university."

Sociopsychological adaptation assesses emotional well-being, peer integration, social comfort, and perceived support [41, 42]. These constructs align with findings that highlight the positive role of self-esteem, social support, and institutional belonging [43]. A sample item is: "Assess your overall emotional state since you started studying at the university."

Academic adaptation refers to adjustment to study demands and learning routines [44, 45]. The scale is designed to assess several aspects of this process. These include students' motivation to learn, the effectiveness of learning, time management skills, presentation skills, comprehension of academic texts, and readiness for future professional growth [28, 46]. A sample item is: "Assess your ability to plan time for independent work during the semester."

The questionnaire makes it possible to analyze each adaptation component separately, as well as calculate an overall adaptation index by averaging scores across the four domains.

2.4 Data collection and analysis

Data were collected in classrooms before the pandemic and online during the pandemic. Students were informed of the study's aims and provided consent prior to participation. Completion of the questionnaire took, on average, 15–20 minutes. All statistical analyses were performed using the STATISTICA software package. Descriptive statistics (means and standard deviations) were first calculated for each of the adaptation components. To compare differences between bachelor's and master's students, independent-samples Student's t-tests were employed. The normality of the distributions was assessed prior to testing, and homogeneity of variances was examined using Levene's test. Statistical significance was set at $p < 0.05$.

3. Results

3.1 Hypothesis 1

Table 1 presents the overall comparison of adaptation between bachelor's and master's students in engineering, based on data collected both before and during the COVID - 19 pandemic. Significant differences were found on several subscales. In sociocultural adaptation, bachelor's students reported a lower mean score (4.67) than master's students (4.97, $p < 0.01$). A similar pattern was observed for physiological adaptation (4.49 vs. 4.80, $p < 0.01$). Differences also emerged on the overall adaptation index, with master's students scoring higher than bachelor's students (5.01 vs. 4.84, $p < 0.05$). By contrast, no significant differences were found in sociopsychological adaptation (4.53 vs. 4.51, $p = 0.853$) or academic adaptation (4.33 vs. 4.46, $p = 0.093$).

The table presents mean scores, standard deviations (SD), t-values, p-values, and Cohen's d effect sizes for four components of adaptation – sociocultural, physiological, sociopsychological, and academic – as well as an overall index. Higher mean values indicate stronger adaptation. Statistically significant differences (*$p < 0.05$; **$p < 0.01$) appeared for master's students in sociocultural, physiological, and overall adaptation levels. The corresponding Cohen's d effect sizes are small ($|d| = 0.26–0.29$) but also indicate the practical significance of differences in the adaptation components of bachelor's and master's students. These findings indicate that while bachelor's and master's students do not differ significantly in academic or sociopsychological adaptation, master's students demonstrate stronger sociocultural and physiological adjustment, as well as higher overall adaptation. Hypothesis H1 is therefore supported.

3.2 Hypothesis 2

As shown in **Table 2**, before the COVID-19 pandemic, there were almost no statistically significant differences in adaptation between bachelor's and master's students. The only exception was the **sociocultural** adaptation subscale, where master's students scored higher (4.95 vs. 4.65, $p < 0.05$). No significant differences emerged in physiological, sociopsychological, or academic adaptation, nor in the

Adaptation	Mean bachelors, n = 450	*SD*	Mean masters, n = 176	*SD*	*t-value*	*p*	*Cohen's d*
Sociocultural	4.67	1.17	4.97	1.14	-2.65	0.008**	0.26
Physiological	4.49	1.06	4.80	1.09	-3.05	0.002**	0.29
Sociopsychological	4.53	0.96	4.51	0.90	0.18	0.853	0.02
Academic	4.33	0.79	4.46	0.86	-1.68	0.093	0.16
Overall	4.84	0.85	5.01	0.86	-2.08	0.038*	0.20

*Note: SD = standard deviation; *statistically significant differences: *p < 0.05; **p < 0.01.*

Table 1.
Comparative analysis of adaptation components among bachelor's (n = 450) and master's (n = 176) engineering students (t-tests).

Adaptation	Mean bachelor's before the Covid-19 pandemic, n = 240	*SD*	Mean masters before the Covid-19 pandemic, n = 96	*SD*	*t-value*	*p*	*Cohen's d*
Sociocultural	4.65	1.18	4.95	0.99	−2.24	0.025*	0.28
Physiological	4.61	0.99	4.57	0.88	0.35	0.722	0.04
Sociopsychological	4.56	0.89	4.44	0.86	1.15	0.252	0.14
Academic	4.23	0.78	4.30	0.68	−0.72	0.469	0.10
Overall	4.85	0.81	4.88	0.69	−0.34	0.730	0.04

*Note: SD = standard deviation; * statistically significant differences; * p < 0.05.*

Table 2.
Comparative analysis of adaptation components among bachelor's (n = 240) and master's (n = 96) engineering students before the COVID-19 pandemic (t-tests).

overall adaptation index, suggesting that, under normal conditions, bachelor's and master's students adapted to university life in broadly similar ways.

Mean scores, SDs, t-values, p-values, and Cohen's d are reported for the four adaptation components and the overall index. Higher means denote stronger adaptation. A statistically significant difference ($p < 0.05$) was observed only in the sociocultural component. The Cohen's d effect size indicates that the difference in sociocultural adaptation between master's and bachelor's students is small ($|d| = 0.28$), while the differences in all other components are negligible.

During the COVID-19 pandemic, however, the distribution of adaptation scores shifted (**Table 3**). On every subscale, master's students reported higher mean values than bachelor's students. Statistically significant differences were found in sociocultural adaptation (4.98 vs. 4.68, $p < 0.05$), physiological adaptation (4.95 vs. 4.42, $p < 0.01$), and the overall adaptation index (5.10 vs. 4.84, $p < 0.05$). In academic adaptation, the difference approached but did not reach significance (4.56 vs. 4.39, $p = 0.090$). These results indicate that during the pandemic, master's

Adaptation	Mean bachelors during the COVID-19 pandemic, n = 210	*SD*	Mean masters during the Covid-19 pandemic, n = 80	*SD*	*t-value*	*p*	*Cohen's d*
Sociocultural	4.68	1.18	4.98	1.23	-2.06	0.040*	0.25
Physiological	4.42	1.10	4.95	1.18	-3.88	0.0001**	0.46
Sociopsychological	4.50	1.01	4.6	0.94	−0.38	0.707	0.10
Academic	4.39	0.80	4.56	0.94	−1.7	0.090	0.19
Overall	4.84	0.88	5.10	0.95	-2.34	0.019*	0.28

*Note: SD = standard deviation; * statistically significant differences * = p < 0.05; ** = p < 0.01.*

Table 3.
Comparative analysis of adaptation components among bachelor's (n = 210) and master's (n = 80) engineering students during the COVID-19 pandemic (t-tests).

students adapted more successfully than bachelor's students, particularly in sociocultural and physiological domains.

The table includes mean scores, SDs, t-values, p-values, and Cohen's d for each adaptation component and the overall index. Higher means indicate stronger adaptation. Significant differences ($p < 0.05$; $^{**}p < 0.01$) were found in the sociocultural, physiological, and overall components. The Cohen's d effect of the difference was small for sociocultural and overall adaptation ($|d| \approx 0.25–0.28$) and approaching medium for physiological adaptation ($|d| \approx 0.46$).

A comparison of the two periods highlights how the adaptation landscape changed. Before COVID-19, bachelor's and master's students were broadly similar, with only a modest sociocultural advantage for master's students. During the pandemic, however, clear differences emerged: master's students consistently reported higher scores across all components, with significant advantages in sociocultural and physiological adaptation as well as in overall adaptation. This shift suggests that, while adaptation was comparable under normal conditions, master's students demonstrated greater resilience in the disruptive context of the pandemic.

An additional observation concerns domains where the relative positions of the groups changed. Before the pandemic, bachelor's students scored marginally higher in sociopsychological adaptation (4.56 vs. 4.44, $p = 0.252$) and in physiological adaptation (4.61 vs. 4.57, $p = 0.722$). During the pandemic, however, the pattern reversed: master's students reported slightly higher scores in sociopsychological adaptation (4.60 vs. 4.50, $p = 0.707$) and substantially higher scores in physiological adaptation (4.95 vs. 4.42, $p < 0.01$). Although the differences in sociopsychological adaptation remained non-significant, the reversal in physiological adaptation reached statistical significance, highlighting this domain as especially sensitive to the pandemic context.

3.3 Hypothesis 3

To examine whether the COVID-19 pandemic altered the relationships between adaptation components, we first considered the relationship of each component with the overall adaptation index. In contrast to Hypotheses 1 and 2, which presented comparisons beginning with bachelor's students, the order here is reversed because the changes in interrelationships were more pronounced among master's students. As shown in **Table 4**, before the pandemic, the overall adaptation of master's students was most strongly associated with physiological ($r = 0.82$) and

Adaptation	Sociocultural	Physiological	Sociopsychological	Academic	Overall
Sociocultural	1.00	0.27	0.42*	0.51*	0.65*
Physiological	0.27	1.00	0.66*	0.47*	0.82*
Sociopsychological	0.42*	0.66*	1.00	0.36*	0.83*
Academic	0.51*	0.47*	0.36*	1.00	0.74*
Overall	0.65*	0.82*	0.83*	0.74*	1.00

Note: – statistically significant correlations; * = $p < 0.05$.*

Table 4.
Correlations among adaptation components in master's students before the COVID-19 pandemic.

sociopsychological adaptation (r = 0.83), while links with sociocultural (r = 0.65) and academic adaptation (r = 0.74) were weaker. This indicates that, under normal conditions, the adaptation of master's students depends mainly on health, well-being, and emotional support.

The table presents Pearson correlation coefficients between the sociocultural, physiological, sociopsychological, and academic components and the overall adaptation index. Most correlations are statistically significant ($p < 0.05$), except for the sociocultural–physiological pair, which is not significant. Statistically significant correlations are marked with an asterisk.

During the pandemic (**Table 5**), overall adaptation became more closely tied to academic adaptation ($r = 0.88$) and sociocultural adaptation ($r = 0.77$), while links with sociopsychological adaptation ($r = 0.88$) and physiological adaptation ($r = 0.80$) remained strong. This suggests that, in crisis conditions, successful adaptation for master's students increasingly depended on academic functioning and integration into the sociocultural environment.

Table 5 presents the Pearson correlation coefficients between the four adaptation components and the overall index. All correlations are positive and statistically significant ($p < 0.05$), with the coefficients generally higher than those observed before the COVID - 19. The change in correlations in the structure of adaptations before and during COVID - 19 among master's students is presented more clearly in **Figure 2**.

For bachelor's students, a similar pattern was observed. Before COVID-19 (**Table 6**), their overall adaptation was most strongly linked to sociopsychological adaptation ($r = 0.85$) and academic adaptation ($r = 0.81$), with weaker ties to sociocultural adaptation ($r = 0.75$) and physiological adaptation ($r = 0.74$). Thus, in normal conditions, bachelor's students' adaptation relied more heavily on emotional stability and academic adjustment.

Table 6 shows Pearson correlation coefficients among the four adaptation components and the overall index, specifically for bachelor's students. All correlations are positive and statistically significant ($p < 0.05$), with moderate to strong coefficients across the adaptation components.

During the pandemic (**Table 7**), these relationships shifted only slightly. The overall index showed strengthened associations with sociocultural (r = 0.77) and physiological adaptation (r = 0.77), while links with sociopsychological (r = 0.84) and academic adaptation (r = 0.79) remained strong but were slightly weaker than before. This indicates that health and cultural integration became somewhat more central to bachelor's students' overall adaptation during the pandemic.

Adaptation	Sociocultural	Physiological	Sociopsychological	Academic	Overall
Sociocultural	1.00	0.49*	0.60*	0.60*	0.77*
Physiological	0.49*	1.00	0.60*	0.57*	0.80*
Sociopsychological	0.60*	0.60*	1.00	0.72*	0.88*
Academic	0.60*	0.57*	0.72*	1.00	0.88*
Overall	0.77*	0.80*	0.88*	0.88*	1.00

*Note: * – statistically significant correlations; * = $p < 0.05$.*

Table 5.
Correlations among adaptation components in master's students during the COVID-19 pandemic.

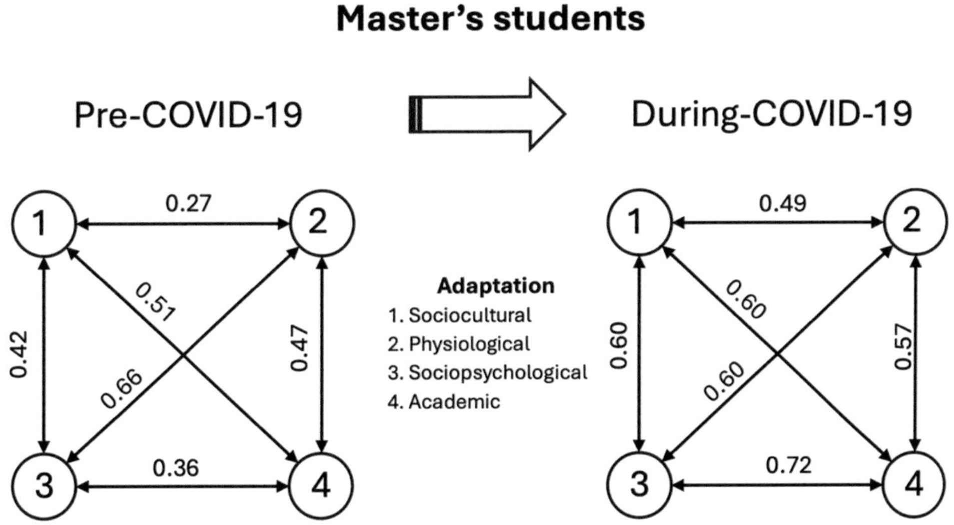

Figure 2.
Correlation graph of the adaptation structure before and during COVID-19 in master's students.

Adaptation	Sociocultural	Physiological	Sociopsychological	Academic	Overall
Sociocultural	1.00	0.34*	0.56*	0.52*	0.75*
Physiological	0.34*	1.00	0.54*	0.51*	0.74*
Sociopsychological	0.56*	0.54*	1.00	0.58*	0.85*
Academic	0.52*	0.51*	0.58*	1.00	0.81*
Overall	0.75*	0.74*	0.85*	0.81*	1.00

*Note: * – statistically significant correlations; * = p < 0.05.*

Table 6.
Correlations among adaptation components in bachelor's students before the COVID-19 pandemic.

Adaptation	Sociocultural	Physiological	Sociopsychological	Academic	Overall
Sociocultural	1.00	0.48*	0.56*	0.53*	0.77*
Physiological	0.48*	1.00	0.54*	0.50*	0.77*
Sociopsychological	0.54*	0.54*	1.00	0.57*	0.84*
Academic	0.53*	0.50*	0.57*	1.00	0.79*
Overall	0.77*	0.77*	0.84*	0.79*	1.00

*Note: * – statistically significant correlations; * = p < 0.05.*

Table 7.
Correlations among adaptation components in bachelor's students during the COVID-19 pandemic.

As in **Table 6**, **Table 7** presents Pearson correlation coefficients among the four adaptation components and the overall index for bachelor's students, but during the pandemic. All correlations are positive and statistically significant ($p < 0.05$), and most coefficients are slightly higher than those observed before the pandemic.

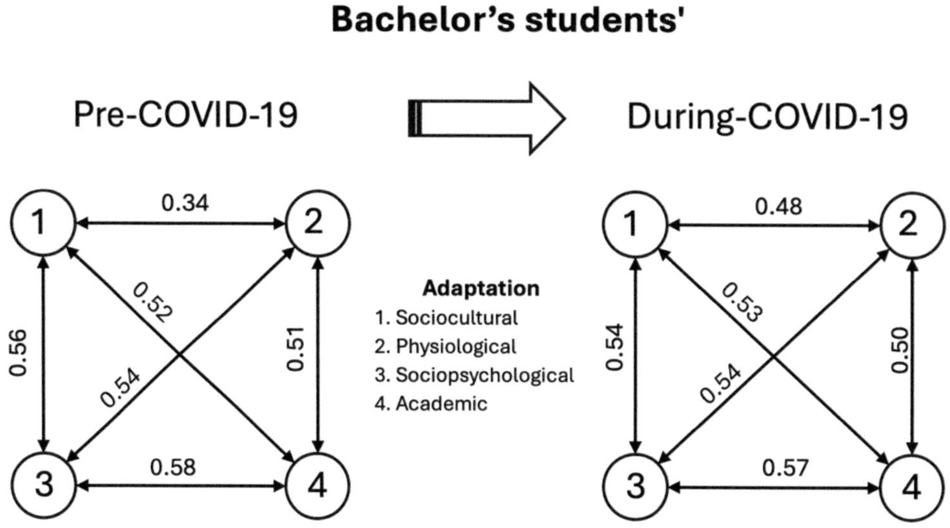

Figure 3.
Correlation graph of the adaptation structure before and during COVID-19 in bachelor's students.

The change in correlations in the structure of adaptations before and during COVID-19 among bachelor's students is presented in **Figure 3**.

A closer look at specific interrelationships highlights how the pandemic reshaped the adaptation system. For master's students, sociocultural adaptation emerged as a central hub: its correlations strengthened with physiological adaptation (0.27 → 0.49, $\Delta = +0.22$), sociopsychological adaptation (0.42 → 0.60, $\Delta = +0.18$), academic adaptation (0.51 → 0.60, $\Delta = +0.09$), and overall adaptation (0.65 → 0.77, $\Delta = +0.12$). This shows that under pandemic conditions, integration into the social and cultural environment became more strongly interwoven with health, well-being, and academic performance. For bachelor's students, correlations among components remained relatively stable, though the link between sociocultural and physiological adaptation grew from 0.34 to 0.48 ($\Delta = +0.14$), suggesting that health and well-being became more connected to social and cultural integration.

Taken together, these results confirm Hypothesis 3: the COVID-19 pandemic altered the relationships between adaptation components for both groups, but the changes were more pronounced for master's students than for bachelor's students.

4. Discussion and conclusions

Taken together, the results of Hypotheses 1–3 show that adaptation to university among engineering students is shaped by the level of study and becomes more differentiated under crisis conditions. Before the pandemic, bachelor's and master's students were broadly similar across most components, with a statistically significant difference observed only in sociocultural adaptation, where master's students scored higher. During the pandemic, however, master's students consistently reported stronger adaptation, especially in sociocultural and physiological domains, while sociocultural adaptation itself emerged as a central hub for overall adjustment. The limited pre-pandemic differences and the subsequent widening of gaps during the pandemic can be explained by the developmental vulnerability of late adolescence: as Nebylitsyn described, this period is characterized by heightened reactivity to external stimuli, and Hagquist emphasized adolescents' increased susceptibility to psychological difficulties [23, 24]. This led to a decrease in physiological adaptation during COVID-19 among undergraduate students [47]. In turn, the sociocultural advantage of master's students reflects their prior experience of social interaction, peer support, and familiarity with university life, which contribute to higher self-ratings of adaptation [48, 49].

Sociocultural adaptation emerges as a pivotal dimension in our study. It was the only domain where master's students consistently outperformed bachelor's students both before and during the pandemic (H1–H2), and it became more tightly interwoven with other domains of adaptation for master's students, where correlations with physiological, sociopsychological, and academic adaptation all strengthened during COVID-19 (H3). For bachelor's students, too, the link between sociocultural and physiological adaptation gained importance, suggesting that cultural integration was increasingly connected to well-being in crisis conditions. This finding is consistent with recent studies confirming the importance of the sociocultural environment and direct interaction with teachers and peers in sustaining student adaptation during the pandemic [50, 51]. "Sociocultural adaptation" appears as one of the keywords in 2025 year's bibliometric research on the topic, linked to concepts such as "psychological adaptation" and "international students" [52]. The authors report an annual publication growth rate of 4.71%. The steady rise in academic attention confirms that sociocultural adaptation is not only a construct of growing theoretical interest but also an empirically central factor in how engineering students adjust to university life, particularly under conditions of uncertainty and disruption.

The pandemic also significantly disrupted the educational experience of bachelor's students, reducing opportunities for collaboration, limiting access to laboratories, and undermining their sense of belonging to university life [53–55]. This highlights the importance of structured mentoring systems, teamwork-based online classes, health promotion initiatives, and resilience training as preventive measures to support vulnerable groups, particularly undergraduates, during social and educational disruptions. The strengthened correlation between sociocultural and physiological adaptation during the pandemic (from 0.27 to 0.49 for master's students and from 0.34 to 0.48 for bachelor's students) indicates that students' health and well-being became more closely tied to their sense of social and cultural integration at the university under crisis conditions.

Building on our previous findings on gender differences, we propose that female undergraduates in engineering may represent the most vulnerable group, whose adaptation is especially at risk under conditions of disruption [56, 57]. This perspective resonates with the wider inequalities revealed by the pandemic: limited digital competencies, unequal access to technology, and socioeconomic disadvantage all hindered students' ability to adapt, underscoring that adaptation is not experienced under uniform conditions [58]. In this sense, the pandemic highlighted both the centrality of resilience in engineering education and the uneven vulnerabilities that shape how students can achieve it [59, 60].

This study demonstrates that adaptation to university life is a multi-component process sensitive to both the educational stage and external disruptions. The COVID-19 pandemic accentuated differences between bachelor's and master's students:

while their adaptation was broadly similar before the crisis, master's students showed clear advantages during it, particularly in sociocultural and physiological adaptation. Moreover, sociocultural adaptation emerged as a central hub, becoming more strongly linked with academic, sociopsychological, and health-related domains under pandemic conditions. For engineering education, these findings underline that resilience cannot be reduced to academic performance alone. Strengthening students' adaptation requires a holistic strategy that integrates cultural belonging, peer networks, well-being, and study skills. In an era of rapid technological and social change, these dimensions are not peripheral but central to preparing engineers whose adaptability directly conditions the workforce's readiness for Industry 4.0 [2].

A major advantage of this study is its survey methodology, which enabled the collection of a wide range of student perspectives across two universities and two time periods. Nevertheless, the study also has limitations. As the instrument relied on self-assessment and reflection, responses may have been shaped by temporary factors such as the respondents' mental or physical state. Although the large sample size mitigates the impact of individual variability, the broader sociopsychological context of the pandemic may still have influenced how students evaluated their adaptation.

Despite these limitations, the findings provide several practical implications. By identifying vulnerable aspects of student adaptation – particularly the central role of sociocultural adjustment and the heightened challenges faced by undergraduates – universities can design targeted interventions to mitigate the negative consequences of large-scale disruptions. Continuous monitoring of adaptation throughout the semester is especially important for bachelor's students, who appear more susceptible to academic and social challenges. Early identification of at-risk individuals would make it possible to provide timely, tailored support, thereby strengthening resilience and improving overall adaptation outcomes.

Table 8 illustrates how the components of university adaptation evolve into key professional competencies required in Industry 4.0 environments. Each component of adaptation aligns with distinct workforce behaviors (see columns 1 and 2 of

Adaptation	Workplace behaviors	Typical stressors	Support actions
Academic	Continuous learning, flexible task management, and adaptation to new technologies.	Rapid skill obsolescence, performance pressure, complex project environments.	Structured onboarding, ongoing professional development opportunities.
Sociocultural	Teamwork, communication across groups, integration into organizational culture.	Cross-functional collaboration challenges, unfamiliar cultural norms, limited social connectedness.	Mentoring networks, inclusive community-building, peer support programs.
Sociopsychological	Motivation, emotional self-regulation, resilience during uncertainty.	High workload variability, performance evaluation stress, feedback sensitivity.	Access to mental health resources, reflective supervision, or coaching.
Physiological	Sustaining well-being and productivity under demanding or hybrid work conditions.	Fatigue, limited recovery opportunities, and ergonomic challenges in remote work.	Workload management policies, health promotion initiatives, ergonomic support.

Table 8.
Workforce bridge: mapping university adaptation to early-career behaviors, stressors, and support actions.

Table 8) and challenges encountered during the early stages of engineering careers. Academic adaptation underpins continuous learning, cognitive flexibility, and the ability to navigate complex technological and organizational systems. Sociocultural adaptation extends beyond interpersonal communication to include cross-cultural competence, collaboration across diverse teams, and integration into multicultural organizational environments – skills increasingly vital in globalized and digitally networked workplaces. Sociopsychological adaptation promotes motivation, emotional regulation, and persistence under uncertainty, supporting professional resilience and effective self-management. Finally, physiological adaptation ensures sustained well-being and productivity under intensive or hybrid work conditions, emphasizing the importance of health-supportive practices and workload management.

Each of these components also corresponds to typical workplace stressors and can be strengthened through targeted organizational support actions (columns 3 and 4) – such as mentoring structures, inclusive community-building, professional development, and health promotion initiatives – thereby facilitating a smoother transition from university to professional life.

Together, these correspondences show that university adaptation forms the basis of professional readiness and long-term well-being. As a dynamic and transferable competence, adaptation can be cultivated through both educational and workplace practices, reinforcing the continuity between learning and professional development.

Other declarations

The authors used ChatGPT (OpenAI, 2025) for language polishing and clarity improvements. All intellectual content, analysis, and interpretations are solely the responsibility of the authors.

Author details

Dzhamilia Nugmanova[1*], Inna Kozlova[1] and Roman Kupriyanov[2]

1 Department of Translation, Interpreting and East Asian Studies, Autonomous University of Barcelona, Barcelona, Spain

2 Kazan (Volga Region) Federal University, Kazan, Russian Federation

*Address all correspondence to: djamilia_ng@list.ru

References

[1] UNESCO, Global Education Monitoring Report Team. Global Education Monitoring Report, 2021/2: Non-state Actors in Education: Who Chooses? Who Loses? Paris: UNESCO, 2021. 573 p. DOI: 10.54676/XJFS2343

[2] Valeyeva NS, Kupriyanov RV, Valeeva E, Kraysman NV. Influence of the Fourth Industrial Revolution (Industry 4.0) on the System of the Engineering Education. In Auer M, Hortsch H, Sethakul P, editors The Impact of the 4th Industrial Revolution on Engineering Education. ICL 2019. Cham: Springer; 2020. p. 316–325. Adv Intell Syst Comput. vol 1135. DOI:10.1007/978-3-030-40271-6_32

[3] Baygin M, Yetis H, Karakose M, Akin E. An effect analysis of industry 4.0 to higher education. Paper presented at: 15th International Conference on Information Technology Based Higher Education and Training (ITHET); 2016 Sep 8–10; Istanbul, Turkey. IEEE; 2016. DOI:10.1109/ ITHET.2016.7760744

[4] Meisenberg G. Cognitive human capital and economic growth in the 21st century. In Abrahams TG, editor. Economic Growth in the 21st Century. Hauppauge, NY: Nova Science Publishers; 2014. p. 49–106

[5] Nugmanova D, Kupriyanov R, Valeyeva NS. Analysis of the differences in adaptation to higher education of the first-year engineering and humanities students. In Auer ME, Rüütmann T, editors Educating Engineers for Future Industrial Revolutions. Cham: Springer; 2021. p. 88–95. DOI: 10.1007/978-3-030 -68201-9_9

[6] Felder R, Brent R. Ch. 6. In: Teaching and Learning STEM: A Practical Guide. San Francisco: Jossey-Bass; 2016

[7] World Economic Forum. The Future of Jobs Report 2025. Geneva: World Economic Forum; 2025. 290 p. Available from: https://www.weforum.org/publi cations/the-future-of-jobs-report-2025/ [Accessed: 2025-October-14]

[8] Tinto V. Reconstructing the first year of college. Planning for Higher Education. 1996;**25**(1):1–6

[9] Baker RW, Siryk B. Student Adaptation to College Questionnaire Manual. Los Angeles: Western Psychological Services; 1989

[10] Kim YY. Becoming Intercultural: An Integrative Theory of Communication and Cross-cultural Adaptation. Thousand Oaks: CA: SAGE Publications Inc.; 2001. DOI: 10.4135/ 9781452233253

[11] Vlasova TA. Социально-психологическая Адаптация Студентов Первого Курса К Условиям Обучения [Social-psychological Adaptation of First-year University Students to Learning Conditions], Vol. **1**. 2009. 13–22.

[12] Morozov V, Mykhailenko L, Falko N, Puente ER, Galaidin A. Socio-psychological adaptation of first-year students at the economic university as a component of individual professional development in the context of cultural and educational space. Periodyk Naukowy Akademii Polonijnej. 2017;**24** (5):98–106. DOI: 10.23856/2411

[13] Rean AA, Kudashev AR, Baranov AA. Психология Адаптации Личности: Анализ, Теория, Практика [Psychology of Personality Adaptation: Analysis, Theory, Practice]. Saint Petersburg: PRIME-EVROZNAK; 2006. p. 479. in Russian

[14] Friedlander LJ, Reid GJ, Shupak N, Cribbie R. Social support, self-esteem, and stress as predictors of adjustment to university among first-year undergraduates. Journal of College Student Development. 2007;**48**:259–274. DOI: 10.1353/csd.2007.0024

[15] Martin F, Ertzberger J. Here and now mobile learning: An experimental study on the use of mobile technology. Computers and Education. 2013;**68**:76–85. DOI: 10.1016/j.compedu.2013.04.021

[16] Nugmanova D, Kozlova I, Kupriyanov R. The peculiarities of adaptation of first-year students to the university during the COVID-19 pandemic in Russia. Revista de Psicología Clínica Con Niños Y Adolescentes. 2022;**9**(2):32–38. DOI: 10.21134/rpcna.2022.09.2.4

[17] Organisation for Economic Cooperation and Development (OECD). Education and COVID-19: Focusing on the Long-Term Impact of School Closures. Paris: OECD; 2020

[18] Crawford J, Butler-Henderson K, Rudolph J, Glowatz M, Burton R, Magni P, Lam S. COVID-19: 20 countries' higher education intra-period digital pedagogy responses. Journal of Applied Learning and Teaching. 2020;**3**(1):1–20. DOI: 10.37074/jalt.2020.3.1.7

[19] Soltovets E, Chigisheva O, Dubover D, Dmitrova A. Russian digital education landscape during the current pandemic: Is the impact felt? E3S Web of Conferences. 2021;**273**:12026. DOI:10.1051/e3sconf/202127312026

[20] Aleshkovski IA, Gasparishvili AT, Krukhmaleva OV, Narbut NP, Savina NE. Студенты России об обучении в период пандемии COVID-19: ресурсы, возможности и оценка учебы в удаленном режиме [Russian students about learning under the COVID-19 pandemic: Resources, opportunities and assessment of the distance learning. RUDN Journal of Sociology. 2021;**21**(2):211–224. in Russian. DOI: 10.22363/2313-2272-2021-21-2-211-224

[21] Bekova S, Terentev E, Maloshonok N. Образовательное неравенство в условиях пандемии COVID-19: связь социально-экономического статуса семьи и опыта дистанционного обучения студентов [Educational inequality and COVID-19 pandemic: Relationship between family socio-economic status and student experience of remote learning]. Voprosy Obrazovaniya/Educational Studies Moscow. 2021;1:74–92. DOI: 10.17323/1814-9545-2021-1-74-92 in Russian

[22] Kalugina TN, Timchenko MV. Цифровизация высшего образования в 2021 году: вызовы для студентов университетов в России [Digitalization of higher education in 2021 – Challenges for university students in Russia]. New Media and Human Communication. 2023;**5**(2):e344. in Russian. DOI: 10.46539/gmd.v5i2.344

[23] Nebylitsyn VD, Gray JA, editors.. Biological Bases of Individual Behavior. New York and London: Academic Press; 1972. 440

[24] Hagquist C. Psychosomatic health problems among adolescents in Sweden–are the time trends gender

related? European Journal of Public Health. 2009;**19**(3):331–336. DOI: 10.1093/eurpub/ckp031

[25] Baker RW, Siryk B. Measuring adjustment to college. Journal of Counseling Psychology. 1984;**31**(2):179–189. DOI: 10.1037/0022-0167.31.2.179

[26] Ananyev BG. О Проблемах Современного Человековедения [On the Problems of Modern Human Studies]. Moscow: Nauka; 1977. p. 379. in Russian

[27] Deniz M, Yılmaz E. Üniversite Öğrencilerinin Duygusal Zeka ve Stresle Başa Çıkma Stilleri Arasındaki İlişkinin İncelenmesi. Turkish Psychological Counseling and Guidance Journal. 2006;**3**(25):17–26. in Turkish

[28] Nugmanova D, Kupriyanov R, Valeyeva NS. The relationship between motivation for studying and academic adaptation levels of first-year students. In Auer ME, Hortsch H, Michler O, Köhler T, editors. Mobility for Smart Cities and Regional Development – Challenges for Higher Education. ICL 2021. Lecture Notes in Networks and Systems. Vol. **390**. Cham: Springer; 2022. p. 531–538. DOI: 10.1007/978-3-030-93907-6_53

[29] Sedgwick P. Cross sectional studies: Advantages and disadvantages. The BMJ. 2014;**348**:g2276. DOI: 10.1136/bmj.g2276

[30] Wang X, Cheng Z. Cross-sectional studies: Strengths, weaknesses, and recommendations. Chest. 2020;**158**(1S):S65–71. DOI: 10.1016/j.chest.2020.03.012

[31] Thompson B, Diamond KE, McWilliam R, Snyder P, SW S. Evaluating the quality of evidence from correlational research for evidence-based practice. Exceptional Children. 2005;**71**(2):181–194. DOI: 10.1177/001440290507100204

[32] Kupriyanov R, Nugmanova D. El cuestionario de evaluación del nivel de adaptación de estudiantes a la universidad. In: V International Congress of Clinical and Health Psychology on Children and Adolescents; Oviedo, Spain. Elche: AITANA-UMH, University of Oviedo; 2019. Available from: https://www.aitanacongress.com/2019/wp-content/uploads/2019/11/abstracts_2019-.pdf [in Spanish] [Accessed: 2025-October-14]

[33] Nugmanova D, Kupriyanov R. Теоретическое обоснование и оценка надежности методики «Оценка уровня адаптации студентов к университету» [Theoretical justification and assessment of the reliability of the questionnaire "assessment of the level of adaptation of students to the university"]. In: Bekhterev and modern personality psychology: collection of articles of the VI All-Russian Scientific and Practical Conference. Kazan; 2020. p. 162–164. [in Russian]

[34] Ergin-Kocaturk H, Tekel E, Su A, Kocaturk M, Karadag E. Acculturation strategies of international higher education students in Türkiye: The role of social support, cultural capital, self-esteem, general trust, and general self-efficacy. Current Psychology. 2025;1–18. DOI: 10.1007/s12144-025-07919-4

[35] Hirai R, Frazier P, Syed M. Psychological and sociocultural adjustment of first-year international students: Trajectories and predictors. Journal of Counseling Psychology. 2015;**62**(3):438–452. DOI: 10.1037/cou0000085

[36] Cohen JR, Pant LW, Sharp DJ. An examination of differences in ethical decision-making between Canadian business students and accounting professionals. Journal of Business Ethics. 2001;**30**:319–336. DOI: 10.1023/A:1010745425675

[37] Yankovsky LV. Адаптация личности к новой социокультурной среде [Adaptation of personality to a new sociocultural environment. In Sonin VA, editor Psychodiagnostic Cognition of Professional Activity. Saint Petersburg: Rech' (Речь); 2004. p. 206–211. in Russian

[38] Ward C, Kennedy A. The measurement of sociocultural adaptation. International Journal of Intercultural Relations. 1999;**23**(4):659–677. DOI: 10.1016/S0147-1767(99)00014-0

[39] Ward C, Rana-Deuba A. Acculturation and adaptation revisited. Journal of Cross-Cultural Psychology. 1999;**30**(4):422–442. DOI: 10.1177/0022022199030004003

[40] Chemers MM, Hu L, Garcia BF. Academic self-efficacy and first-year college student performance and adjustment. Journal of Educational Psychology. 2001;**93**(1):55–64. DOI: 10.1037/0022-0663.93.1.55

[41] Osnitskiy AK. Определение характеристик социальной адаптации [Definition of characteristics of social adaptation]. Journal of Psychology and School. 2004;**1**:43–56. in Russian

[42] Pilugina SA, Taranenko OG. Особенности социально-психологической адаптации студентов первого курса [Specific features of socio-psychological adaptation of first-year students]. Vestnik Tomskogo Gosudarstvennogo Pedagogicheskogo Universiteta. 2016;**8**(173):104–108. in Russian

[43] Wu S, Liu H, Li Y, Teng Y. The influence of self-esteem on sociocultural adaptation of college students of Hong Kong, Macao and Taiwan: The chain mediating role of social support and school belonging. Psychology Research and Behavior Management. 2024;**17**:905–915. DOI: 10.2147/PRBM.S445042

[44] Baeva IS, Gayazova AV, Kondakova IV, Laktionova EB. Психологическая безопасность личности и ценности подростков и молодежи [Psychological Security and Values in Adolescents and Young People]. Psychological Science and Education. 2020;**25**(6):5–18. in Russian. DOI: 10.17759/pse.2020250601

[45] Jardim ME, Soares AB. Academic adaptation, stress, self-efficacy and social skills in college students from public and private institutions. Estudios de Psicologia (Campinas). 2025;**42**:e220088. DOI: 10.1590/1982-0275202542e220088

[46] Kozlova IV. Академическая адаптация первокурсников: психологические детерминанты [Academic adaptation of first-year students: Psychological determinants]. Bulletin of Kazan University. 2010;**152**(3):45–52. in Russian

[47] Orgilés M, Espada JP, Delvecchio E, Francisco R, Mazzeschi C, Pedro M, Morales A. Anxiety and depressive symptoms in children and adolescents during COVID-19 pandemic: A transcultural approach. Psicothema. 2021;**33**(1):125–130. DOI: 10.7334/psicothema2020.287

[48] Ward C, Bochner S, Furnham A. The Psychology of Culture Shock. 2nd ed. London: Routledge; 2001

[49] Poyrazli S, Arbona C, Nora A, McPherson R, Pisecco S. Relation between assertiveness, academic self-efficacy, and psychosocial adjustment among international graduate students. Journal of College Student Development. 2002;**43**(5): 632–642

[50] Kubikova K, Bohacova A, Slowik J, Pavelkova I. Student adaptation to distance learning: An analysis of the effectiveness, benefits and risks of distance education from the perspective of university students. Social Sciences & Humanities Open. 2024; **9**:100875. DOI: 10.1016/j.ssaho.2024.100875

[51] Labbaf S, Abbasian M, Nguyen B, Lucero M, Ahmed MS, Yunusova A, Rivera A, Jain R, Borelli JL, Dutt N, Rahmani AM. Physiological and emotional assessment of college students using wearable and mobile devices during the 2020 COVID-19 lockdown: An intensive, longitudinal dataset. Data in Brief. 2024;**54**:110228. DOI: 10.1016/j.dib.2024.110228

[52] Soheili F, Lanz M. Psychological adaptation of international students in higher education: A bibliometric analysis. Current Psychology. 2025;**44**:10670–10678. DOI: 10.1007/s12144-025-07926-5

[53] Paul A, Lewis RS, DeWitt NJ, Bork SJ. Adapting to alternative learning: Insights from engineering graduate students during the COVID-19 pandemic. Presented at: 2025 ASEE Annual Conference & Exposition; 2025 Jun; Montreal, Quebec, Canada. DOI:10.18260/1-2–57563

[54] Nagel R, Aleman M, Sadel K. Features of continuity and change through COVID-19 in an undergraduate engineering program. In: Proceedings of the 2022 ASEE Annual Conference & Exposition; 2022 Jun 26–29; Minneapolis, MN. DOI:10.18260/1-2–41501

[55] Ruiz-Robledillo N, Vela-Bermejo J, Clement-Carbonell V, Ferrer-Cascales R, Alcocer-Bruno C, Albaladejo-Blázquez N. Impact of COVID-19 pandemic on academic stress and perceived classroom climate in Spanish university students. International Journal of Environmental Research and Public Health. 2022;**19**(7):4398. DOI: 10.3390/ijerph19074398

[56] Nugmanova D, Kozlova I, Kupriyanov R. Gender-specific response to stress in master's adaptation to university in Spain and Russia. Journal of Social Studies Education Research. 2024;**15**(5):196–225. Available from: https://jsser.org/index.php/jsser/article/view/5702/714 [Accessed: 2025-October-14]

[57] García-Fernández L, Romero-Ferreiro V, Padilla S, López-Roldán PD, Monzó-García M, Rodriguez-Jimenez R. Gender differences in emotional response to the COVID-19 outbreak in Spain. Brain and Behavior. 2021;**11**(1):e01934. DOI: 10.1002/brb3.1934

[58] Marler E, Bruce MJ, Abaoud AF, Henrichsen C. The impact of COVID-19 on university students' academic motivation, social connection, and psychological well-being. Scholarship of Teaching and Learning in Psychology. 2024;**10**(3):320–330. DOI: 10.1037/stl0000294

[59] Nesmith JE, Hickey JW, Haase E. Improving biomedical engineering undergraduate learning through use of online graduate engineering courses during the COVID-19 pandemic. Biomedical Engineering Education. 2021;**1**(2): 317–324. DOI: 10.1007/s43683-020-00041-w

[60] Al-Khatib M, Alkhatib A, Talhami M, Kashem AHM, Ayari MA, Choe P. Enhancing engineering students' satisfaction with online learning: Factors, framework, and strategies. Frontiers in Education. 2024;**9**:1445885. DOI: 10.3389/feduc.2024.1445885

Chapter 9

Human-Centered Digitalization in Training Language Industry Specialists

Inna Kozlova Mikurova

Abstract

This study examines how the initial training of language specialists can effectively incorporate online instruction. Grounded in Transactional Distance Theory and the Zone of Proximal Development, it compares synchronous online and on-campus classes to identify how dialog and learner autonomy were affected by shifts in delivery mode. The analysis is based on a stable cohort of 25 first-year translation students who experienced both instructional formats during consecutive semesters. Data included attendance records, partial exam grades, homework submissions, and an anonymous questionnaire combining closed and open questions. Attendance strongly correlated with exam performance ($r = 0.64$ and $r = 0.62$, $p < 0.01$), showing that regular participation – and thus opportunity for dialog – was key to achievement. Attendance remained stable across semesters, but students reported that online dialog was disrupted by technical issues and reduced immediacy, while on-campus classes felt more motivating and interactive. The online format, however, fostered greater flexibility and higher homework completion (82% vs. 72%), indicating improved autonomy, whereas on-campus classes saw a decline as fixed schedules limited self-regulation. Results highlight the need for human-centered digitalization in higher education: synchronous videoconferencing can effectively preserve dialog when supported by institutional infrastructure and teacher training. The study's small, context-specific cohort limits generalization but offers transferable insight into the dialogic dynamics of synchronous hybrid learning.

Keywords: language industry, digitalization, synchronicity, sustainability, human-centered

1. Introduction

The training of language specialists today is confronted with both technological and social challenges. Professionals such as translators, reviewers, localization engineers, and project managers increasingly operate within human–machine collaborative environments, with many working remotely or in hybrid settings. These professions, while part of the broader service sector with a significant information technology component, tend not to be highly paid. However, they offer considerable flexibility, enabling practitioners to balance work and personal life, often from locations with lower living costs or as digital nomads.

This emerging professional context raises a paradox for the training of language industry professionals. Although their future work environments are likely to be remote and digitally mediated, current educational models still require students' physical presence on campus. This chapter explores how blended learning, particularly the use of synchronous, computer-mediated face-to-face teaching (i.e., videoconferencing), can provide a viable pedagogical alternative. Videoconferencing has become an increasingly accepted and effective substitute for in-person attendance in higher education [1, 2].

The training of translators involves the development of several subcompetences, including bilingual, extralinguistic, strategic, and instrumental components, as well as general knowledge about translation [3]. This chapter does not focus on translators' instrumental subcompetence – such as the use of translation technologies, machine translation systems, term bases, or quality assurance tools – which are better addressed in more advanced stages of translator training (see [4, 5]). Instead, our focus lies on the development of bilingual competence, a foundational step not only for translators and interpreters but also for social mediators and many other language professionals.

Acquiring reading, speaking, and writing skills in a foreign language requires significant effort and time. Given that class time is inherently limited, autonomous learning outside of the classroom becomes essential for student progress. Homework – defined as tasks assigned by teachers for completion outside regular class hours [6] – has long played a role in supporting language acquisition, dating back to memory research in the late 19th century that highlighted the value of practice [7]. Even prior to digitalization, language education already incorporated a blended model by combining synchronous face-to-face instruction with asynchronous, autonomous practice.

The process of digitalization in higher education has been underway for decades, mirroring broader societal trends. Universities have increasingly adopted digital systems for administration (e.g., online registration, electronic grade submission, and thesis archiving) and pedagogical delivery (e.g., virtual campuses, especially through platforms like Moodle). As a result, student–teacher communication has progressively shifted online, with email, digital feedback, and virtual tutorials becoming routine. Despite these advancements, in-person instruction on campus has remained the defining feature of traditional education, distinguishing it from fully remote or distance learning.

The outbreak of the COVID-19 pandemic accelerated the shift to distance education, prompting an urgent transition to Emergency Remote Teaching (ERT). Numerous studies have documented the difficulties faced by both educators and students during this transition [8–15]. Torres-Martín et al. [15] highlight the lack of preparedness among educators for effective online instruction. However, these findings must be understood in context: as Hodges et al. note, the standard development cycle for an online course is typically six to nine months, which was unavailable during the emergency pivot [10].

In theory, synchronous online classes could have substituted in-person sessions, assuming the necessary technological infrastructure was in place. In fact, synchronous online learning has been widely explored in the literature since at least 2005 [16–27]. Yet, several practical barriers limited this transition. These included limited familiarity with videoconferencing platforms among faculty, unequal access to stable internet connections and equipment, a phenomenon widely referred to as the digital divide [8, 9, 12, 13, 28]. For instance, Pérez-López et al. report that only 79% of students in their study had access to a dedicated device, while 19% had to share, and 2% relied solely on mobile phones [14]. Further challenges included financial constraints [8], emotional distress from social isolation [9], and physical fatigue from prolonged screen exposure [12, 13].

Although there is extensive research on student and teacher perceptions of the shift from face-to-face to online education during the pandemic, there remains a lack of systematic investigation into what exactly changes when synchronous instruction simply moves online. To address this gap, the chapter poses two guiding research questions:

How do students perceive class dialog in synchronous online instruction?

Does the change in class delivery mode affect students' autonomous learning practices and overall academic outcomes?

The chapter is structured into two major sections. The first presents the theoretical background, outlining relevant concepts from the fields of ERT and Distance Education (DE). It places particular emphasis on the concepts of dialog and autonomy, which are central to the pedagogical frameworks under discussion. Special attention is given to the role of synchronicity in DE, particularly its capacity to sustain continuous teacher–student interaction [29–31]. The role of dialog is further examined through the lens of Vygotsky's Zone of Proximal Development [32], which emphasizes the importance of interaction with More Knowledgeable Others in the learning process. However, Vygotsky's theory does not explicitly address the role of mediated resources. This gap is addressed by Moore's Transactional Distance Theory [29], which conceptualizes learning as occurring beyond immediate, real-time human interaction, thereby providing a theoretical bridge for understanding dialog in digital learning contexts.

1.1 Synchronous communication in DE and ERT

The concept of synchronicity in DE has undergone significant changes in recent years, largely due to technological advancements. Moore identified teleconferencing as transformative to the very notion of DE [29], while Keegan envisioned virtual classrooms where instructors and students are electronically connected – an idea made plausible by the elite status of videoconferencing at that time [33]. Simonson [34] even questioned whether "virtual education (interactive, live televised instruction) [is] a subset of distance education or to be regarded as a separate field of educational endeavour" [34, p. 7].

However, broadband internet and audio communication via Skype only became widely accessible in 2003. Pan and Sullivan reported the success of small-group class sessions using Skype [18], while Godwin-Jones referred to Skype as a "disruptive technology" [16]. Videoconferencing capabilities were not introduced on Skype until 2006. Prior to this, synchronicity in DE was generally limited to live text chat during class sessions and occasional telephone use.

Despite its availability, videoconferencing remained peripheral in educational settings for many years. Watts demonstrated that synchronous communication tools continued to be used sporadically, with Skype often cited as the most popular platform [26]. Yet, the courses discussed in such studies typically involved telecollaboration projects – virtual exchanges [17, 20–22], small-group instruction, or individual tutorials [24–27] – rather than standard university courses.

The COVID-19 pandemic introduced a radically different context for synchronous online instruction and new terminology, such as ERT [10] or Emergency Distance Education [28]. These terms refer to the urgent shift from physical classrooms to remote learning environments. Hodges et al. used the term "remote teaching" specifically to describe synchronous online classes that temporarily replaced face-to-face instruction [10]. According to these authors, ERT is not equivalent to traditional online learning but rather a rapid, makeshift adaptation that lacks the instructional design typically required for quality distance learning.

When lockdowns began, course content was not redesigned, but synchronous videoconferencing was there as a technology, able to substitute for face-to-face onsite classes. Why, then, was the transition not straightforward?

The early months of 2020 saw an unprecedented demand for free group videoconferencing tools. Skype, available since 2003, was initially the go-to platform for telecollaboration, but its free version was limited to small groups. Microsoft, which acquired Skype in 2011, had already launched Microsoft Teams in 2017 to replace Skype for Business – a paid service supporting group calls of up to 50 people [35]. Zoom, launched in 2011, began offering free services for larger groups (initially up to 25 users) as early as 2013. This feature gave Zoom a significant advantage in early 2020, helping the company enter the NASDAQ-100 in April of that year. However, many organizations restricted or prohibited Zoom due to privacy and security concerns. Google responded by launching a free version of Google Meet in May 2020 [36], and open-source platforms like Jitsi Meet gained traction through word of mouth.

Still, the initial lack of a familiar, reliable, and free videoconferencing platform significantly hindered the swift transition to synchronous online communication, which was essential to maintaining the class dialog.

1.2 Dialogue in ZPD theory

The role of teacher–student dialog in education can be illuminated through Vygotsky's Zone of Proximal Development (ZPD) theory [32]. ZPD refers to the cognitive space between what a learner can achieve independently and what they can accomplish with the support of More Knowledgeable Others (MKOs). In higher education, MKOs typically include both instructors and peers.

The teacher is usually responsible for assessing each student's prior knowledge and providing tailored scaffolding to facilitate progress within their individual ZPD [37]. However, in group-based learning environments – common in on-campus university classes – students also learn from their peers. Consequently, the teacher–student interaction is better conceptualized not as a dialog, but as a "multilogue" [38]. Dennen and Wieland [39] assert that learning "is not narrowly defined as a one or two-way dialog between individual student and teacher," but instead depends on shared meaning-making and intersubjectivity [39, p. 283]. Stahl et al. [40] similarly argue that "the learning takes place largely through interactions among students" [40, p. 410].

Interestingly, peer interaction was found to have little impact on students' perceptions of online courses [41]. It appears that students primarily value "teacher's presence" and "teacher's immediacy" [42, p. 8]. It is the instructor who bridges the gap between current knowledge and the next developmental stage [43], especially when students' ZPDs differ widely.

Vygotsky's theory was developed based on his work with children – learners with limited autonomy and no access to textual resources. Consequently, he emphasized the role of MKOs without addressing independent resource use, which requires greater autonomy. Vygotsky, according to Wood [44], believed that the ability to learn through instruction was a "fundamental aspect of human intelligence and the main vehicle for knowledge transmission" [44, p. 25]. In higher education, however, instruction is inseparable from books, articles, dictionaries, and multimedia resources. Dennen and Wieland emphasize the role of such artifacts in supporting situated meaning-making [39]. As students are increasingly expected to act autonomously, with less direct guidance, they are encouraged to rely more on documentation. The pandemic accelerated this shift toward greater learner autonomy.

1.3 Dialog and autonomy in TDT

Moore's Transactional Distance Theory (TDT), first proposed in 1972, explores the relationship between the instructor, course structure, and learner within DE environments [29]. Focusing on transactional distance, Moore identifies the "interactive nature of the medium of communication" as "a major determinant of dialog in the teaching-learning environment." [29, pp. 24–25]. He notes that teleconferencing is "more dialogic and less structured compared with broadcast, recorded, or correspondence media" [29, p. 38].

Yet, many factors beyond media may affect dialog: class size, which can reduce opportunities for individual interaction; the physical or emotional environments of teachers and students; and personal characteristics of both. In fact, as Rumble [45] notes, sometimes students attending in-person lectures may experience greater transactional distance than distance learners who "regularly meet, correspond with or telephone his tutor" [45, p. 8]. Some learners may choose not to engage in interaction, opting instead for independent study. In such cases, student-content interaction replaces student-teacher dialog [30], signaling a deliberate shift toward autonomy. Even for students who do engage in dialog, autonomy is exercised between sessions, as they manage their time to stay on track with course requirements.

Apart from dialog and autonomy, Moore presents the concept of course structure, which encompasses the flexibility and adaptability of teaching. A textbook, by itself, is low in structure [29, p. 35], offering students the freedom to navigate content at their own pace. However, when integrated into a rigid course schedule with assigned readings and exercises, it turns into a highly structured resource. According to Moore, the structure of a course reflects its responsiveness to learners' individual needs [46]. A course has high structure when its "objectives, teaching methods, and evaluation" [46, p. 10] are predetermined by the instructor. Conversely, when these elements are defined by the learner, we speak of learner autonomy. In blended learning environments – where face-to-face teaching (whether on-campus or via videoconferencing) is combined with independent homework – both modes coexist, one characterized by high dialog, the other by high structure. Dialog in this context, represents a middle ground where the instructor and learner negotiate course content and expectations.

Figure 1. demonstrates the role of dialog as central to both theories and highlights how it connects TDT and ZPD.

In TDT, dialog interacts with structure and autonomy to shape the pedagogical relationship, whereas in the ZPD, dialog alone defines the space where learning and

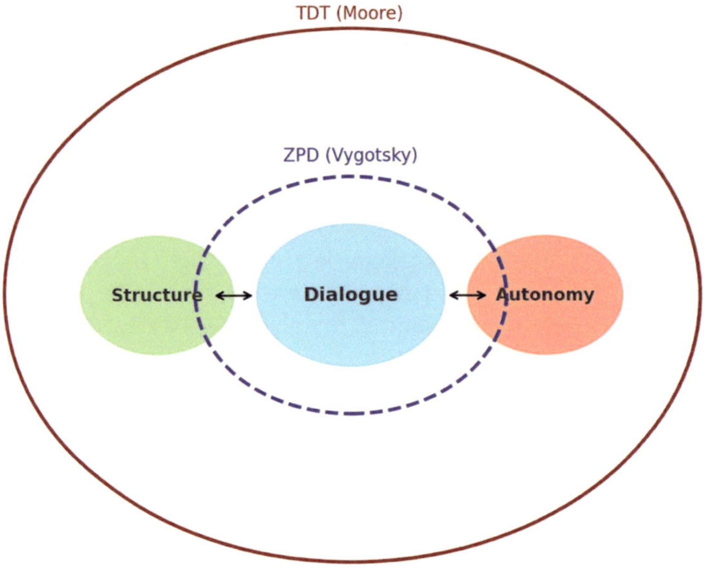

Figure 1.
Dialog as a central construct in both ZPD and TDT.

development occur. The figure illustrates this overlap, positioning dialog as the common ground between the two frameworks.

This leads to the following questions, reinterpreted considering the conceptual connection between the two theories:

> Research Question 1: To what extent was opportunity to maintain dialogue related to learning outcomes?
>
> Research Question 2: How effectively was dialogue maintained in synchronous online classes, and how did this compare to on-campus classes?
>
> Research Question 3: To what extent did students engage in autonomous work during on-campus classes, and how did this compare to synchronous online classes?
>
> Research Question 4: What were the main challenges encountered in synchronous online and on-campus learning?

2. Methodology

2.1 Participants

The participants in this exploratory case study were first-year students enrolled in the Faculty of Translation and Interpreting at a public Spanish university. The sample size was 25 students: in the first semester, the group under study comprised 25 students; in the second semester, the same 25 students continued. There was a new, 26th student who registered for the second semester, attended class just once, and did not submit any assignments.

Approximately two-thirds of the participants were women. Their ages ranged from 19 to 23, with an average age of 20. These students were enrolled in the course Foreign Language C for Translators and Interpreters 1 during the first semester, and its continuation, Foreign Language C for Translators and Interpreters 2, in the second semester. The designation "C" refers to a second foreign language, which in this case was Russian. At the outset of the first semester, most students had no prior knowledge of Russian. By the start of the second semester, the majority had acquired knowledge roughly equivalent to A1.1 on the CEFR scale. While there were some variations in proficiency levels at the beginning of the second semester, likely due to differences in individual learning progress, it is worth noting that two false beginners were already present in the first semester. Thus, while a degree of unevenness existed, it was not entirely new to the group in the second semester, which reduced the potential semester effects.

Out of the overall sample of 25 students, 21 participants responded to the questionnaire anonymously, providing their consent for their responses to be used in the study. To preserve the personal data and privacy of the participants, as they requested, these responses were analyzed as a group survey and not individually, which would have allowed us to triangulate them with personal attendance data, final grades, and Moodle assignment statistics.

2.2 Research design

This study adopts an exploratory case study design aimed at comparing the performance of a group of participants under different instructional conditions. The participant group remained stable throughout both semesters, and the students' workload in the two semesters was intentionally kept consistent, which ensured cohort comparability. The independent variable in this study was the mode of class delivery: during the first semester, classes were delivered primarily through computer-mediated face-to-face instruction, whereas in the second semester, instruction occurred on campus in a traditional classroom setting.

It was expected that the change in class delivery would have an effect on class dialog and, possibly, on students' autonomy. A group survey was designed to capture these potential effects. Participants' responses were analyzed through summative content analysis. The uniqueness of the context – specifically, the shift in delivery mode under relatively stable conditions – qualifies this as what Yin [47] calls a "revelatory case" [47, p. 42], where the researcher has an opportunity to observe and analyze a phenomenon that was previously inaccessible to scientific investigation.

2.3 Case description

The cohort for this study faced unique circumstances that made it particularly suitable for examining the dynamics of class dialog in two different settings: synchronous online via Teams in the 1st semester and on-campus in the 2nd semester. At the start of the academic year on September 14, 2020, both students and teachers were aware of the increasing likelihood of another lockdown due to rising COVID-19 cases. The experience of the previous academic year (2019–2020) with ERT had already familiarized both groups with various videoconferencing tools and online learning platforms. This background allowed for a smoother transition when the expected lockdown was announced in early October, and on-campus classes were canceled.

Rather than fully transitioning to asynchronous learning or other alternatives, the instructors of this student cohort chose a computer-mediated synchronous approach. This decision was based on a desire to maintain the highest level of interaction and dialog with students. The synchronous online format mirrored the traditional on-campus experience in terms of class schedules, with students completing homework assignments between classes and engaging in group discussions during the virtual meetings. This approach aimed to retain a sense of connection and motivation despite physical separation.

When the epidemiological situation improved in February 2021, the second semester of 2020–2021 saw a return to traditional on-campus classes with the required safety measures.

2.3.1 Course objectives and teacher's challenges

The course targeted foreign language proficiency at level A1.1 in the first semester and A1.2 in the second semester, and students were expected to engage with both the textbook and the exercise book. The teacher's primary role was to monitor students' progress, establish class pacing, provide explanations, conduct interactive class activities, and post assignments on Moodle for students to complete independently. Both semesters' classes were held at the same time slots, which included five hours a week distributed between three sessions (1.5 h sessions on Mondays and Wednesdays and 2 h session on Fridays). Feedback on assignments was given at least once a week, and answer sheets were also provided for students to self-assess their work.

Given the described learning environments and course structure, maintaining dialog during reading, speaking, and writing tasks became particularly challenging in a computer-mediated format. The challenges faced in these areas can be linked directly to the format of teaching, student engagement, and the technology used.

2.3.2 Tools and technical solutions

Microsoft Teams was used for synchronous computer-mediated sessions in the first semester, when students were expected to master reading in the Cyrillic alphabet. Unlike typical online settings, where microphones are muted to avoid background noise, in this case, students' microphones were left unmuted precisely to encourage a fluent and interactive dialog. This approach helped maintain the natural flow of turn-taking in reading exercises, but it also presented challenges in managing transitions between students smoothly.

For students to read aloud one after the other, the teacher needed to actively monitor the flow of the class and ensure that each student had a turn to read. This became particularly challenging in the computer-mediated format, where the usual non-verbal cues (such as eye contact or body language) that help facilitate turn-taking were absent. The teacher had to pay careful attention to students' readiness to speak and guide the reading process, which required more active management compared to an in-person setting.

The reading material was projected on the screen for students, which helped streamline the process and allowed all students to follow along. For students with vision problems, this fit better as physical distance to the screen in a traditional class no longer represented a problem. However, it still required constant attention from the teacher to keep students engaged. In class, teachers would use a whiteboard to maintain students' attention. Due to technological unpreparedness and the unavailability of a proper tool, chat was used instead.

Writing in a foreign language often requires familiarity with tools and formats that students may not have encountered before. In this case, while the traditional handwriting exercises were still part of the course at the very beginning of the first semester, the computer-mediated format represented a shift to typing, which required learning a Cyrillic keyboard layout. Many students had not yet installed the Russian keyboard or were unfamiliar with the layout at the beginning of the semester. This created an initial gap in their ability to write efficiently and participate in writing tasks. However, the ability to type in Russian was also seen as a critical skill for students' professional futures. As the semester progressed, students overcame this initial difficulty, and the skill of typing in Russian became an asset in the long run.

Speaking tasks, typically carried out in small groups or pairs, also presented a set of issues. The online platform allowed for the creation of breakout rooms, but managing these virtual spaces and ensuring that all students were participating actively was difficult. Dividing the cohort of 25 students into just four groups for the speaking activities helped manage the logistics, but the teacher could only listen to one group at a time. This became particularly challenging when the teacher was trying to monitor group discussions and provide feedback on how students were applying useful expressions in their conversations. In an in-person class, the teacher could circulate among the groups and give more direct, immediate feedback.

2.4 Data collection

To assess how dialog and autonomy were affected by the change in class delivery, the study used multiple data sources. Data were collected at the end of the instructional period, providing a cross-sectional view of the students' experiences. The data collection sources included:

- Class attendance statistics for each semester measure student participation and engagement during classes. Class attendance is not mandatory in Spanish universities, but failure to attend directly reduces opportunities for classroom dialog and this, if not compensated by autonomous work, could affect the final grade.

- Two partial exams of the second semester, to triangulate with individual class attendance.

- Moodle assignment statistics capture students' completion of homework and the extent to which they engage with out-of-class tasks. Failure to complete the work, especially when combined with failure to participate in the class dialog, could affect the final grade.

- A questionnaire was designed by the instructors, which incorporated both closed and open-ended questions, allowing students to reflect on their experiences with class communication, course structure, workload, teacher feedback, and obstacles faced in the learning process.

The questionnaire, administered on the last day of class, included eight questions used for the purposes of this study. Questions 1–4 of the questionnaire were open-ended and sought to identify students' perceptions of the advantages and disadvantages of the two class delivery formats, providing insight into how they viewed the dialog and their interaction with the teacher in each setting. Question 5 was closed and asked students to rate the workload and timing on a Likert scale, and Question 6 was open-ended and allowed students to elaborate on the previous one. Closed Questions 7 and 8 addressed the challenges students faced, asking them to prioritize difficulties in each semester, specifically lack of motivation, lack of time, lack of speaking opportunities, and lack of resources. Given the small number of closed questions and the predominance of open-ended items, internal reliability coefficients were not calculated.

Representative items from the two second-semester partial exams are provided in Appendix A, illustrating the controlled vocabulary, grammar, and short reading/writing tasks used for the quantitative analysis. These examples show the standardized structure of the assessments and clarify why total exam scores could be used as a reliable proxy for lexico-grammatical performance. The complete student questionnaire is reproduced in Appendix B, which includes all open and closed items used to explore students' perceptions of dialog, autonomy, and obstacles to learning across the two delivery modes.

Data were analyzed using descriptive statistics, correlations, and paired-sample comparisons. Missing data were treated by listwise exclusion for correlational analyses.

3. Results

3.1 Research question 1

Our first research question asked: To what extent was the opportunity to maintain dialog related to learning outcomes?

According to ZPD theory, dialog constitutes the essential mechanism of learning, whereas TPD theory acknowledges its importance but allows for the possibility of autonomous learning as a substitute. To assess the role of classroom dialog within traditional on-campus teaching, we examined the correlation between students' class attendance and their exam performance. Attending class provided students with opportunities to engage in dialog, while non-attendance deprived them of this possibility.

In the second semester, examinations were conducted on campus, offering more reliable measures of academic achievement. Two partial exams were administered, enabling us to calculate correlations separately for the periods leading up to each partial exam (Feb 22–Apr 12 and Apr 14–May 24).

Figure 2. shows the relationship between attendance and partial exam grades in the second semester. Students who did not sit for the corresponding exam were excluded from the analysis. Only second-semester data were considered, as first-semester exam results – conducted remotely – offered limited reliability for

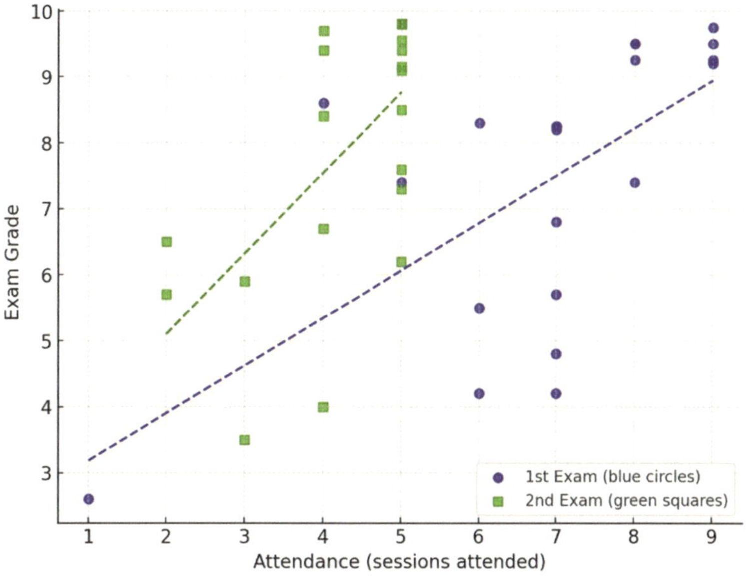

Figure 2.
Partial exam grades in relation to individual attendance (second semester only).

comparison, as explained in the contributions and limitations of the study section. Attendance before the first exam is plotted against first exam grades (blue circles), while attendance before the second exam is plotted against second exam grades (green squares). The dashed trend lines indicate linear regression fits.

Exam 1 (first partial, 2nd semester): Correlation (r): 0.64, $p = 0.002$ → strong positive association, Cohen's d: 0.94 → large effect.

Exam 2 (second partial, 2nd semester): Correlation (r): 0.62, $p = 0.003$ → strong positive association, Cohen's d: 1.30 → very large effect.

These findings indicate that students who attended more classes achieved substantially higher partial exam grades. Since the p-values were effectively zero in both analyses, the relationships can be considered statistically highly significant.

To complement the correlation analysis, a paired comparison was conducted between the two partial exams administered in the second semester. Students' mean performance showed a slight increase from the first to the second partial exam (7.44 → 7.90), but this difference was not statistically significant, $t(20) = 1.48$, $p = 0.15$. The paired effect size was small ($d_z = 0.32$, 95% CI [−0.19, 1.11]), indicating modest progress rather than a substantial shift in proficiency. This pattern is consistent with the consecutive nature of the assessments, which evaluated related lexico-grammatical content within the same learning phase. The stability of results across the two partials suggests that students maintained their performance throughout the semester, while differences in attendance continued to account for much of the variance in individual outcomes.

3.2 Research question 2

Our second research question was: How effectively was dialog maintained in synchronous online classes, and how did this compare to on-campus classes?

Sustaining dialog was a central objective in both semesters. Attendance remained stable across semesters (15–21 students), with initial peaks of up to 24 students due to the looming lockdown in the first semester and the enthusiasm to return in the second. These peaks quickly subsided, and no substantial differences were observed between on-campus and online participation.

Survey responses highlight the motivational and pedagogical value of on-campus classes. Students emphasized that these sessions promoted interaction, improved peer relations, and supported the clarification of doubts: "*better learning, more didactic,*" "*helps to improve the relation between the students and to resolve doubts,*" "*no distractions, more possibility to resolve doubts*" and "*active and dynamic classes adapted to everybody.*"

By contrast, online classes were often hindered by technological issues such as poor internet connections ("*bad quality of the call,*" "*connection problems,*" "*problems with the computer and Wi–Fi,*" "*interferences*") as well as by coordination difficulties ("*delays in response caused by muted micros,*" "*difficulty to understand classes,*" "*misunderstanding,*" "*bad communication in oral exercises*"). Students also pointed to the absence of usual visual support, such as the whiteboard, and reported fatigue, concentration problems, and a loss of classroom dynamics, describing online sessions as less "*fluent*" and "*entertaining.*"

Nevertheless, both modes presented limitations. In on-campus classes, students noted that "*those in the last rows could become disengaged*" and that "*we work faster, sometimes it's confusing.*"

3.3 Research question 3

Our third research question was: To what extent did students engage in autonomous work during on-campus classes, and how did this compare to synchronous online classes?

The course was highly structured, with a defined textbook and a clear pacing plan that required students to manage their time for homework and remain aligned with class progress. Engagement data from the Virtual Campus showed frequent daily logins, with activity declining predictably between 3 a.m. and 7 a.m.

Figure 3. shows the distribution of homework submissions in the first semester. Of all assignments, 61% were submitted on time, 21% were submitted late, and 18% were not submitted. The overall submission ratio (on-time + late) was, therefore, 82%.

Figure 4. presents the outcomes for the second semester. On-time submissions accounted for 47%, late submissions for 25%, and missing submissions for 28%. The overall submission ratio was thus 72%.

Assignment submissions revealed higher on-time completion in the first semester (0.61) than in the second (0.47), and a drop in overall submission ratios from 0.82 in the first semester to 0.72 in the second.

This pattern suggests a shift in time management and engagement between semesters. In the first semester, when instruction was computer-mediated, students benefited from flexible scheduling. Many used deadline extensions, non-teaching weeks, and holidays to catch up on assignments, indicating that autonomy and adaptable timelines supported sustained coursework management.

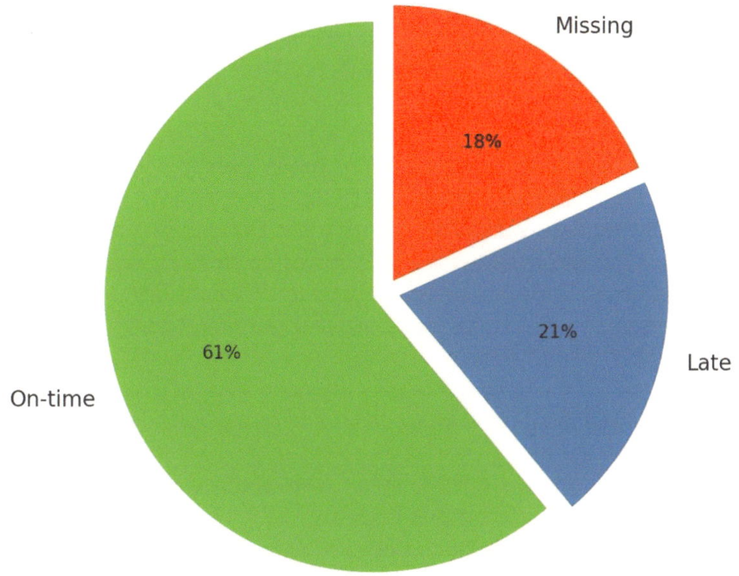

Figure 3.
Assignment submissions of the 1st semester with online synchronous classes.

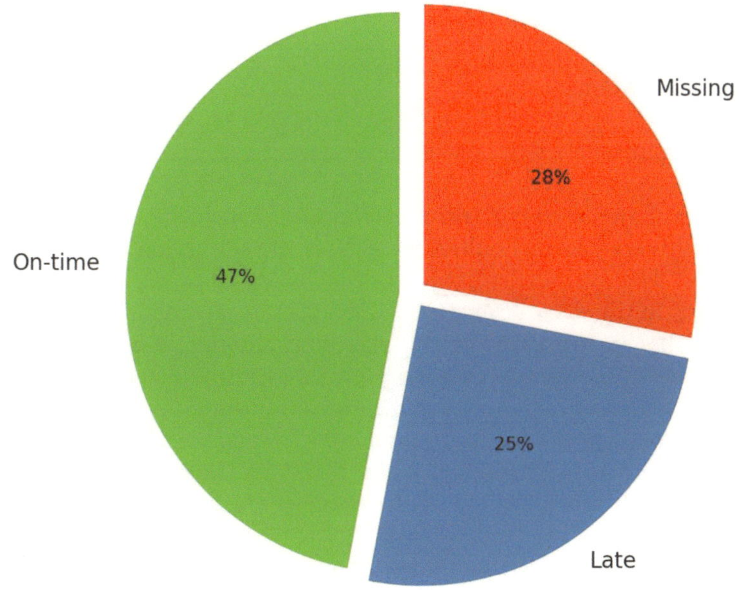

Figure 4.
Assignment submissions of the 2nd semester with on-campus classes.

By contrast, the second semester's return to on-campus instruction imposed a rigid schedule and removed the recovery periods that had previously allowed students to improve submission records. This contributed to lower homework completion, as students struggled to balance face-to-face classes with continuous assignments. To counter this, the instructor extended deadlines, which maintained relatively high submission rates but reduced consistency in pacing.

Survey feedback confirmed this trade-off. Students valued the flexibility but also described the workload as excessive, particularly during exam weeks. While course pacing was rated positively (7.71/10), many reported difficulties managing the assignment volume within the constraints of the in-person schedule: *"I had a lot to do, I couldn´t do homework every two days."*

Overall, the findings show that the structure of the learning environment shaped students' capacity for autonomous work. The online format enabled greater self-regulation, while the on-campus format constrained flexibility, leading students to complete assignments more strategically around exams rather than as part of ongoing learning.

3.4 Research question 4

Our fourth research question read: What were the main challenges encountered in synchronous online and on-campus learning?

Questions 7 and 8 asked students about the main problems affecting their learning progress in the first and second semesters, respectively. In both cases, students were asked to rank the following four obstacles in order of importance:

- Lack of time
- Lack of people to speak to
- Lack of motivation
- Lack of resources

As **Figure 5** shows, during the first semester of synchronous online classes, the most frequently reported difficulties were "lack of motivation" and "lack of people to speak to."

By contrast, in the second semester, when classes returned to campus, **Figure 6** indicates that the "lack of time" became the predominant perceived obstacle.

A comparison of **Figures 5** and **6** highlights a striking shift: while the lack of time was virtually absent as a main concern during online learning, it became the primary obstacle to learning once students resumed on-campus classes. Conversely, the lack of social interaction and motivation, which had been the most pressing difficulties during lockdown, dropped to the lowest positions after the return to face-to-face onsite learning.

This pattern suggests an inverse relationship between motivation reinforced by social contacts and the time available for study. Students' responses to the four open-ended questions of the questionnaire further illustrate this dynamic. Students reported that during the lockdown, they had "*more time to do homework,*" a "*more flexible timetable,*" the possibility to "*attend classes in a passive way when one has to go to the doctor's*" and even "*more hours for sleep.*" In contrast, the majority emphasized the time "*wasted*" on commuting once on-campus classes resumed, particularly for those living far from the university. Descriptions such as "*convenient*" and "*comfortable*" were commonly used in reference to study from home during the first semester of computer-mediated classes.

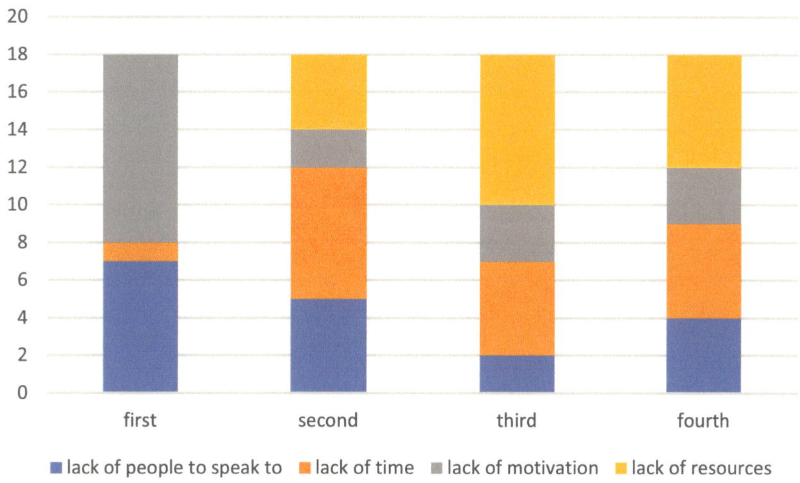

Figure 5.
Perceived obstacles to learning in the 1st semester, ordered by reported importance. Note: Three respondents failed to complete this part of the survey.

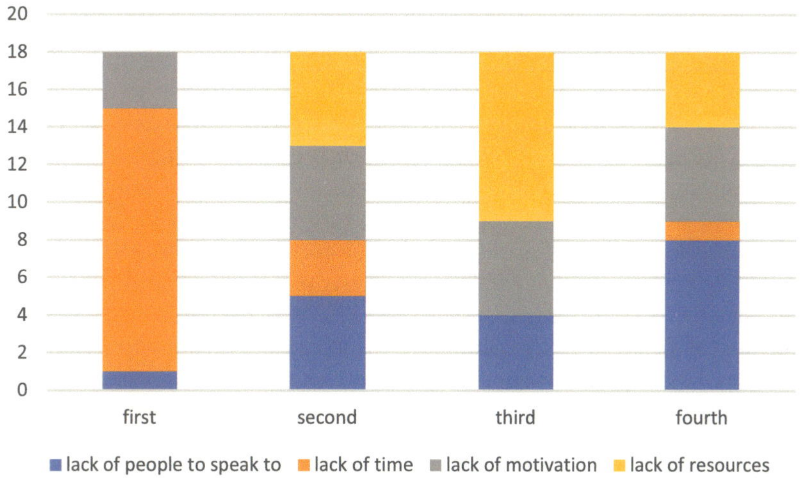

Figure 6.
Perceived obstacles to learning in the 2nd semester, ordered by reported importance. Note: Three respondents failed to complete this part of the survey.

4. Conclusions

This study contributes to the ongoing debate initiated in The Manifesto for Teaching Online [48]. First, it challenges the conventional ontology of online education as contrasted with traditional onsite teaching. Instead of reproducing this binary, we emphasize synchronicity as a defining feature: synchronous distance education is pedagogically closer to onsite instruction than to asynchronous online learning. In this respect, we build on the arguments of Simonson [34] and Midkiff and DaSilva [49], who proposed modeling synchronous online classes on F2F settings.

Second, the study highlights the central role of dialog. While some scholars underscore student autonomy and the teacher's role as a facilitator [50], we align with the view that dialog with a human teacher is essential for motivation and for extending students' ZPD. Autonomous learning can complement but cannot fully replace this interaction.

Our findings further distinguish between synchronous and asynchronous dimensions of a university course: classroom sessions provide opportunities for dialog, while homework fosters autonomy. The COVID-19 lockdown created a unique situation for comparing delivery modes of the synchronous component. In our blended course, we observed which elements of transactional distance theory were affected when classes moved online and when students later returned to campus.

Attendance rates were similar across both formats, broadly supporting Radford et al., who found that the delivery mode did not significantly affect engagement or persistence [51]. Consistent with findings on ERT [9, 13], students expressed gratitude for the possibility of continuing their education safely. At the same time, our results align with evidence that synchronous communication with instructors was crucial in mitigating isolation during lockdown [9]. In line with [23, 52], our case supports the value of combining synchronous and asynchronous elements in blended learning.

Nevertheless, students identified important differences between formats. In the synchronous online mode, transactional distance was most evident in dialog, often due to technical issues or the absence of familiar tools such as the whiteboard. Students' views should be complemented by the teacher's technological challenges described in the Case Description subsection of the Methodology. As students' microphones remained mostly unmuted to facilitate dialog, this created the potential for background noise or overlapping speech at times. Still, the open-microphone policy allowed the teacher to provide real-time feedback on students' reading and speaking, helping them improve their pronunciation and comprehension on the spot. Students' view of synchronous distance classes as less dynamic should be placed in the context of a computer-mediated environment making group dynamics harder to control and limiting the teacher's ability to ensure that all students were actively engaging in group work through pre-established channels. Students expressed relief at returning to on-campus teaching, which they found more motivating, less distracting, and more conducive to understanding. However, consistent with [45], students also acknowledged that on-campus instruction carries its own forms of transactional distance. Further research is therefore needed to better capture how dialog and interaction evolve in synchronous online classes, especially in medium-sized groups.

With respect to autonomous work, our findings resonate with [13, 53], which report improved self-directed learning and time management during the pandemic. Interestingly, our results indicate that students' autonomy management worsened when classes returned to campus. The resumption of social life and a more rigid schedule left students with less time for homework, which coincided with higher perceived workload pressure. In this sense, the balance between study and social activity emerged as a critical factor shaping engagement.

Finally, our analysis underscores the role of attendance as a condition for maintaining dialog. Irregular attendance disrupts the ZPD dynamic: teachers struggle to gauge students' current learning stage, and students lose clarity about class objectives. Our correlation analysis confirmed a strong link between attendance and academic performance. Thus, ensuring students' ability to attend – whether in person or online – is essential for sustaining learning continuity.

Videoconferencing, when integrated into blended formats, can serve two functions: enabling individual students in exceptional circumstances to attend remotely and allowing teachers to connect with students when they cannot be physically present. For this to become a viable long-term option, institutional support and adequate infrastructure are indispensable. Universities, however, face the tension between responding to student expectations for flexible, technology-supported learning and preserving the traditional campus model. Addressing this tension will be central to shaping the post-pandemic future of higher education.

4.1 Contribution and limitations of the study

The main contribution and novelty of this study lie in its focus on class delivery mode as the sole variable, while other elements of the course and the cohort remained constant. The data analyzed refer to the academic year 2020–21, while this chapter is being published in 2025. Although the time gap might reduce the immediacy of the findings, the pandemic created a unique and unrepeatable set of conditions that remain highly relevant for understanding the evolution of hybrid learning. The value of the dataset lies precisely in its documentation of practices and student responses during that exceptional period. Moreover, subsequent publications have rarely addressed synchronicity and dialog as central variables. Even the most recent work within the Transactional Distance Theory paradigm [54] continues to treat onsite and online participation in binary terms, rather than examining the dialogic dynamics of synchronous hybrid learning. This underlines the continuing relevance of the present study.

A further limitation is the small cohort size, which constrains the generalizability of the findings. Although further research is needed to provide stronger evidence for the claims presented here, the unique conditions created by the COVID-19 pandemic – unlikely and undesirable to replicate – nonetheless allowed for this exploratory comparison.

Another limitation is that we took into consideration exam grades of the second semester only. The exams in the first semester were conducted in a distance format, which makes their reliability somewhat questionable when compared with on-campus assessments in the second semester. Students with valid attendance but no exam grade (0) were excluded, as their inclusion would have artificially suppressed the correlation between attendance and performance. A further limitation concerns the measurement of attendance. Before the first partial exam, attendance was marked twice per week, while before the second partial exam, it was recorded only once per week. This reduced frequency may have made the second period's attendance measure somewhat less precise. Nonetheless, the correlations between attendance and exam performance remained strong and highly significant in both cases, suggesting that the overall findings are robust despite this difference in measurement.

Finally, the questionnaire administered to students was completed anonymously, as participants requested, which prevented linking responses to individual exam performance and assignment statistics. While this anonymity encouraged candid feedback, it limited the possibility of contrasting students' perceptions with their actual outcomes.

We therefore suggest applying the preliminary findings of this study to current higher education contexts, particularly in the design of synchronous hybrid learning environments [55]. Each mode of class delivery – synchronous computer-mediated and on-campus – shows both advantages and disadvantages. Our contribution has been to make these more visible and to call for further research. As higher education institutions face increasing pressure to respond to both society's sustainability challenges and students' expectations that technology should support their learning, lifestyles, and individual circumstances, they will need to make corresponding decisions regarding the future of course delivery.

The dialogic and autonomous learning habits observed in this study resonate with the interactional competencies required in contemporary language industry workflows. In translation communities enhanced by computer-assisted and AI-supported translation environments, language professionals engage in a form of "prompt dialog", in line with the dialogic dimension of Transactional Distance Theory, with their MKOs, among them more expert human colleagues, and with large language model systems, while using machine translation and consulting terminology databases, interpreting and refining their output through iterative feedback. The capacity for self-regulation observed in students' autonomous work parallels the professional need for critical evaluation, post-editing discipline, and quality assurance within human-in-the-loop production. Cultivating these habits during initial training helps prepare future practitioners to integrate technological tools effectively while maintaining human oversight, trust calibration, and accountability in digitally mediated translation processes.

Conflict of Interest

The author declares no conflict of interest.

Other declarations

The author used ChatGPT (OpenAI, 2025) for language polishing and clarity improvements. All intellectual content, analysis, and interpretations are solely the responsibility of the author.

Appendices

Appendix A. Sample items from the second-semester partial exams

Purpose:

To illustrate the type of controlled lexico-grammatical and short reading/translation tasks used for quantitative analysis (Partial 1–26 March; Partial 2–14 May). Both exams were graded on a 0–10 scale and shared the same structure, emphasizing vocabulary range, grammar accuracy, and comprehension of short texts.

Partial 1 (26 March) – examples

1. Vocabulary completion
Fill in the missing Russian word in context:
Он … в университет каждый день.
(a) ходит (b) пишет (c) читает (d) ждёт

2. Reading micro-task
Read the short text and answer the question in Spanish.
Анна живёт в Барселоне. Она изучает русский язык.
→ ¿Dónde vive Anna y qué estudia?

3. Controlled translation
Translate into Russian: "My friend works in Madrid."
→ ____________________________

Partial 2 (14 May) – examples

1. Grammar transformation
Complete the sentence with the correct verb form (aspect and tense):
Вчера мы (писать) тест в аудитории.

2. Functional lexicon
Choose the correct preposition.
Он говорит … своей мамой по телефону.

3. Short written production
Write 4–5 sentences describing your daily routine using at least five verbs in the correct aspect.

Scoring:
Each section contributed equally to the overall exam score (total = 10 points). The tasks were designed to assess recognition and controlled use of vocabulary and grammar rather than open writing skills, allowing direct comparison between the two partials as indicators of lexico-grammatical progress.

Appendix B. Student Questionnaire on Synchronous Online and On-Campus Learning

Purpose:
Students completed this questionnaire at the end of the academic year to compare their experiences of synchronous online and on-campus instruction. It addressed three areas: (1) teacher-led group classes in the two modes of class delivery, (2) autonomous work, and (3) obstacles to learning.

Synchronous online classes

1. Mention one advantage: ______________________________

2. Mention one disadvantage: ____________________________

On-campus classes

3. Mention one advantage: ___________________________

4. Mention one disadvantage: _________________________

Timing and workload

5. Was the pacing of work adequate (homework after each class and flexibility for late submission)?

☐ 0 ☐ 1 ☐ 2 ☐ 3 ☐ 4 ☐ 5 ☐ 6 ☐ 7 ☐ 8 ☐ 9

6. Additional comments: _________________________

Obstacles to learning

What has most limited your progress in Russian?
1. Lack of people to speak to 2. Lack of time 3. Lack of motivation 4. Lack of resources

7. Rank these obstacles for synchronous online classes period___________________

8. Rank these obstacles for on-campus classes period _________________________

Note: Given the small number of closed questions and the predominance of open-ended
responses, internal reliability coefficients (e.g., Cronbach's α) were not calculated.

Author details

Inna Kozlova Mikurova
Autonomous University of Barcelona, Barcelona, Spain

*Address all correspondence to: inna.kozlova@uab.cat

References

[1] Al-Maroof R, Al-Qaysi N, Salloum SA, Al-Emran M. Blended learning acceptance: A systematic review of information systems models. Technology, Knowledge and Learning. 2022;**27**:891–926. DOI: 10.1007/s10758-021-09519-0

[2] Camilleri MA, Camilleri AC. The acceptance of learning management systems and video conferencing technologies: Lessons learned from COVID-19. Technology, Knowledge and Learning. 2021;**26**:1069–1093. DOI: 10.1007/s10758-021-09561-y

[3] PACTE Group. Investigating translation competence: Conceptual and methodological issues. Meta. 2005;**50** (2):609–619

[4] Chan V. MOOC-based blended learning: A new paradigm in translation technology education. Interactive Technology and Smart Education. 2025;**22**(3):481–505. DOI: 10.1108/ITSE-12-2024-0325

[5] Peng W, Hu M, Bi P. Investigating blended learning mode in translation competence development. SAGE Open. 2023;**13**(4). DOI: 10.1177/21582440231218628

[6] Cooper H. Homework. White Plains: Longman; 1989

[7] Ebbinghaus H. Memory: Investigations in Experimental Psychology [In German]. Leipzig: Duncker & Humblot; 1885

[8] Alvarez Jr A. The phenomenon of learning at a distance through emergency remote teaching amidst the pandemic crisis. Asian Journal of Distance Education. 2020;**15**(1):144–153. DOI: 10.5281/zenodo.3881529

[9] Ghazi-Saidi L, Criffield A, Kracl CL, McKelvey M, Obasi SN, Vu P. Moving from face-to-face to remote instruction in a higher education institution during a pandemic: Multiple case studies. International Journal of Technology in Education and Science. 2020;**4**(4):370–383. DOI: 10.46328/ijtes.v4i4.169

[10] Hodges C, Moore S, Lockee B, Trust T, Bond A. The difference between emergency remote teaching and distance learning [Internet]. EDUCAUSE Review. 2020 March. Available from: https://er.educause.edu/articles/2020/3/the-difference-between-emergency-remote-teaching-and-online-learning [Accessed: 2025-September-19]

[11] Klyagin AV, Abalmasova ES, Garev KV, et al. The storm of the first weeks: How higher education stepped into the reality of a pandemic. Modern Analytics of Education Series. 2020;**6** (36)

[12] Bergdahl N, Nouri J. Covid-19 and crisis-prompted distance education in Sweden. Technology, Knowledge and Learning. 2021;**26**:443–459. DOI: 10.1007/s10758-020-09470-6

[13] Cloete M, Ellington J, Jansen van Vuuren A, Marais EA, Masinga P. Migrating from face-to-face to online learning during the COVID-19 pandemic: The experiences of psychology students. Central European Journal of Educational Research. 2021;**3**(3):11–21. DOI: 10.37441/cejer/2021/3/3/10003

[14] Pérez-López E, Vázquez Atochero A, Cambero Rivero S. Educación a distancia en tiempos de COVID-19. Revista Iberoamericana de Educación a Distancia. 2021;**24** (1):331–350. DOI: 10.5944/ried.24.1.27855

[15] Torres-Martín C, Acal C, El Homrani M, Mingorance Estrada AC. Impact on the virtual learning environment due to COVID-19. Sustainability. 2021;**13**(2):582. DOI: 10.3390/su13020582

[16] Godwin-Jones R. Emerging technologies: Skype and podcasting. Language Learning & Technology. 2005;**9**(3):9–12. DOI: 10.64152/10125/44026

[17] Godwin-Jones R. Telecollaboration as an approach to developing intercultural communication competence. Language Learning & Technology. 2019;**23**(3):8–28. DOI: 10.10125/44691

[18] Pan CC, Sullivan M. Promoting synchronous interaction in an eLearning environment. T.H.E. Journal. 2005;**33** (2):27–30

[19] Mabrito M. A study of synchronous versus asynchronous collaboration in an online business writing class. American Journal of Distance Education. 2006;**20**(2): 93–107

[20] Bower M. Synchronous collaboration competencies in web-conferencing environments. Distance Education. 2011;**32**(1):63–83. DOI: 10.1080/01587919.2011.565502

[21] Kern R. Technology as pharmakon: The promise and perils of the Internet for language education. The Modern Language Journal. 2014;**98**(1):340–357. DOI: 10.1111/j.1540-4781.2014.12065.x

[22] Helm F. The practices and challenges of telecollaboration in higher education in Europe. Language Learning & Technology. 2015;**19**(2):197–217

[23] Moallem M. The impact of synchronous and asynchronous communication tools. The Online Journal of Distance Education and e-Learning. 2015;**3**(3):55–77

[24] Salbego NN, Tumolo CHS. Skype™ classes: Teachers' and students' perceptions. International Journal of Language and Applied Linguistics. 2015;**1**(3):36–45

[25] Rehn N, Maor D, McConney A. Investigating teacher presence in courses using synchronous videoconferencing. Distance Education. 2016;**37**(3):302–316. DOI: 10.1080/01587919.2016.1232157

[26] Watts L. Synchronous and asynchronous communication in distance learning. Quarterly Review of Distance Education. 2016;**17** (1):23–32

[27] Rennar-Potacco D, Orellana A. Supporting STEM students from a distance through videoconferencing. American Journal of Distance Education. 2018;**32**(2):131–149. DOI: 10.1080/08923647.2018.1446121

[28] Correia A-P, Liu C, Xu F. Evaluating videoconferencing systems. Distance Education. 2020;**41** (4):429–452. DOI: 10.1080/01587919.2020.1821607

[29] Moore MG. Theory of transactional distance. In: Keegan D, editor. Theoretical Principles of Distance

Education. London: Routledge; 1993. p. 22–38

[30] Zhang M. Transactional Distance in Web-based Learning. [dissertation]. Richmond: Virginia Commonwealth University; 2003

[31] Lo HF. Learning from A Transactional Distance: A Systematic Review [dissertation]. Oxford: University of Oxford; 2021

[32] Vygotsky LS. Thought and Speech. Russian. Moscow, Leningrad: State Educational pedagogical Publishing House; 1934

[33] Keegan D. Distance Education Technology for the New Millennium: Compressed Video Teaching. Hagen: Central Institute for Distance Education Research (ZIFF); 1995

[34] Simonson M. Equivalency theory and distance education. TechTrends. 1999;**43**(5):5–8. DOI: 10.1007/BF02818157

[35] Warren T. Microsoft Teams jumps 12 million daily users to 44 million [Internet]. The Verge. 2020 March 19. Available from: https://www.theverge.com/2020/4/29/21241972/microsoft-teams-75-million-daily-active-users-stats [Accessed: 2025-September-19]

[36] Schroeder S. Google takes on Zoom by going completely free for everyone [Internet]. Mashable. 2020. Available from: https://mashable.com/article/google-meet-goes-free [Accessed: 2025-September-19]

[37] Kozlova I. Ellis' corrective feedback in a problem-solving context. ELT Journal. 2010;**64**(1):95–97. DOI: 10.1093/elt/ccp064

[38] Schwab G. From dialogue to multilogue. Classroom Discourse. 2011;**2**(1):3–19. DOI: 10.1080/19463014.2011.562654

[39] Dennen VP, Wieland K. From interaction to intersubjectivity. Distance Education. 2007;**28**(3):281–297. DOI: 10.1080/01587910701611328

[40] Stahl G, Koschmann T, Suthers D. Computer-supported collaborative learning: A historical perspective. In: Sawyer RK, editor. Cambridge Handbook of the Learning Sciences. Cambridge: Cambridge University Press; 2006. p. 409–426

[41] Kuo Y-C, Walker AE, Schroder KEE, Belland BR. Interaction, Internet self-efficacy, and self-regulated learning as predictors of student satisfaction. The Internet and Higher Education. 2014;**20**:35–50. DOI: 10.1016/j.iheduc.2013.10.001

[42] Chakraborty M, Nafukho FM. Strategies for virtual learning environments: Focusing on teaching presence and teaching immediacy. Journal of Online Learning Research and Practice. 2015;**4**(1):8–37

[43] Mishra RK. Vygotskian perspective of teaching/learning. Innovation: International Journal of Applied Research. 2013;**1**(1):23–30

[44] Wood D. How Children Think and Learn: The Social Contexts of Cognitive Development. 2nd ed. Oxford: Blackwell; 1998

[45] Rumble G. The Planning and Management of Distance Education. Kent: Croom Helm; 1986

[46] Moore M. Editorial: The evolution of distance education. Open Learning:

The Journal of Open, Distance and e-Learning. 1990;**5**(3):10–15. DOI: 10.1080/0268051900050303

[47] Yin R. Case Study Research: Designs and Methods. 3rd ed. Thousand Oaks: Sage; 2003

[48] Bayne S, Evans P, Ewins R, Knox J, Lamb J. The Manifesto for Teaching Online. Cambridge (MA): MIT Press; 2020

[49] Midkiff SF, DaSilva LA. Leveraging the Web for synchronous versus asynchronous distance learning. In: Proceedings of the International Conference on Engineering Education (ICEE 2000); Taiwan: 2000. p. 14–18

[50] MacKenzie A, Bacalja A, Annamali D, et al. Dissolving the dichotomies between online and campus-based teaching: A collective response to The Manifesto for Teaching Online (Bayne et al. 2020). Postdigital Science and Education. 2021;**4**:271–329

[51] Radford K, Stewart H, Sarvaiya H, et al. Teaching presence in online business education: A framework for designing synchronous interactions. In: Allen S. editor. Handbook of Teaching with Technology in Management, Leadership, and Business. Cheltenham: Edward Elgar; 2020. p. 406–417

[52] Giesbers SJH, Rienties BC, Tempelaar DT, Gijselaers WH. A dynamic analysis of the interplay between asynchronous and synchronous communication in online learning: The impact of motivation. Journal of Computer Assisted Learning. 2014;**30**(1):30–50. DOI: 10.1111/jcal.12020

[53] Foung D, Chen J, Lin L. When "blended" becomes "online": A data-driven study on the change of self-directed engagement during COVID-19. CALICO Journal. 2022;**39**(1):1–25. DOI: 10.1558/cj.19666

[54] Ottenheim VL, Meulenbroeks R, Drijvers P. Transactional distance and interaction in hybrid education: A case study. European Journal of Open, Distance and E-Learning. 2025;**27**(1):44–62. DOI: 10.2478/eurodl-2025-0003

[55] Bülow MW. Designing synchronous hybrid learning spaces: Challenges and opportunities. In: Gil I, Mor Y, Dimitriadis Y, Köppe C, editors. Hybrid Learning Spaces. Cham: Springer; 2022. p. 135–163. DOI: 10.1007/978-3-030-88520-5_9

Chapter 10

Industry 4.0 in Logistics Higher Education

Brigita Gajšek and Beata Mrugalska

Abstract

Logistics encompasses transport, warehousing, inventory, packaging, and information management, with strong links to economic growth. Driven by Industry 4.0 and sustainability demands, the sector is rapidly evolving toward digital, smart logistics. As a result, a skilled, adaptable workforce is increasingly vital. However, gaps remain between required and existing competencies, especially in Industry 4.0, sustainability, and digitalization. Modern logistics roles demand technical, behavioral, and organizational skills. This study examines the readiness of logistics students and professionals, comparing self-assessments with expert evaluations and the needs of the logistics sector. It aims to inform workforce development by identifying skill gaps and offering recommendations for aligning education and training with emerging logistics industry demands.

Keywords: logistics, Industry 4.0, higher education, curriculum, learning outcomes

1. Introduction

The logistics industry encompasses a broad spectrum of operations, offering extensive employment opportunities. According to Council of Supply Chain Management Professionals' (CSCMP's) definition, logistics management personnel should be able to plan, implement, and control the efficient, effective forward and reverse flow and storage of goods, services, and related information between the point of origin and the point of consumption in order to meet customers' requirements [1]. From almost exclusively being about cost savings and reducing tied-up capital, logistics management today is also associated with terms like sustainability, value creation, revenue streams, competitiveness, and strategy [2]. Logistics is more relevant than ever as societal trends reshape patterns of production, distribution, and consumption. One of the most powerful among them is technological advancements, widely recognized by the public in the context of digitalization, Industry 4.0, and sustainability. While the definition of logistics and logistics management has remained relatively unchanged for years, the skills required to plan and adapt logistics systems in accordance with contemporary developments and competitive pressures are evolving significantly.

Just a few years ago, logistics expertise relied heavily on operational efficiency and task-driven execution. Today, however, professionals must also be equipped to implement technological advancements associated with digitalization and Industry 4.0. This includes competencies in data analytics, process automation, digital platform management, and the integration of technologies such as the Internet of Things (IoT), artificial intelligence (AI), and advanced robotics. As highlighted in recent industry insights from CSCMP, the future logistics manager is expected to be an information-driven decision-maker and a multi-skilled professional, capable of combining analytical, operational, and project management expertise to navigate the increasing complexity of modern supply chains. To achieve this, it will be important to possess not only technical but also soft competencies such as communication, adaptability, collaborative leadership, problem-solving, project management, conflict resolution, and stress management [3].

Education is often visionarily positioned as a catalyst for positive global transformation, regardless of the industry for which it is educating. Within its inherent opportunities and constraints, logistics higher education contributes significantly to the development of managerial and analytical capabilities of future logistics and supply chain experts, which are very likely necessary to address the complexities of modern logistics and supply chains. Recently, this mission has also required greater agility in adapting study content due to the fastest technological developments in human history.

Technological advancements are progressing at such a rapid pace that logistics employees, without continuous education, risk becoming less competent, slower to adapt, and more resistant to change. One such transformative innovation steadily permeating the industry is AI. The integration of AI has reignited attention toward technologically enhanced workplaces, including those already upgraded with collaborative robotics or IoT systems. Across the entire logistics value chain, from operational roles to strategic decision-making, workplaces are undergoing significant technological transformation. All of this consequently requires upgrading the competencies of the employees who manage and operate these systems.

The maturity of logistics competencies can be assessed using the Logistics Competence Maturity Matrix, a model that distinguishes between three stages of development [4]. At the basic level, countries or institutions are just beginning to build structured logistics education, often relying on external support, and systematic curriculum review is rare. At the intermediate level, training programs are more established and increasingly shaped by national industry needs. In this case, curricula may be revised occasionally but not yet in a consistent, forward-looking manner. At the advanced level, logistics education systems are highly developed, with curricula that are regularly updated to reflect technological changes, global supply chain integration, and international best practices. The analysis described by McKinnon and coauthors [4] clearly shows that the more mature the logistics competency level, the stronger the emphasis on keeping curricula up to date. Regular updates are not just an academic exercise but a hallmark of advanced logistics education systems, ensuring alignment with global trends and industry needs.

As logistics increasingly evolves from a cost-oriented function into a strategic driver of sustainability, value creation, and competitiveness, and as rapid technological advances such as Industry 4.0 reshape production and consumption patterns, there is a clear need to critically examine existing curricula and the competencies they produce. Previous analyses have shown that only education systems that regularly review and update their curricula are able to progress to higher levels of logistics competence maturity, enabling them to align with the demands of Industry 4.0 and advance toward Industry 5.0. To address this need, we conducted a comprehensive study, the outcomes of which are presented in the following chapters: the research methodology in Section 3, the results in Section 4, and the guidelines for logistics higher education in Section 5.

2. Literature review

Competency cannot be conceptualized as the passive reception of knowledge but rather as the integration of concepts, problem-solving methods, and practical skills developed through learning [5]. Competencies delineate desired educational outcomes by uniting cognitive, functional, ethical, and personal dimensions. Competency-based education thus seeks to prepare learners with the capacities needed to address complex personal and professional challenges, standing in contrast to traditional models that emphasize rote memorization and routine skill acquisition [6].

Research on logistics competencies has developed significantly over the past decades, laying the groundwork for understanding the skills required in this rapidly evolving sector. Early models, such as the one presented by Boyatzis [7] and the BLM framework described by Murphy and Poist [8], emphasized the importance of business, logistics, and management skills as a foundation for effective performance. Subsequent studies by Gammelgaard and Larson [9], Mangan and Christopher [10], and Derwik and Hellström [11] expanded these ideas, identifying sets of individual and professional competencies critical for logistics and supply chain managers. Research consistently highlights that competency requirements vary across countries due to differences in regulations, technological maturity, market demands, and cultural factors [12–15]. Industry interviews and surveys confirm the demand for cross-functional expertise, strong leadership and communication skills, analytical and IT skills, and the ability to operate effectively in multicultural environments [16, 17].

Kotzab et al. [18], further systematized competencies into professional, methodological, social, and personal categories, emphasizing adaptability and self-organization as key traits. At the same time, newer studies note a growing gap between the academic focus on social skills and industry's urgent need for digital, cognitive, and meta-competencies [19, 20]. Global research, such as the World Bank and KLU study [21], has documented persistent shortages of qualified personnel and misalignments between workforce skills and technological developments, particularly in the context of rapid digitalization and the increasing reliance on third-party logistics (3PL) providers. The COVID-19 pandemic accelerated these trends, highlighting the need for digital skills, resilience, and flexibility in supply chain operations [22].

More recent studies link the transformation of logistics competencies to Industry 4.0 and 5.0, highlighting the growing importance of digital literacy, big data analytics, cognitive agility, and a sustainability-oriented mindset [23–26]. The integration of AI into supply chain management has been shown to improve forecasting, decision-making, and operational efficiency, while also demanding new digital and analytical competencies from logistics professionals to manage increasingly complex, data-driven supply chains [27]. Pacher, Woschank, and Zunk [28] proposed competency frameworks for industrial logistics education aligned with the principles of Industry 5.0, while Nayeri et al. [29] explored adaptive Supply Chain 5.0 models within the healthcare sector.

Ścibiorek [30] emphasized that rapidly changing organizational conditions driven by globalization, Industry 4.0, and societal transformations require the education system, particularly in management and logistics, to shift from theory-heavy instruction toward flexible, practice-oriented learning that integrates scientific knowledge, fosters continuous skill development, and prepares future entrepreneurs to adapt quickly to uncertainty and change.

Overall, the literature shows broad agreement that existing competency models must be updated to reflect digitalization, Industry 4.0 and 5.0, sustainability imperatives, and many other contemporary studies are contributing to this discussion, underscoring how current and dynamic this field of research has become.

3. Methodology

The study employed a multi-phase mixed-method approach to examine the current offerings of logistics-related study programs in higher education institutions (HEIs), with particular emphasis on Industry 4.0, digitalization, and sustainability. The objective was to (1) determine the extent to which existing curricula incorporate these competencies, (2) identify gaps between educational provisions and the needs of the logistics labor market, and (3) translate these insights into design principles for advanced curricula.

The research was conducted in 2023 by a multidisciplinary consortium comprising the University of Maribor (Slovenia), Universidade de Aveiro (Portugal), Politechnika Poznanska (Poland), Gaziantep Üniversitesi (Turkey), ECQA (Austria), ValueDO (Italy), and Zerynth (Italy). This diverse partnership ensured both academic rigor and practical relevance by combining higher education expertise, industry perspectives, and professional certification experience.

The research was conducted in four stages.

3.1 Stage 1: Content analysis of existing logistics study programs

A state-of-the-art analysis was carried out to map the current logistics education landscape in Europe and selected non-European countries. The sample comprised 50 higher education study courses in logistics, supply chain management, and related fields, offered by HEIs across 17 countries. The largest shares originated from Sweden (11), United Kingdom (7), Slovenia (7), and Germany (7), with additional study courses from Netherlands (3), Norway (3), Poland (2), and several other countries represented by one study course each. Regarding study level, the sample included 34 master's courses, 11 bachelor's, 1 doctoral, and 1 post-graduate course. Most courses (40) were delivered in English, with the remainder taught in Slovenian (5), German (3), Polish (1), and Portuguese (1).

For each study course, data were collected on the HEI, country, department or study program, degree level, disciplinary focus, ECTS (European Credit Transfer and Accumulation System) credits, language of instruction, course content, stated objectives and expected competencies, learning outcomes, innovative aspects, and teaching methods. This diverse composition provides a comprehensive overview of contemporary logistics-oriented study courses and serves as a basis for analyzing their structure, pedagogical approaches, and alignment with emerging competency requirements associated with Industry 4.0, digitalization, and sustainability.

3.2 Stage 2: Content analysis of existing logistics study programs

A structured online questionnaire was designed to explore competency gaps from the perspective of key stakeholders: students, logistics employees in logistics companies, representatives of national logistics associations, and organizations recognized for best practices in sustainability, digitalization, and Industry 4.0. This allowed the consortium to identify gaps between existing study programs and the skills required in professional practice, particularly in relation to sustainability, digitalization, and Industry 4.0.

The questionnaire consisted of five thematic sections: demographics, current knowledge and skills in logistics-related domains, the importance of various competencies for future logistics professionals, preferred teaching and learning methods, and perceived gaps between higher education and workplace requirements.

In total, the questionnaire contained 42 questions, primarily using a 1–4 scale (1 – definitely not able, 4 – definitely able), complemented by multiple-choice and open-ended questions. The instrument was validated through internal review, pilot testing with a small group of students and professionals, and then translated into the national languages of participating countries to ensure accessibility and cultural relevance.

The survey collected 425 valid responses from target groups: 344 students, 59 logistics company employees, 6 representatives of national logistics associations, and 16 representatives of organizations recognized for best practices in sustainability and Industry 4.0. Respondents came from the project partners' countries – Slovenia (162 respondents, 38.1%), Poland (123 respondents, 28.9%), Türkiye (82 respondents, 19.3%), and Portugal (58 respondents, 13.6%) – to provide a diverse cross-section of the European logistics education and labor market landscape.

3.3 Stage 3: External expert evaluation of Industry 4.0 competencies

In parallel with Stage 2 activities, 22 representatives of national logistics associations and companies recognized for advanced Industry 4.0 implementations were invited to assess how capable current students and employees are of implementing a set of specific Industry 4.0-related activities. They rated each activity on a 1–4 scale using the same criteria as the self-assessment (1 – definitely not able, 4 – definitely able). This allowed for a comparison between self-perceived readiness and external judgment.

3.4 Stage 4: Triangulation and synthesis

Finally, results from the content analysis, the training needs analysis, and the external expert evaluation were triangulated to identify critical competency gaps and to formulate guidelines for the design of an advanced curriculum.

To quantify the magnitude of the difference between self-assessment and expert ratings, effect sizes (Cohen's d) were calculated as the difference between group means divided by their pooled standard deviations.

4. Results

The results provide an integrated overview of the current state of competency development in logistics higher education, highlighting existing strengths, critical gaps, and priority areas for curriculum improvement in the context of Industry 4.0, digitalization, and sustainability.

4.1 State-of-the-art training in the logistics sector

Despite the sector's significant economic importance, relatively few HEIs offer dedicated logistics or supply chain management programs, and many of those that do exist remain limited in size, scope, or industry relevance.

Curricula often lack coherence and balance, with some programs focusing heavily on quantitative methods, while others narrow their scope to operational logistics, leading to inconsistent graduate preparedness. A shortage of qualified faculty, combined with limited student awareness of logistics as a dynamic career path, further constrains the expansion and modernization of these programs.

Research strongly supports the incorporation of experiential learning approaches, such as internships, industry-led projects, and simulation-based training, to close the gap between academic preparation and labor market expectations.

Kolb's experiential learning theory [31] provides the foundation for these approaches, emphasizing that learning through experience fosters deeper understanding and skill acquisition. Empirical studies confirm that structured internships and work-integrated learning programs enhance both technical and soft skills, such as problem-solving, communication, and adaptability, while facilitating smoother transitions to professional roles [32, 33]. Similarly, simulation-based methods, including serious games in logistics education, have been shown to improve decision-making, analytical thinking, and collaboration in realistic operational scenarios [34].

To meet the growing demand for digitally literate, multi-skilled professionals, logistics education must integrate digitalization, data analytics, and sustainability competencies, supported by stronger collaboration between academia, industry, and policymakers. Embedding experimental and practice-oriented models through internships, live case studies, and advanced simulations can ensure graduates are better equipped to navigate the demands of Industry 4.0 and 5.0 while contributing to more resilient and sustainable supply chains.

In our study, the analyzed 50 higher education courses are predominantly offered at the Master's level (68%), indicating that logistics education is still largely positioned as an advanced specialization rather than as entry-level training. Most courses award approximately 6.5 ECTS credits per course and are primarily taught in English (80%). This distribution in our sample does not necessarily reflect the actual situation within individual countries or across the European Union, as assessing national differences and conducting cross-country comparisons was not the aim of this research. The disciplinary focus spans logistics, supply chain management, operations, and industrial engineering, with many courses situated within business schools, engineering faculties, or specialized logistics departments, reflecting a broad yet fragmented institutional anchoring.

The study courses in our sample typically combine lectures, seminars, project-based work, and case studies, with varying degrees of experiential and practice-oriented components. Specifically, capstone projects or theses are included in 22 courses (44%), problem-based or case-based learning in 19 (38%), industry collaboration in 18 (36%), internships or placements in 15 (30%), and simulations, labs, or serious games in 9 (18%). However, only a minority describe mandatory internships or industry-led projects as core components of the curriculum. While several courses incorporate interdisciplinary content and digital learning tools as innovative elements, traditional didactic approaches continue to predominate, suggesting that the transition toward practice-based and student-centered pedagogies is gradual rather than systemic.

Although nearly all courses refer to analytical, managerial, and cross-functional skills, explicit integration of digitalization, Industry 4.0 technologies, and sustainability remains limited. Only 14 courses (28%) explicitly reference digitalization or Industry 4.0, 11 (22%) mention AI, machine learning, or data analytics, and 21 (42%) address sustainability or ESG-related (environmental, social, and governance) content. Where present, these topics are often embedded within broader strategy or operations modules rather than offered as stand-alone subjects. This pattern suggests a partial and uneven alignment with evolving labor market requirements for digital literacy, cognitive agility, and sustainability-oriented thinking. Furthermore, course descriptions frequently rely on generic competency formulations (e.g., "critical thinking" and "teamwork"), indicating that many curricula have yet to be systematically updated to reflect the digital transformation and the competencies envisioned under Industry 4.0.

Table 1 summarizes the key concepts and their frequencies appearing in descriptions of course study content, grouped broadly into themes such as sustainability, technology/innovation, and digitalization/data. Terms related to sustainability are the most dominant, with sustainability itself appearing 48 times,

Sustainability-related	Count	Technology-related	Count	Digital-related	Count
Sustainability	48	Project	27	Digital, data	8
Traffic	10	Technology	26	Communication	4
Environmental	24	Smart/Intelligent	15	Software, programming	3
Climate	12	Simulation	11	Cloud computing, AI, machine learning	1
Energy	9	Economic	10	Twin	2
Society	8	Industrial, modelling	9		
Green/Greener	7	Optimization	8		
Emissions	5	Forecasting	7		
Ethics, health	3	Uncertainty, lean, trend	6		
Hydrogen	2	Robots, IoT, autonomous	5		
		Mobile, automated, machine, flexible, innovation, responsiveness	4		
		Standards, routing	3		
		AMR	2		

Note: Shows how often sustainability-, technology-, and digitalization-related terms appear in analyzed curricula, highlighting the underrepresentation of core Industry 4.0 topics.

Table 1.
Keywords and frequencies in the reviewed logistics education context.

followed by other environmentally oriented concepts such as environmental (24), climate (12), traffic (10), and energy (9).

Technology-related terms also feature prominently, indicating the increasing role of technological innovation in logistics. High-frequency terms in this category include project (27), technology (26), smart/intelligent (15), simulation (11), IoT (5), and economics (10), alongside a cluster of emerging topics like robots, autonomous systems, automation, and innovation.

An analysis of the course content reveals that explicit references to core Industry 4.0 technologies at the beginning of 2024 were largely absent from logistics education courses. While the IoT (five), AI (one), machine learning (one), and cloud computing (one) are mentioned occasionally, cyber-physical systems (CPS) (zero), big data analytics (zero), and additive manufacturing (zero) are not mentioned at all. These findings underscore that most key Industry 4.0 technologies are either underrepresented or entirely missing, highlighting the need to more systematically embed technological and digital competencies into logistics and supply chain management (SCM) curricula to align education with the demands of future supply chain environments.

4.2 Training needs analysis: Gaps and readiness in logistics education

The training needs analysis revealed significant gaps between current logistics higher education curricula and the skills required in professional practice, particularly in the domains of digitalization, Industry 4.0 technologies, and sustainability. While respondents rated their soft skills (team work, communication, problem-solving, adaptability) relatively high (mean 3.2/80% "able"), most technical competencies scored below 3.0 on a 1–4 scale, indicating only moderate readiness for emerging sectoral requirements.

Regarding digitalization-related competencies, respondents reported only moderate ability in core digitalization skills, including data analysis programs (2.97/67% "able"), big data collection and analysis (2.59/53% "able"), digitalization of document flows (2.71/59% "able"), and digital twins (2.73/59% "able"). These findings indicate a strong need to enhance digital literacy, analytical thinking, and data-driven decision-making within logistics education. Although basic information and communications technology (ICT) use is established, advanced digital tools and data-based approaches are not yet systematically embedded in curricula or workplace practices.

Competencies directly connected to Industry 4.0 technologies and processes received some of the lowest readiness scores overall, reflecting a clear readiness gap among respondents. At the strategic level, the ability to explain Industry 4.0 was rated at a mean of 2.7 (SD = 0.91), with only 17.4% of respondents definitely able to implement it and 42.2% probably able, while 32.2% considered themselves not able (11.3% definitely not, 20.9% probably not). Similarly, the ability to introduce Logistics 4.0 into a company's environment received a mean of 2.59 (SD = 0.94), with only 15.3% definitely able, 37.9% probably able, and 38.9% not able (14.4% definitely not, 24.5% probably not). Related strategic competencies, such as assessing the level of a company's Industry 4.0 maturity, scored 2.56, with 50.4% of respondents considering themselves able. These results suggest that basic conceptual knowledge exists, but practical and strategic application competencies remain limited. Even lower readiness was reported for the implementation of concrete Industry 4.0 technologies. Respondents assessed their ability to implement IoT resource tracking (2.43/45.2% "able"), optimization of logistics processes using advanced technologies (2.72/59.5% "able"), planning warehouse automation (2.49/47.5% "able"), AI solutions design (2.67/68% "able"), cybersecurity and blockchain (2.40/43% "able"), collaborative robots (2.52/49.4% "able"), autonomous warehouse equipment (2.59/52.9% "able"), drones (2.18/23% "able"), augmented/virtual/mixed reality (2.45/44.5% "able"), machine learning (2.52/51.1% "able"), and computer vision systems (2.12/24% "able"). In nearly all cases, less than half of respondents considered themselves probably or definitely able to implement the technology, and mean ratings stayed below 3.0 on the 1–4 scale. This indicates that while awareness of emerging technologies is present, respondents lack the applied technical and managerial competencies to deploy them in an operational logistics context. Combined with the content analysis, which found no references to many of these technologies, such as collaborative robots, drones, or augmented/virtual/mixed reality, in current curricula, these results confirm a critical gap between technological developments and logistics higher education. Addressing this gap will require embedding applied technological training, simulation-based learning, and real-life project work into logistics study programs to prepare graduates for the digital transformation of supply chains.

To gain deeper insight into respondents' perceived confidence levels, **Figure 1** presents their readiness to implement specific Industry 4.0 technologies, showing

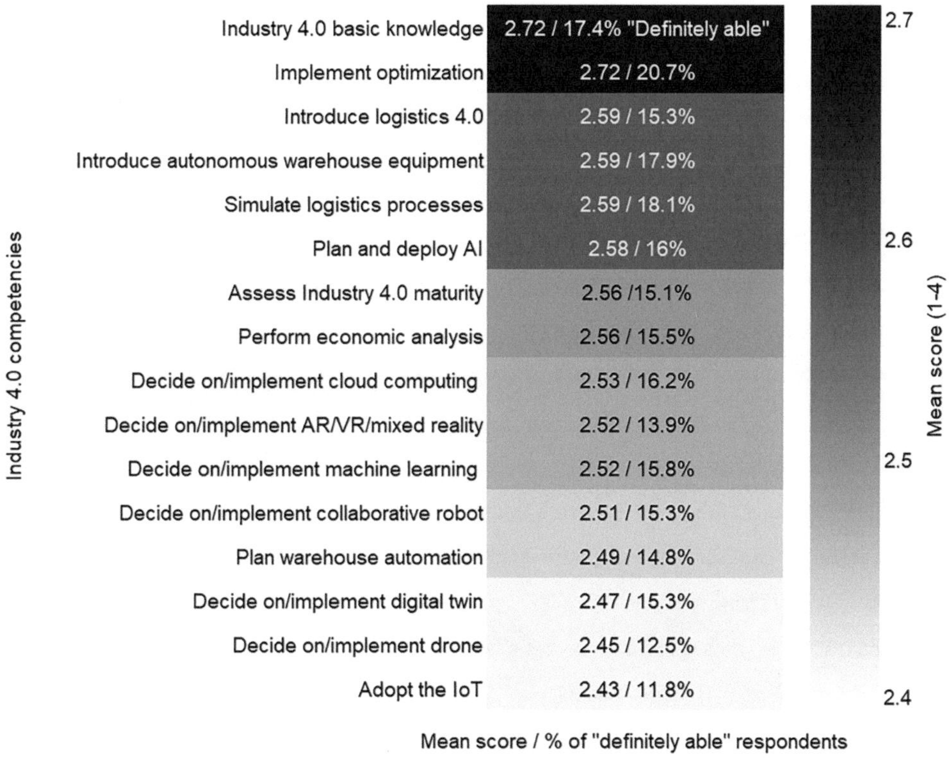

Figure 1.
Industry 4.0 implementation readiness (self-assessment). Note: This shows the mean self-assessment scores and the percentage of respondents who reported being "definitely able" to perform specific Industry 4.0 activities, revealing low practical confidence despite moderate familiarity.

mean self-assessment scores and the percentage of respondents who reported being definitely able to perform each competency. While the mean values indicate overall moderate familiarity, the proportion of respondents who feel fully confident is markedly low, often below 20%, which highlights the substantial gap between conceptual awareness and the practical ability to implement advanced technological solutions in logistics settings.

Sustainability emerged as a strongly demanded but underdeveloped competency area. Respondents reported moderate ability in green/sustainability competencies (2.6–2.8/ ~60% "able"), economic analysis of sustainability investments (2.95/77% "able"), and optimization tools for sustainable routing (~2.8/65% "able"). Respondents emphasized the importance of environmental and social responsibility, circular economy principles, and ethical decision-making, yet felt their current study programs do not systematically integrate these topics. Overall, the results demonstrate that while basic operational and communication skills are well represented, digitalization, Industry 4.0, and sustainability competencies are insufficiently covered. This highlights an urgent need to embed these competencies as core components of logistics higher education curricula to prepare graduates for rapidly evolving supply chain environments.

The following heat map (see **Figure 2**) visually summarizes the mean self-assessed competency scores and the percentage of respondents who reported being probably or definitely able to perform each activity.

Figure 2 illustrates the relative strength of assessed competencies, showing that soft skills score highest, while digitalization and Industry 4.0-related competencies remain notably weaker, indicating clear areas for curriculum development.

Preliminary subgroup analyses indicated that competency gaps were more pronounced among students than among employees, particularly in Industry 4.0 and

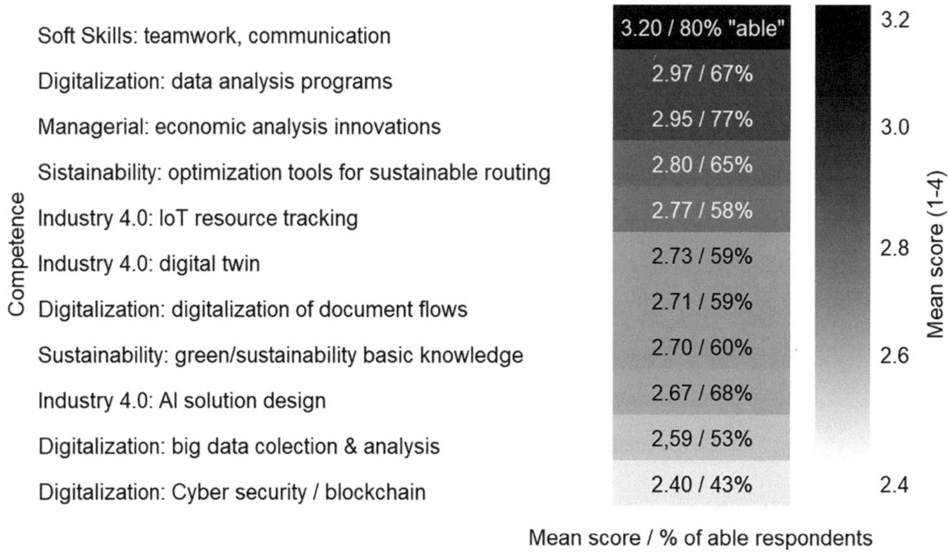

Figure 2.
Key competencies in sustainability, Industry 4.0, and digitalization (self-assessment). Note: Heat map of mean competency scores (color) and the percentage of respondents who reported being able or definitely able to perform each competency (labels), showing strong soft skills but weak Industry 4.0 and digitalization competencies.

digitalization competencies, while sustainability competencies showed smaller differences. Minor cross-country variations also emerged, suggesting that regional context influences exposure to advanced technologies. These patterns highlight the importance of adapting curriculum design to different learner profiles and regional contexts to ensure inclusiveness and relevance.

4.3 Training needs analysis: External evaluation of Industry 4.0 readiness

In addition to self-assessment, the study gathered evaluations from representatives of national logistics associations and organizations recognized for advanced implementations of Industry 4.0 solutions ($N = 22$). These experts were asked to assess the ability of logistics students and employees to implement a range of Industry 4.0-related activities in practice. Their ratings indicate that overall readiness remains moderate and that fully confident ("definitely able") performance is still rare across most competencies.

Experts perceived strategic competencies as relatively more developed. The ability to introduce Logistics 4.0 into a company's logistics strategy received the highest rating (mean 3.10), with over two-thirds of respondents describing students and employees as probably or definitely able, and more than one-third judging them as definitely able. Similarly, the ability to explain Industry 4.0 concepts and developments (mean 2.81) and to assess a company's Industry 4.0 maturity level (mean 2.76) were viewed as moderately present, though only around a quarter of respondents considered current learners definitely capable of these tasks.

When it comes to deciding on and adopting specific technologies, readiness was more uneven. Experts rated readiness for cloud computing (mean 2.95) as relatively stronger, yet only one-third judged current students and employees as definitely able to implement it. Machine learning (mean 2.73), digital twins (mean 2.76), and especially virtual/augmented/mixed reality (mean 2.59) received lower ratings, with fewer than one in five respondents perceiving definite readiness.

Assessments of operational technologies and automation also indicated only moderate readiness. Abilities to introduce autonomous warehouse equipment (mean 2.86), plan warehouse automation up to dark warehouse level (mean 2.76), adopt IoT (mean 2.80), or implement unmanned aerial vehicles (drones) for logistics purposes (mean 2.73) were seen as still emerging, with most respondents describing current learners as only probably, but not yet definitely, able to perform them.

Interestingly, competencies involving analytics and economic decision-making were viewed more positively. Respondents rated the ability to implement optimization of logistics processes, routes, and supply chains using advanced tools (mean 2.95) and carry out economic analyses of Industry 4.0 interventions (mean 2.95) as among the strongest competencies, with nearly one-third judging current learners as definitely able.

Overall, these external assessments presented a picture of growing conceptual awareness but limited confident practical ability. Stakeholders acknowledge that students and employees are starting to understand and engage with Industry 4.0, yet they rarely perceive them as fully capable of independent implementation, particularly in newer technological areas such as immersive reality, drones, machine learning, and robotics. This underlines the need for more applied and hands-on learning opportunities in logistics higher education, ensuring graduates can move beyond theoretical knowledge to practical implementation of emerging Industry 4.0 solutions.

While **Figure 3** provides a detailed overview of competency gaps at the level of individual skills, it is also important to examine how these gaps appear when aggregated into broader competency families. To anchor the magnitude of the observed differences, **Table 2** summarizes the mean self-assessment and expert ratings for the three main competency families (Industry 4.0, digitalization, and sustainability), together with their standard deviations and an indicative effect-size measure.

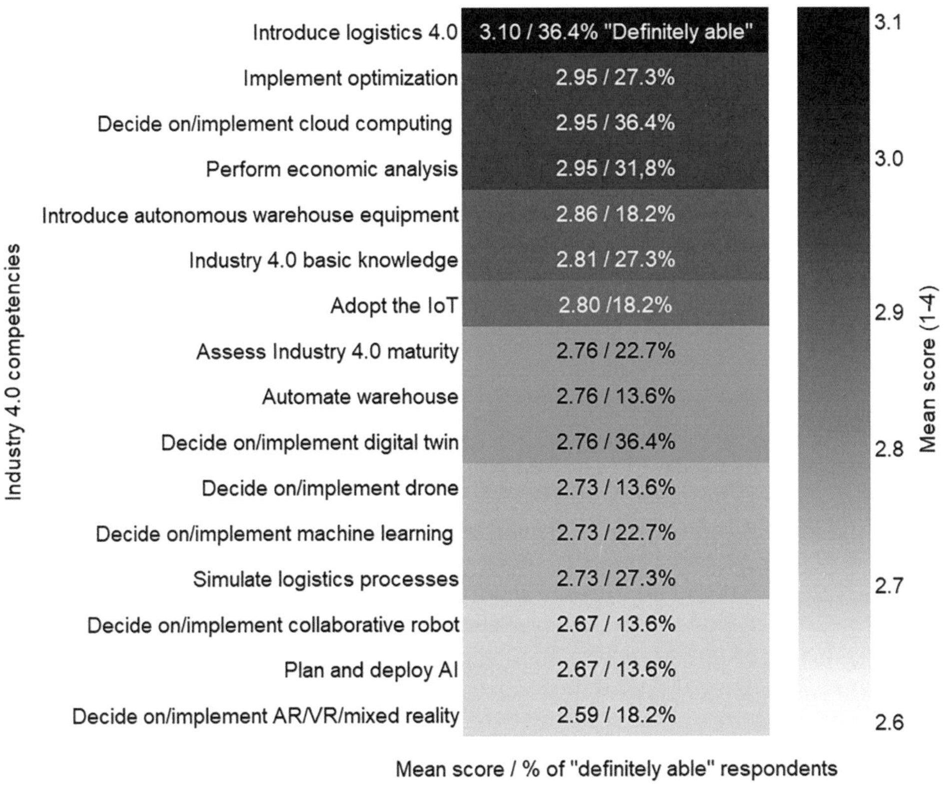

Figure 3.
External evaluation of readiness to implement Industry 4.0 activities. Note: Expert ratings show moderate means but a very low percentage judged as "definitely able," confirming the need for more applied training.

Competency family	Self-assessment (M ± SD)	Expert rating (M ± SD)	Δ (Cohen's *d*)	Observed gap
Industry 4.0	2.61 ± 0.64	1.98 ± 0.52	1.07	Large
Digitalization	2.84 ± 0.58	2.21 ± 0.49	1.13	Large
Sustainability	2.97 ± 0.55	2.68 ± 0.5	0.54	Medium

Note: Presents mean scores (±SD) and effect sizes (Cohen's d*), revealing large gaps in sustainability, indicating misalignment with labor market needs.*

Table 2.
Comparison of self-assessment and expert ratings for selected competency families.

5. Guidelines for logistics higher education

Based on the evidence collected through the literature review, the content analysis of existing study courses, the training needs analysis of logistics students and logistics employees, and the external evaluation by logistics associations and best-practice organizations, several key directions emerge for the development of future-oriented logistics higher education curricula. These guidelines aim to address identified competency gaps, particularly in the domains of digitalization, Industry 4.0 technologies, sustainability, managerial-analytical skills, and pedagogical approaches, to better prepare graduates and already employed individuals for the dynamic and technologically evolving logistics environment.

Strengthening competencies in logistics higher education can directly enhance the resilience of supply chains, understood as the capacity to anticipate, absorb, adapt to, and recover from disruptions. While resilience is often discussed as an abstract system property, it can be operationalized by linking specific competency domains to concrete resilience mechanisms and measurable performance indicators. **Table 3** maps selected competency areas identified in this study to the resilience mechanisms they support and suggests indicative key performance indicators that could be used to monitor their contribution in practice. This framework illustrates how investing in target competencies, such as anomaly detection, digital twin modeling, and agile project management, can translate into measurable improvements in system adaptability and recovery capacity, thereby bridging the gap between education outcomes and operational resilience.

The mapping (see **Table 3**) illustrates that competency development is not only a response to emerging technological trends but also a strategic lever for building

Competency domain	Examples of specific competencies	Resilience mechanisms	KPI
Digitalization and data analytics	Data literacy, anomaly detection, and predictive analytics	Early disruption detection, response coordination	Order recovery lead time
Digital twin and simulation	Digital twin modeling, scenario simulation, and system optimization	Proactive risk assessment, scenario planning	Time-to-reconfiguration
Technology integration	IoT systems, autonomous equipment, cobots	Process continuity under stress through automation	Uptime ratio during disruption [%]
Project and change management	Agile project methods, change leadership	organizational adaptability	Change implementation cycle time
Sustainability and circular logistics	LCA, reverse logistics, resource efficiency	Resource buffering and redundancy	Inventory buffer coverage days
Communication and collaboration	Cross-functional coordination, stakeholder management	Relational resilience and information transparency	Information latency
Ethics and risk-aware decision-making	Risk assessment, ethical reasoning, compliance	Decision quality under uncertainty	Decision lead time variance

Note: Links core competencies to resilience mechanisms and measurable KPIs, illustrating how target competency development can strengthen logistics system resilience.

Table 3.
Competency domains linked to resilience mechanisms and KPIs.

resilient logistics systems. Each competency area contributes to specific resilience mechanisms, highlighting the need for intentional curriculum design that aligns learning outcomes with system-level robustness. The following subsections outline key competency domains, starting with digitalization, whose integration into logistics higher education can drive both innovation and resilience.

5.1 Integration of digitalization competencies

Given the low self-reported readiness in core digital competencies, logistics curricula should place a stronger emphasis on data-driven decision-making and digital process management. This includes embedding content on big data analytics, data visualization, and digital supply chain platforms, as well as the digitalization of document flows and real-time information sharing. Students should be trained to work with actual logistics datasets to develop data literacy and analytical reasoning. Additionally, curricula should incorporate digital ethics, data security, and data governance to ensure responsible use of digital tools.

5.2 Strengthening Industry 4.0 and emerging technologies

A major gap identified concerns the readiness to implement advanced technologies associated with Industry 4.0 and Logistics 4.0. To address this, study programs should provide students with both theoretical understanding and practical exposure to key technologies such as the IoT, AI, machine learning, collaborative robots, automated warehouse systems, and autonomous mobile robots. Students should also gain experience with digital twin simulations and AR/VR/mixed reality to build familiarity with immersive and simulation-based logistics tools. HEIs should establish labs or simulation platforms for experimenting with these technologies and embed modules on strategic adoption, change management, and technology integration to prepare students for their deployment in real logistics settings.

5.3 Embedding sustainability and ethics

Although sustainability is widely recognized as a critical trend, it remains insufficiently embedded in most logistics curricula. Study programs should integrate the environmental, social, and economic dimensions of sustainability across courses, including content on life-cycle assessment, green logistics, carbon accounting, energy efficiency, and circular economy models. Students should be trained to perform environmental impact analyses and to evaluate the trade-offs between operational efficiency and sustainability objectives. Furthermore, curricula should strengthen ethical reasoning, corporate social responsibility, and stakeholder impact analysis to build responsible decision-making competencies. This content should be linked to real-world case studies, industry projects, and policy frameworks to ensure relevance and applicability.

5.4 Enhancing managerial and analytical skills

Results from both self-assessment and external evaluations indicate that students require stronger project management, process improvement, and change leadership competencies to successfully drive logistics transformation. Programs should combine operational and technological content with analytical thinking, problem-solving, system modeling, and decision analysis. Interdisciplinary projects that integrate financial, economic, operational, and technological perspectives should be introduced to reflect the complex nature of real logistics challenges. Furthermore, elective courses or minors could be offered in business analytics, innovation management, and technology strategy to develop students' ability to connect technical solutions with strategic and economic decision-making.

5.5 Innovating pedagogy and collaboration with industry

To effectively build these competencies, logistics higher education must also innovate its pedagogical approaches and strengthen collaboration with industry. Study programs should shift from predominantly theoretical delivery to experiential learning methods, including problem-based learning, case studies, simulations, and logistics games. Stronger partnerships with logistics companies and professional associations are needed to provide guest lectures, industry challenges, and co-supervised thesis projects. Embedding mandatory internships or practice placements would give students direct exposure to technological and organizational realities. Finally, curricula should be regularly updated through structured feedback loops involving alumni, employers, and professional associations to maintain alignment with rapidly evolving industry requirements.

Importantly, it is not sufficient merely to increase the number of guest lectures, industry challenges, or case studies. Such activities must be deliberately planned with clear alignment to the target competency profile of graduates, specifying which competencies each activity aims to strengthen and how they address identified weaknesses. At present, some HEIs report high quantitative indicators of industry involvement but demonstrate limited focused impact on specific competency gaps. A more strategic and outcome-oriented integration of industry collaboration is, therefore, essential to ensure that experiential learning activities directly contribute to closing the documented competency gaps. Achieving such targeted impact requires not only better-designed collaboration activities but also a more agile and strategically coordinated approach to curriculum development, which ensures that competency gaps are addressed systematically and continuously, an aspect further elaborated in the following section.

5.6 Fostering agility, lifelong learning, and industry cooperation

To remain relevant in a rapidly transforming logistics landscape, higher education must become more agile and responsive to changing competency demands. The current slow pace of curriculum renewal, combined with the long lag between evaluation and the incorporation of updates, contributes to the observed gaps between industry requirements and educational outcomes. Research on logistics competency maturity highlights that some countries, such as Germany, which is widely recognized as one of the most advanced in implementing Industry 4.0 principles, achieve and sustain high maturity through continuous curriculum upgrading, strong industry–academia linkages, and early integration of emerging technologies into formal education.

To emulate this model, logistics study programs should establish regular curriculum review cycles (e.g., every two to three years) based on structured feedback from industry partners, alumni, and professional associations. At the same time, HEIs should strengthen international orientation through activities such as student mobility, industry placements in high-maturity logistics ecosystems, and offering foreign language modules (e.g., German for business and technical contexts) to enhance graduates' access to advanced logistics environments and best practices.

Moreover, the findings reveal a pressing need to support the upskilling and reskilling of already employed logistics personnel, who often lack opportunities to systematically update their competencies in line with technological change. HEIs should, therefore, build flexible, modular training offers and micro-credential pathways, enabling professionals to acquire targeted skills without enrolling in full-degree study programs. Such micro-credentials could be aligned with established international certification schemes, such as the European Logistics Association certification of competencies. While such certificates are widely recognized within the industry, they remain insufficient on their own to meet the rapidly evolving competency demands.

To support this shift, cooperation between HEIs and logistics companies should extend beyond internships to long-term partnerships, involving joint problem-solving projects, mentoring schemes, and continuous professional education programs. Such collaboration would allow companies to contribute real-world challenges to the curriculum while benefiting from the innovative capacity of students and academic staff. In turn, it would help ensure that both students and current employees can continuously develop the technological, analytical, and managerial competencies required in a rapidly changing environment.

5.7 Curriculum starter map

To support the practical adoption of these guidelines, **Table 4** presents a curriculum starter map that aligns modules with target competencies, teaching modes, assessment approaches, and micro-credentials with indicative ECTS allocations. The suggested ECTS allocations are intentionally minimal, as the program is designed for participants who are already enrolled in logistics-related education or employed in the logistics industry and possess prior professional experience. For more details, please follow the Slog4.0 webpage [35].

5.8 Limitations and future research

While this study provides a comprehensive overview of competency gaps and readiness levels in logistics higher education, several limitations should be acknowledged. First, the content analysis of 50 study courses was based on publicly available curriculum documents, which may not fully reflect the actual teaching practices or recent informal updates.

Second, the training needs analysis relied on self-assessments from students and employees, which can be subject to response bias, such as overestimation or underestimation of personal competencies. Similarly, the external evaluation involved a relatively small sample ($n = 22$) of experts from selected countries, limiting the generalizability of their assessments.

Given that some regions already display higher levels of logistics competence maturity, future efforts should focus less on expanding the geographic sample and more on transferring good practices from high-maturity regions. However, the education system should balance its development according to the logistics maturity

Module	Core competencies developed	Teaching and pedagogical modes	Assessment methods	Micro-credential and ECTS
Module 1: Logistics 4.0 in the smart society	Industry 4.0 and Logistics 4.0 fundamentals, Industry 4.0 maturity, AR/VR/mixed reality, (collaborative) robotics, AGV/AMR, material flow platform, simulation	Student-centered approach, e-lectures, asynchronous remote learning in a virtual classroom, mind and concept maps, videos, problem-based learning, hands-on activities, case studies	Self-assessment with online quizzes, final examinations conducted by an independent organization (ECQA)	Industry 4.0 and digitalization fundamentals (0.7 ECTS)
Module 2: Sustainability in Logistics 4.0	Regulatory frameworks, sustainability terminology, sustainability maturity, sustainability KPIs, sustainability methods/tools, CO_2 emissions, social, economic, and environmental sustainability			Sustainability fundamentals (0.6 ECTS)
Module 3: IoT: adopting 4.0 for green logistics	IoT integration, sensor networks, data-driven decision-making, and automation			IoT fundamentals (0.7 ECTS)
Module 4: Green challenge	problem solving, cross-functional project work, innovation, systems thinking	seminar, mentoring sessions, industry challenge	capstone project evaluation, oral defence	Logistics 4.0 innovator (1 ECTS)

Note: Aligns modules with target competencies, teaching modes, assessment methods, and micro-credentials (ECTS), offering a structured starting point for curriculum adoption.

Table 4.
Curriculum starter map for sustainable logistics 4.0.

in the region and its goals. Future research should emphasize longitudinal tracking of competency development and use mixed methods, such as structured interviews, classroom observations, and pre- and post-assessments of competency growth to capture how students' skills evolve over time.

Despite these limitations, this study offers unique and timely insights into the alignment between logistics higher education and the rapidly changing competency demands in the sector. By combining content analysis, self-assessment, and external evaluation, it provides one of the first integrated, multi-perspective mappings of logistics competency gaps at the European level, thereby offering a valuable evidence base for the design of future-oriented, agile curricula in the field.

6. Conclusion

The findings of this study underscore that the transformation of logistics under the pressures of Industry 4.0, digitalization, and sustainability is outpacing the ability of current higher education systems to adapt. While numerous logistics study programs provide solid foundations in operational and managerial domains, they remain only partially aligned with the emerging technological, analytical, and ethical competencies required in modern supply chains. The study revealed that although students and employees are conceptually aware of digital and Industry 4.0 developments, they often lack the confidence and applied capability to implement them in practice.

Addressing this challenge will require more than incremental curriculum adjustments. HEIs should adopt a systemic approach to competency development, one that is agile, forward-looking, and embedded in close collaboration with industry. This includes frequent curriculum updates, integration of applied technological training, sustainability principles, and data-driven decision-making, as well as stronger use of experiential learning, micro-credentials, and lifelong education pathways to support both students and already-employed professionals.

By repositioning logistics education as a dynamic, innovation-oriented, and continuously evolving system, higher education can not only reduce the current competency gaps but also actively drive the sector's digital and sustainable transformation, ensuring a workforce capable of thriving in complex and rapidly changing global logistics networks. Only by embracing agility, interdisciplinarity, and industry collaboration can logistics higher education keep pace with the sector's rapid transformation.

Beyond addressing immediate competency gaps, strengthening higher education also contributes to the systemic resilience of supply chains by preparing a workforce capable of adapting to technological disruptions, market volatility, and sustainability mandates. Embedding human-centered and inclusive learning approaches, particularly those that support learners from underrepresented regions and groups, can help ensure that the digital transformation of logistics is socially sustainable and broadly accessible. Achieving this requires not only action from individual HEIs but also coordinated efforts by universities, government agencies, and industry associations to define competency standards, fund curriculum innovation, and support micro-credentials and lifelong learning pathways. Such multi-level collaboration is essential for aligning logistics education with the evolving demands of Industry 4.0 while safeguarding its societal relevance and resilience.

Acknowledgments

The chapter is co-funded by the Erasmus+ project no. 2022-1-PL01-KA220-HED-000086366 (Sustainable Logistics 4.0: Digital and green skills for boosting innovation and sustainability of the logistics sector).

Conflict of interest

The authors declare no conflicts of interest.

Appendix

Survey questionnaire is available at: https://drive.google.com/file/d/1R33iIbLql38EHNSJtwfdLO_n4iAPKpfy/view?usp=sharing

Author details

Brigita Gajšek[1]* and Beata Mrugalska[2]

1 University of Maribor, Faculty of Logistics, Celje, Slovenia

2 Poznan University of Technology, Faculty of Engineering Management, Poznan, Poland

*Address all correspondence to: brigita.gajsek@um.si

References

[1] Vitasek K. Supply chain terms and glossary [Internet]. 2013. Available from: https://cscmp.org/CSCMP/CSCMP/Educate/SCM_Definitions_and_Glossary_of_Terms.aspx?hkey=60879588-f65f-4ab5-8c4b-6878815ef921 [Accessed: 2025-September-10]

[2] Sandberg E. Strategic Logistics Management - Contemporary Principles and Practice. London: KoganPage; 2025. p. 232

[3] Cayone BM. 10 Skills for logistics managers in 2025 [Internet]. 2025. Available from: https://training.safetyculture.com/blog/skills-for-logistics/ [Accessed: 2025-September-10]

[4] McKinnon A, Flöthmann C, Hoberg K, Busch C. Logistics Competencies, Skills, and Training: A Global Overview. Washington DC: World Bank Studies; 2017. DOI: 10.1596/978-1-4648-1140-1

[5] Sleurs W, editor. Competences for ESD (Education for Sustainable Development) Teachers. In: A Framework to Integrate ESD in the Curriculum of Teacher Training Institutes. Hochschule Zürich, Switzerland: Brussels; 2008. p. 90. Available from: https://unece.org/fileadmin/DAM/env/esd/inf.meeting.docs/EGonInd/8mtg/CSCT%20Handbook_Extract.pdf [Accessed: 2025-November-25]

[6] Lozano R, Merrill MY, Sammalisto K, Ceulemans K, Lozano FJ. Connecting competences and pedagogical approaches for sustainable development in higher education: A literature review and framework proposal. Sustainability. 2017;**9**(10):1889

[7] Boyatzis RE. The Competent Manager: A Model for Effective Performance. New York: John Wiley & Sons; 1991. p. 328

[8] Murphy PR, Poist RF. Skill requirements of senior-level logisticians: Practitioner perspectives. International Journal of Physical Distribution & Logistics Management. 1998;**28**(4):284–293

[9] Gammelgaard B, Larson PD. Logistics skills and competencies for supply chain management. Journal of Business Logistics. 2001;**22**(2):27–50. DOI: 10.1002/j.2158-1592.2001.tb00002.x

[10] Mangan J, Christopher M. Management Development and the Supply Chain Manager of the Future. International Journal of Logistics Management. 2005;**16**(2):178–191

[11] Derwik P, Hellström D. Competence in supply chain management: A systematic review. Supply Chain Management: An International Journal. 2017;**22**(2):200–218. DOI: 10.1108/SCM-09-2016-0324

[12] Čuček M, Kač SM. Projected and expected competences in logistics and supply chain in slovenia. In: Proceedings of the 23rd International Scientific Conference Business Logistics in Modern Management; 5-6 October 2023; Osijek. Osijek: Josip Juraj Strossmayer University of Osijek, Faculty of Economics; 2023. p. 79–91.

[13] Cvetić B, Vasiljević D, Danilović M. Competence requirements for logistics managers in the republic of Serbia. Management: Journal of Sustainable Business and Management Solutions in Emerging Economies. 2017;**22**(2):37–46

[14] Vinh T, Ibrahim KB, Vidya R, Huang HY. Competency profile of managers in the Singapore logistics industry. The Asian Journal of Shipping and Logistics. 2012;**28**(2):161–182. DOI: 10.1016/j.ajsl.2012.08.002

[15] Wahab SN, Tan A, Roche O. Identifying supply chain manager leadership skills and competencies gaps in Malaysia. Higher Education, Skills and Work-Based Learning. 2024;**14**(5):921–937. DOI: 10.1108/HESWBL-07-2023-0179

[16] Derwik P, Hellström D, Karlsson S. Manager competences in logistics and supply chain practice. Journal of Business Research. 2016;**69**(11): 4820–4825. DOI: 10.1016/j.jbusres.2016.04.037

[17] Clarke B 12 Essential Logistics Skills for an Effective Supply Chain Manager [internet]. 2022. Available from: https://www.logisticsit.com/articles/2022/12/16/12-essential-logistics-kills-for-an-effective-supply-chain-manager [Accessed: 2025-September-10]

[18] Kotzab H, Teller C, Bourlakis M, Wünsche S. Key competences of logistics and SCM professionals – The lifelong learning perspective. Supply Chain Management: An International Journal. 2018;**23**(1):50–64. DOI: 10.1108/SCM-02-2017-0079

[19] Katiniene A, Jezerske Ž, Vaičiute K. Research on competencies of logistics specialists in transport organisations. Journal of Business Economics and Management. 2021;**22**(5):1308–1322. DOI: 10.3846/jbem.2021.15299

[20] Sapper S, Kohl M, Fottner J. Future competency requirements in logistics due to industry 4.0: A systematic literature review. In: Proceedings of 10th International Conference on Industrial Technology and Management (ICITM); 26-28 March 2021. Cambridge, United Kingdom; p. 94–105

[21] McKinnon A, Flöthmann C, Hoberg K, Busch C. Logistics Competencies, Skills, and Training: A Global Overview. Washington: World Bank Publications; 2017. DOI: 10.1596/978-1-4648-1140-1

[22] Future-proofing the supply chain [internet]. Available from: https://www.mckinsey.com/capabilities/operations/our-insights/future-proofing-the-supply-chain [Accessed: 2025-September-10]

[23] Wahab SN, Rajendran SD, Yeap SP. Upskilling and reskilling requirement in logistics and supply chain industry for the fourth industrial revolution. LogForum. 2021;**17**(3):399–410. DOI: 10.17270/j.log.2021.606

[24] Wamba SF, Akter S. Understanding supply chain analytics capabilities and agility for data-rich environments. International Journal of Operations and Production Management. 2019;**39**(6/7/8):887–912. DOI: 10.1108/ijopm-01-2019-0025

[25] Jafari N, Azarian M, Yu H. Moving from Industry 4.0 to Industry 5.0: What Are the Implications for Smart Logistics? Logistics. 2022;**6**(2):26. DOI: 10.3390/logistics6020026

[26] Garrido S, Muniz J, Batista Ribeiro V. Operations management, sustainability & industry 5.0: A critical analysis and future agenda. Cleaner Logistics and Supply Chain. 2014;**10**:100141. DOI: 10.1016/j.clscn.2024.100141

[27] Hryhorak MY, Harmash OM, Popkowski T. Artificial intelligence in supply chain management: Opportunities and threats for professional competence. Intellectualization of Logistics and Supply Chain Management. 2023;**19**: 24–44. DOI: 10.46783/smart-scm/2023-19-3

[28] Pacher C, Woschank M, Zunk BM. The role of competence profiles in industry 5.0-related vocational education and training: exemplary development of a competence profile for industrial logistics engineering education. Applied Sciences. 2023;13. DOI: 10.3390/app13053280

[29] Nayeri S, Sazvar Z, Heydari J. Designing an IoT-enabled supply chain network considering the perspective of the fifth industrial revolution: Application in the medical devices industry. Engineering Applications of Artificial Intelligence. 2023; **122**:106–113. DOI: 10.1016/j.engappai.2023.106113

[30] Ścibiorek Z. Educational challenges in reference to management. Logistics and Transport. 2021;**1-2** (49-50):5–16. DOI: 10.26411/83-1734-2015-1-49-1-21

[31] Kolb DA. Experiential Learning: Experience as the Source of Learning and Development, 2nd ed. Upper Saddle River, New Yersey: FT Press; 2014. 146 p

[32] Moqaddam AF. The impact of experiential learning on the development of practical skills in internships. Journal of Study and Innovation in Education and Development. 2023;**3**:31–40. DOI: 10.61838/jsied.3.2.5

[33] Cooper L, Orrell J, Bowden M. Work Integrated Learning: A Guide to Effective Practice. United Kingdom: Taylor & Francis; 2010. p. 240

[34] Rzeczycki A, Chrzastek G, Niemcewicz P. Serious games in logistics education – Analysis of effectiveness and students' decision-making processes. European Research Studies Journal. 2024;**XXVI** I :952–967. DOI: 10.35808/ersj/3758

[35] Sustainable Logistics 4.0. Available from: https://slog4.put.poznan.pl/index.php [Accessed: 2025-September-19]

Section 3

Regional, Sectorial and Public-Sector Transformations beyond Manufacturing

Chapter 11

Perspective Chapter: Transforming Public Infrastructure through Industry 4.0 – Systems Integration and Governance in the MENA and EU Contexts

Ashok Chopra and Anita Rani Chopra

Abstract

This chapter critically examines the role of Industry 4.0 technologies in transforming public infrastructure systems, with a focus on the Middle East and North Africa (MENA) and European Union (EU) regions. Traditionally associated with manufacturing, Industry 4.0 is increasingly redefining public service delivery through the integration of cyber-physical systems, cloud computing, artificial intelligence (AI), and the Internet of Things (IoT). The chapter explores how these technologies are being institutionalized within national digital transformation agendas and regulatory ecosystems, shifting public infrastructure management from static, fragmented models to integrated, data-driven frameworks. The analysis provides a comparative lens, highlighting the EU's emphasis on regulatory harmonization, data sovereignty, and federated architectures – exemplified by frameworks such as the General Data Protection Regulation (GDPR), European Interoperability Framework (EIF), and GAIA-X. Conversely, MENA countries, notably the United Arab Emirates and Saudi Arabia, are pursuing rapid digitalization through centralized governance structures and strategic visions, emphasizing national identity platforms, hybrid cloud models, and evolving cybersecurity regimes. The chapter discusses the critical role of interoperability, cloud infrastructure, and digital identity in enabling seamless service delivery, while also addressing challenges such as legacy systems, institutional capacity constraints, and policy fragmentation. Governance frameworks in both regions are transitioning from reactive compliance models to anticipatory regulatory strategies that support innovation, resilience, and cross-border cooperation. Case studies from Estonia, UAE, Saudi Arabia, and Germany illustrate varying approaches to implementing Industry 4.0 within public sector infrastructure. These examples underscore the importance of aligning technology deployment with coherent governance models and sectoral policy priorities. The chapter concludes by emphasizing the necessity for continuous regulatory foresight, agile institutional reform, and investment in digital skills to ensure sustainable and inclusive public infrastructure development in the Industry 4.0 era.

Keywords: Industry 4.0, governance, system integration, public infrastructure, MENA, EU, context

1. Introduction

Industry 4.0, traditionally associated with the manufacturing sector, has increasingly extended into public infrastructure systems. Governments are now re-thinking the design and operation of essential services such as transportation, utilities, health, and education through the integration of digital technologies. The convergence of cyber-physical systems with data-driven governance mechanisms has led to the emergence of smart cities and adaptive urban ecosystems. Public infrastructure is being redefined as a dynamic, sensor-enabled environment that can collect, process, and respond to real-time data. This shift is not merely technological – it involves institutional reforms, legal re-alignments, and capacity-building to support agile and interoperable digital services. The interplay between digital identity, cloud computing, and regulatory oversight is critical in ensuring the secure and efficient delivery of services. The modernization of public infrastructure through Industry 4.0 technologies is also increasingly shaped by global pressures for resilience, sustainability, and inclusive governance. Driven by digital technologies such as cloud computing, artificial intelligence (AI), and the Internet of Things (IoT), this transformation is redefining how governments in the Middle East and North Africa (MENA) and European Union (EU) regions plan, manage, and deliver services. This chapter examines how public infrastructure is evolving under Industry 4.0, focusing on systems integration, governance mechanisms, and regional implementation (**Figure 1**).

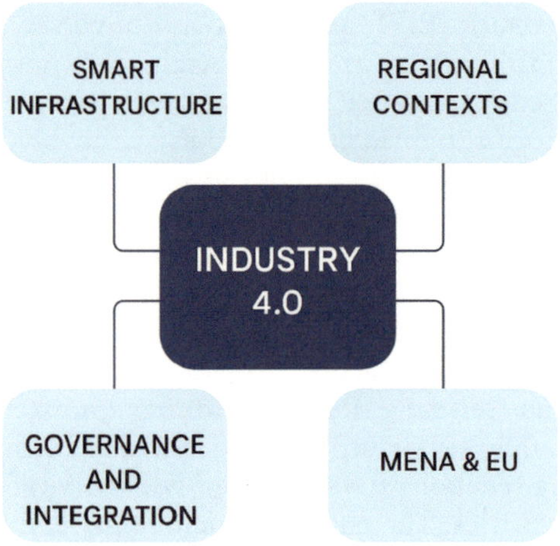

Figure 1.
Industry 4.0 in public infrastructure: systems, integration, and Governance in MENA and EU contexts.

2. E-government and smart infrastructure: Foundational concepts

E-government services represent the digitization of public services, enabling functions such as licensing, healthcare, and taxation through online portals. These platforms reduce bureaucracy, improve efficiency, and enhance transparency. Smart infrastructure, by extension, refers to the embedding of digital systems into physical assets like roads, energy grids, and water systems to facilitate automation and real-time monitoring.

Governments in both regions are shifting from static, paper-based systems to dynamic, digital-first operations. In the EU, this transition is supported by regulatory frameworks such as the European Interoperability Framework (EIF), which provides common models, standards, and recommendations for cross-border and cross-sector public service interoperability [1] (European Commission, 2017). The EIF aims to ensure that digital public services across the EU are compatible and effectively coordinated. In contrast, MENA countries are aligning with broader national digital transformation agendas. For example, UAE Vision 2031 emphasizes the importance of digital infrastructure and smart services to build a globally competitive economy [2] (UAE Government, 2022), while Saudi Vision 2030 prioritizes e-government, digital identity, and data-driven governance as foundational pillars for public sector reform [3] (Kingdom of Saudi Arabia, 2021).

3. Cloud computing as an enabler of integration

Cloud computing provides scalable, secure, and accessible infrastructure to host government services and facilitate inter-agency collaboration. EU countries utilize sovereign cloud strategies to ensure data compliance with GDPR, notably through projects like GAIA-X, which promote secure, federated cloud infrastructure with European control and standards. The initiative responds to growing concerns over data sovereignty, supporting the EU's digital autonomy strategy [4] (European Commission, 2020).

In the MENA region, hybrid cloud models are common. The UAE's Federal Government Cloud (FedNet) exemplifies a centralized, government-managed platform for hosting e-services, while still partnering with global providers under national data residency requirements. Saudi Arabia's Cloud First Policy mandates that public institutions prioritize cloud adoption, with oversight by the Communications and Information Technology Commission (CITC) to ensure national security and economic alignment [5], (CITC, 2021).

4. Systems integration and interoperability challenges

Interoperability is critical for real-time data exchange across departments. However, legacy systems, fragmented governance, and inconsistent data standards hinder integration. EU initiatives, such as the Digital Europe Program, address these issues through standardized protocols and funding mechanisms that promote cross-border digital public services [4] (European Commission, 2020). Projects such as e-SENS and ISA^2 have contributed reusable components and frameworks to enhance interoperability.

In the MENA region, digital interoperability faces additional constraints, including limited open data frameworks and diverse regulatory landscapes across countries. The UAE's Digital Government Strategy promotes a unified digital identity and service integration framework [6] (UAE Government, 2021), while Saudi Arabia's Yesser Program facilitates interoperability through the government service bus (GSB), a middleware platform that enables cross-entity integration. GSB functions as a centralized integration hub that connects disparate government entities, enabling them to exchange data and interact seamlessly through standardized interfaces and protocols. It acts as a secure channel through which electronic services (e-services) are delivered, orchestrated, and monitored. GSB eliminates the need for point-to-point integration by enabling agencies to publish and consume services via a shared infrastructure, reducing duplication and improving scalability. This architecture supports real-time communication, simplifies legacy system integration, and promotes data consistency across ministries and public agencies. Moreover, GSB incorporates robust authentication, logging, and service-level management features, aligning with Saudi Arabia's Vision 2030 to build a unified and digitally capable public sector.

5. Governance frameworks and regulatory compliance

Digital governance involves legal, institutional, and technical systems that guide the deployment of Industry 4.0 technologies. In the EU, the Digital Services Act (DSA) and General Data Protection Regulation (GDPR) establish frameworks for data protection, content moderation, and accountability in digital service provision. The NIS2 Directive further mandates risk management practices and reporting obligations for operators of essential services.

The EIF ensures that digital public services across member states are designed to work seamlessly across borders and sectors. It promotes legal, organizational, semantic, and technical interoperability. Additionally, the EU Cybersecurity Act empowers ENISA (European Union Agency for Cybersecurity) to support a unified certification framework for ICT products and services, which strengthens trust in cloud and AI-enabled systems.

In the MENA region, digital governance frameworks are evolving to accommodate rapid digital transformation. The UAE's Digital Government Strategy 2025 focuses on unified standards for digital service delivery, enterprise architecture, and open data policies. Its National Cybersecurity Strategy addresses risks linked to cloud adoption, IoT, and critical infrastructure vulnerabilities. Regulatory agencies, such as the Telecommunications and Digital Government Regulatory Authority (TDRA) and the Digital Dubai Authority, enforce compliance and coordinate cross-agency digital policies.

Saudi Arabia has established multiple governance layers, including the Digital Government Authority (DGA), the National Data Management Office (NDMO), and the Saudi Data and AI Authority [7] (SDAIA). These institutions implement sector-specific regulatory requirements, including data classification, usage licenses, and cloud security controls, aligned with Vision 2030. The National Digital Government Strategy emphasizes efficiency, transparency, and citizen-centricity, supported by legal frameworks for e-signatures, digital identity, and cloud certification.

Moreover, both regions are increasingly adopting regulatory sandboxes to test and refine emerging technologies such as blockchain, autonomous systems, and AI-powered governance tools before they are fully legislated. These experimental environments are critical for maintaining technological neutrality while ensuring regulatory alignment.

Overall, governance frameworks in both the EU and MENA are shifting from reactive to proactive models – building regulatory foresight, cross-border legal interoperability, and institutional resilience into the foundation of digital public service ecosystems.

6. Security, privacy, and data sovereignty

Governments must manage growing volumes of sensitive citizen data, necessitating robust cybersecurity protocols. In the EU, data sovereignty is enforced by the GDPR and supported by initiatives like the European Cloud Rulebook, ensuring that data remains within controlled jurisdictions and complies with privacy legislation [8]. GAIA-X, a European data infrastructure focused on security, transparency, and data protection, was initiated by the German and French governments and enhances trust by defining transparent governance models and compliance standards [4] (European Commission, 2020).

In the MENA region, sovereign control over public data is a top priority. The UAE mandates that government and critical sector data be hosted locally unless approved exceptions apply. Similarly, Saudi Arabia enforces stringent data residency rules, emphasizing alignment with Vision 2030's national security objectives. The use of encryption, identity-based access control, and continuous monitoring systems is essential to mitigate breaches and maintain service continuity.

7. Regional case studies

Estonia (EU): Estonia has established itself as one of the world's most advanced digital societies. The X-Road platform, a decentralized data exchange layer, enables secure interoperability between over 1,000 public and private institutions. Citizens access services – such as voting, tax filing, prescriptions, and medical records – using a single digital ID authenticated through multifactor encryption. Estonia's use of blockchain technology ensures data integrity and traceability in critical areas such as health and land registries. The nation's e-Residency program also allows global citizens to establish EU-based digital businesses remotely [8] (e-Estonia, 2023).

UAE (MENA): The Smart Dubai initiative leverages AI, IoT, and blockchain to optimize urban operations and citizen services. The UAE Pass serves as a unified digital identity that allows secure access to more than 6,000 services from local and federal government entities, as well as private providers. Services range from renewing trade licenses and accessing healthcare records to signing legal contracts digitally. Integrated with biometric verification and blockchain-backed authentication, the UAE Pass aims to eliminate paper-based processes entirely [9] (Smart Dubai, 2021).

Saudi Arabia (MENA): The NDMO, part of the Saudi Data and Artificial Intelligence Authority (SDAIA), oversees unified data policies across the kingdom. Through the DGA and the National Digital Transformation Unit, Saudi Arabia coordinates standards and compliance for cloud adoption, cybersecurity, and interoperability. Key platforms include the Tawakkalna app and the Absher portal, which integrate identity, health, and civil service functionalities. These initiatives align with Vision 2030's goal of becoming a digitally advanced nation [7] (SDAIA, 2023).

Germany (EU): Germany's GAIA-X project provides a decentralized framework that interlinks cloud service providers while maintaining European values of transparency, portability, and data sovereignty. The initiative includes participation from industries, governments, and research institutions across EU member states. GAIA-X supports cloud-based public services by offering interoperable nodes that respect national compliance rules while enabling cross-border cooperation. It forms part of Germany's strategy to reduce dependency on non-European providers and safeguard digital infrastructure [10] (GAIA-X, 2022).

8. Technical capacity and organizational readiness

A major barrier to effective Industry 4.0 implementation in public infrastructure is the shortage of skilled personnel and institutional agility. In the EU, programs such as the Digital Skills and Jobs Platform, co-funded by the European Commission, support capacity-building by linking stakeholders, offering training tools, and funding digital competency projects [11] (European Commission, 2021).

MENA countries are investing in similar strategies. The UAE's One Million Arab Coders initiative aims to equip citizens with digital and programming skills. In Saudi Arabia, the Future Skills initiative under Vision 2030 targets digital proficiency development across the civil service. Regional disparities in infrastructure, however, create uneven readiness levels, particularly in lower-income MENA states.

9. Future outlook and risks

Integration of AI, IoT, and cloud into public infrastructure enhances responsiveness but also raises dependency risks. In the EU, concerns over vendor lock-in and regulatory lag are addressed through public procurement frameworks and open-source platforms that reduce reliance on single providers. MENA countries, while embracing rapid adoption, must ensure that their governance and legal systems evolve to handle cross-border data flows, cyber resilience, and ethical AI deployment.

Both regions must focus on evolving governance and procurement models to support adaptive, resilient digital ecosystems that balance innovation with accountability and security.

10. Conclusion

Industry 4.0 is reshaping public infrastructure by merging physical assets with digital intelligence. This transformation is not confined to technological upgrades but involves a fundamental reengineering of how public services are conceived, delivered, and governed. Case studies from the EU and MENA regions demonstrate varied but converging approaches to leveraging cloud computing, AI, IoT, and blockchain within robust governance frameworks.

In the EU, the emphasis lies on regulatory harmonization, digital sovereignty, and open, federated architectures such as GAIA-X. In contrast, MENA countries are advancing rapidly through centralized transformation agendas like Saudi Vision 2030 and UAE Vision 2031, focusing on national digital identity systems and hybrid cloud strategies.

Despite regional variations, common enablers include the prioritization of interoperability, the use of secure digital identity mechanisms, and an expanding emphasis on cybersecurity and data sovereignty. Institutional readiness, workforce capacity, and cross-sector coordination remain critical challenges across both contexts.

Looking ahead, continued investment in digital literacy, agile policy-making, and cross-border cooperation will determine the pace and inclusiveness of digital public infrastructure. As the boundaries between physical and digital spaces blur, governance systems must evolve to ensure resilience, transparency, and equitable access to public services.

Thus, Industry 4.0 in public infrastructure is no longer optional; it is central to the future of responsive, secure, and citizen-centered government.

Author details

Ashok Chopra[1*] and Anita Rani Chopra[2]

1 Amity University Dubai, Oman, UAE

2 FIT Institute Dubai, Oman, UAE

*Address all correspondence to: achopra@amityuniversity.ae

References

[1] Union E. New European interoperability framework promoting seamless services and data flows for European public administration.

[2] UAE Government. We the UAE 2031. 2022. https://u.ae/en/about-the-uae/strategies-initiatives-and-awards/federal-governments-strategies/we-the-UAE-2031

[3] Kingdom of Saudi Arabia. Saudi Vision 2030. 2021. https://www.vision2030.gov.sa/

[4] European Commission. Shaping Europe's Digital Future. 2020. https://ec.europa.eu/digital-strategy

[5] CITC. Cloud computing regulatory framework. Communications and Information Technology Commission. 2021;https://www.citc.gov.sa/en/services/CloudComputing/

[6] UAE Government. Digital Government Strategy 2025. 2021. https://u.ae/en/about-the-uae/digital-uae/uae-government-strategy

[7] SDAIA. National Data Management Office. Saudi Data and Artificial Intelligence Authority. 2023. https://sdaia.gov.sa/

[8] e-Estonia. The Digital Society. 2023. https://e-estonia.com/solutions/interoperability-services/x-road/

[9] Dubai S. UAE Pass: One Digital Identity for All. 2021. https://smartdubai.ae/

[10] GAIA-X. GAIA-X: A Federated Data Infrastructure for Europe. 2022. https://www.gaia-x.eu

[11] European Commission. Digital Skills and Jobs Platform. 2021. https://digital-skills-jobs.europa.eu/en

Chapter 12

Establishing a Regional Industry 4.0 Electronics Hub: The GreenUpPCB Model

Rogério Dionísio and Irem Ünal

Abstract

This chapter presents GreenUpPCB as a regional Industry 4.0 hub for electronics repair and retrofitting, based in Castelo Branco, Portugal. The initiative combines artificial intelligence, machine vision, 3D printing, and automation to extend the life cycle of obsolete printed circuit boards, particularly in renewable energy systems and railway rolling stock. By promoting circular economy practices and reducing electronic waste, GreenUpPCB supports sustainable and resilient industrial operations. The hub also strengthens regional development by linking academic research with industrial needs, offering training, innovation, and high-tech services to local small- and medium-sized enterprises. Its model enables smart manufacturing adoption in low-density territories without dependence on urban clusters. Aligned with the European Chips Act and national strategies for microelectronics, GreenUpPCB demonstrates how regional hubs can contribute to the green and digital transitions across multiple sectors, including energy and transport.

Keywords: Industry 4.0, circular economy, electronics retrofitting, artificial intelligence, regional innovation hub

1. Introduction

The rapid evolution of Industry 4.0 technologies is transforming the way industrial production and service provision are organised, placing automation, digitalisation, and sustainability at the core of innovation strategies. At the intersection of these trends lies GreenUpPCB, a regional laboratory based in Castelo Branco, Portugal, dedicated to extending the life cycle of electronic components through testing, retrofitting, and reverse engineering of printed circuit boards (PCBs). Its activities focus particularly on renewable energy systems – most notably wind and solar power – and railway infrastructure, two strategic sectors in which electronics play a critical role in ensuring operational reliability.

Portugal has emerged as one of Europe's leaders in renewable electricity generation. In 2023, renewable sources supplied 70.7% of the national electricity mix, with wind power accounting for 29.3% and solar photovoltaics 8.2%, the latter representing the fastest-growing segment [1]. The Central Region of Portugal, encompassing districts such as Castelo Branco, Coimbra, and Leiria, hosts a significant share of this capacity through large-scale onshore wind farms [2] and an expanding base of solar PV installations [3]. This rapid growth brings new challenges for the maintenance of electronic control systems, inverters, and monitoring devices, many of which are exposed to harsh operating conditions and face risks of obsolescence.

The country's railway sector, operated primarily by Comboios de Portugal (CP), also plays a pivotal role in the national mobility system and in supporting regional connectivity. Like many European operators, CP faces the dual challenge of accommodating increasing passenger demand while managing an ageing fleet. Maintenance, refurbishment, and modernisation of rolling stock are therefore essential to sustaining service quality until new units are fully procured and integrated into operations [4].

In this context, GreenUpPCB emerges as an initiative that directly addresses sectoral needs by providing advanced repair and retrofitting services for high-value electronics. By aligning circular economy principles with the technical requirements of both the renewable energy and railway industries, the laboratory contributes not only to extending equipment life cycles but also to advancing broader policy goals of sustainability and resilience. The following section outlines the context and motivation behind the establishment of this regional hub.

2. Context and motivation

Although wind farms, solar plants, and railway networks operate in different domains, they face similar challenges related to the reliability of complex electronic systems exposed to long service life cycles, high utilisation rates, and demanding environmental conditions. Delays in fleet or asset renewal place further emphasis on maintenance and refurbishment, making local technical capacity a crucial factor for sustaining both performance and safety standards.

In the central region of Portugal, wind and solar power represent the backbone of renewable generation. Extensive wind farms operate in areas such as Serra do Açor, Serra da Lousã, and Penamacor [2], feeding into a national grid that reached 79.3% renewable electricity in the first half of 2025 [1]. In parallel, solar PV capacity has expanded rapidly, supported by both large utility-scale projects and distributed generation on rooftops and industrial facilities. Lightsource BP is advancing a portfolio of five large solar farms totalling approximately 1.35 GW, which includes sites in Castelo Branco, among others [5]. Between late 2024 and mid-2025, solar installations grew by nearly 500 MW, with the central region playing a prominent role thanks to its high irradiation levels and availability of land [3]. This concentration of renewable assets creates a sustained demand for specialised maintenance of electronics, such as turbine controllers, pitch systems, inverter boards, and monitoring modules. In many cases, these devices rely on imported PCBs with limited local support, exposing operators to costly delays whenever faults occur.

The railway sector illustrates a parallel challenge. CP operates with an ageing fleet that requires continuous refurbishment. Its Sustainability Report 2021–2023 highlights the importance of nationwide workshops located in Contumil, Guifões, Sernada do Vouga, Entroncamento, Figueira da Foz, Santa Apolónia, Campolide, Oeiras, Barreiro, and Vila Real de Santo António, which collectively sustain operations through repair, modernisation, and rehabilitation. In 2023, CP carried 172.6 million passengers (24 million more than the previous year), and this upward trend continued into 2024 and 2025 [4]. While new rolling stock is being procured from Alstom and Stadler companies, the procurement and homologation cycles mean that legacy trains will remain in service for several more years. Keeping these assets operational requires targeted electronic maintenance and retrofitting, often under tight schedules.

Beyond the technical dimension, the Portuguese central region faces socio-economic and infrastructural challenges that accentuate the need for a local electronics hub. Low population density and limited industrial infrastructure often oblige operators to send faulty boards to Lisbon, Porto, or abroad, increasing both costs and downtime [6–8]. Establishing a regional facility capable of delivering advanced diagnostic and repair services, therefore, represents not only an industrial necessity but also an opportunity to strengthen technological autonomy.

GreenUpPCB was envisioned in direct response to these challenges. By combining advanced diagnostic equipment, reverse engineering systems, and modular prototyping facilities, the laboratory provides the missing link between the rapid deployment of renewable assets, delayed fleet renewal in railways, and the pressing need for sustainable electronics management.

Its creation aligns with major strategic frameworks: at the European level, it aligns with the EU Chips Act [9] and the European Green Deal [10], both of which emphasise the importance of technological sovereignty and circular economy principles; at the national level, it aligns with the Portuguese National Energy and Climate Plan 2030 [11] and the National Strategy for Semiconductors [12]; and at the regional level, it aligns with the Agenda for Microelectronics e Semiconductors [13].

In addition, the Portuguese central region has consolidated collaborative platforms such as INOVC +, a strategic project bringing together 23 universities, polytechnics, technology centres, and municipalities to reinforce the regional innovation ecosystem. The initiative focuses on technology transfer, entrepreneurship, and knowledge valorisation across multiple domains, including digitalisation, Industry 4.0, and sustainability. While not dedicated solely to microelectronics, INOVC + provides the collaborative framework in which specialised initiatives such as GreenUpPCB can succeed, ensuring that regional innovation hubs are embedded in a wider ecosystem of research, education, and industrial partnerships [14].

2.1 Problem statement

The combined realities of rapid asset deployment in the renewable sector and delayed fleet renewal in the railway sector have created a critical operational gap: essential high-value electronics must remain functional far beyond their original design life. Both industries face mounting pressure to sustain service quality, meet environmental targets, and control costs, all while reducing downtime. Traditional approaches based on full equipment replacement are increasingly impractical due to long procurement cycles, supply chain constraints, and the environmental burden of manufacturing new components. In this context, GreenUpPCB emerges as a strategic solution, providing specialised local capacity to test, diagnose, repair, and retrofit electronic systems, thereby extending asset life cycles, reducing waste, and aligning with circular economy principles. By targeting two of Portugal's most strategically important sectors, the initiative addresses not only technical and economic challenges but also national and European policy priorities for sustainability, resilience, and technological sovereignty.

Addressing these sector-wide challenges requires more than incremental maintenance improvements; it demands a dedicated technological infrastructure capable of combining advanced diagnostics, precision repair, and rapid prototyping with the flexibility to serve multiple industries. The GreenUpPCB laboratory was conceived to fulfil this role, operating as both a technical service hub and a knowledge-transfer platform in the central region of Portugal. By integrating Industry 4.0 tools, such as computer-aided design and manufacturing (CAD/CAM), automated test systems, and data-driven fault analysis, the laboratory can diagnose and restore high-value electronic assemblies to operational standards that meet or exceed their original specifications. In the following section, the operational framework, technological resources, and collaborative strategies of GreenUpPCB are detailed, illustrating how the model supports both renewable energy and railway clients while remaining scalable to other industrial sectors.

3. GreenUpPCB laboratory: Vision, scope, and technology

GreenUpPCB was established with the vision of becoming a regional reference in sustainable PCB reconditioning and reverse engineering. The laboratory's strategic objectives extend beyond technical repair, aiming to position itself as a knowledge centre for electronic circularity and to enable the local industry with diagnostic and retrofit solutions. By connecting education, research, and small- and medium-sized enterprises (SMEs), the initiative seeks to create a dynamic innovation ecosystem that promotes both technological and regional development.

The target sectors served by GreenUpPCB are diverse but strategically chosen. In the renewable energy field, the laboratory supports both wind and solar power operators, whose assets require continuous monitoring and refurbishment of electronic systems. In rail transportation, GreenUpPCB works closely with CP to extend the service life of essential train electronics, such as Eurocard circuit boards, which are often affected by component obsolescence. Beyond these two sectors, the laboratory also collaborates with companies in industrial automation, telecommunications, and the electronics repair market, providing a critical service base for both established SMEs and emerging technology start-ups.

To achieve these goals, GreenUpPCB integrates advanced technological resources into its operations. Diagnostic platforms from ABI Electronics [15] form the backbone of its testing and reverse-engineering activities, while artificial-intelligence-assisted fault detection enhances the precision of data logging and repair. Precision soldering is supported through JBC rework stations [16], enabling the handling of high-density components. The laboratory also employs rapid prototyping tools such as 3D printers and maintains digital repositories and modular component libraries that ensure long-term reproducibility of schematics and designs.

3.1 Artificial intelligence (AI) and machine vision in PCB diagnostics

The GreenUpPCB laboratory integrates AI and machine vision tools to enhance fault detection and accelerate repair processes. Traditional diagnostic methods, while reliable, are often time-consuming and highly dependent on operator expertise. AI-driven approaches enable faster, scalable, and more consistent analysis of PCB defects by leveraging large datasets of annotated images.

Several benchmark datasets have been used to train and validate machine learning algorithms for PCB diagnostics:

- The AI Challenger PCB Dataset contains thousands of labelled PCB images with annotated defects, supporting robust supervised learning methods [17].

- The DeepPCB Dataset provides 1,500 pairs of reference and tested PCB images with defect annotations, designed specifically for image alignment and defect detection tasks [18].

- The Kaggle PCB Dataset collections aggregate multiple community-contributed datasets, enabling the training of models for defect classification and object detection [19].

Machine vision systems, when coupled with these datasets, are capable of detecting shorts, opens, spurious copper, missing components, and misalignments with higher precision than manual inspection.

The integration of AI-assisted diagnosis with reverse engineering tools and precision rework systems [16] creates a hybrid repair process that reduces downtime, improves repair yield, and supports the scalability of circular electronics practices. At GreenUpPCB, such approaches are being developed to complement traditional ABI diagnostic platforms [15, 20], aiming to automate part of the diagnostic workflow. Furthermore, by embedding these techniques into training and micro-credential programmes, GreenUpPCB ensures that technicians acquire cutting-edge skills aligned with Industry 4.0 requirements.

4. Infrastructure and workflow

The GreenUpPCB laboratory (**Figure 1**) has been designed as a modular and expandable facility, capable of adapting to the evolving demands of electronics repair and innovation. The physical set-up initially occupies 25 square meters but is being scaled up to 150 square meters as projects and partnerships grow. This space is complemented by dedicated training laboratories that provide 20 workbenches equipped with oscilloscopes, signal generators, logic analysers, and ABI diagnostic systems [20]. Controlled rework environments ensure that sensitive PCB operations can be performed under safe and reliable conditions, while future expansion plans include the installation of cleanroom infrastructure to support high-precision prototyping and component handling.

The operational framework of GreenUpPCB is structured around three interlinked functions: technical services, collaboration with industry, and knowledge transfer. Technical services include PCB testing, retrofitting, and reverse engineering of obsolete electronic boards – activities that are central to extending the life cycle of equipment in the renewable energy and railway sectors. Collaboration with companies such as CP and HFA [21] ensures that the laboratory's services

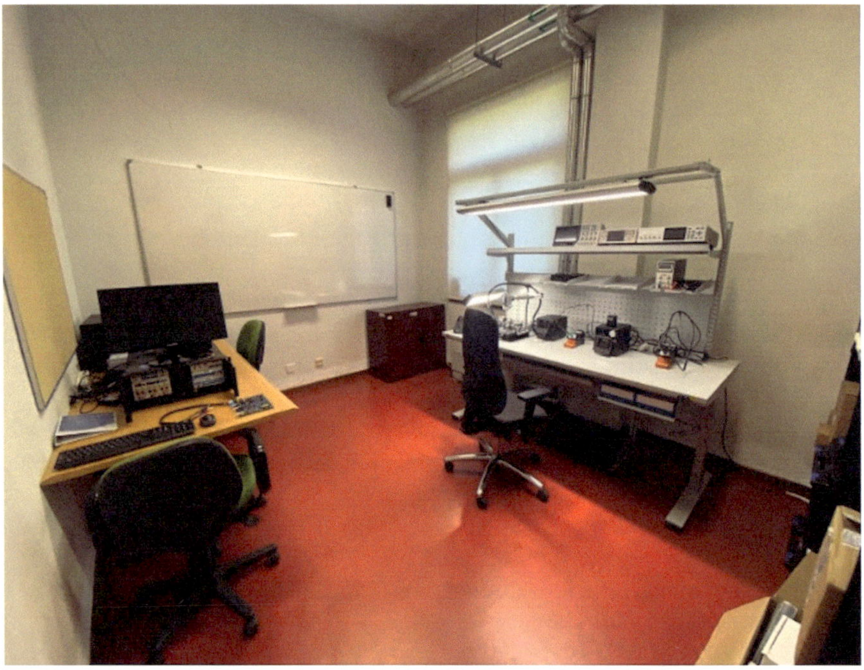

Figure 1.
GreenUpPCB laboratory facilities.

a) b)

Figure 2.
Meeting with partners from (a) CP and (b) HFA.

respond directly to the needs of industrial stakeholders, while prototyping and iterative repair cycles are conducted under strict quality protocols.

A meeting with partners from CP and HFA is shown in **Figure 2**. The workflow is designed to be flexible, allowing the laboratory to address urgent repair requests while simultaneously engaging in long-term development projects.

Technology transfer is another key dimension of the laboratory's mission. By creating demonstrator projects and disseminating repair case studies, GreenUpPCB not only provides direct solutions to its partners but also contributes to broader knowledge sharing within the electronics repair ecosystem. The integration of advanced diagnostics, automated test systems, and CAD/CAM-based design tools further enables the laboratory to diagnose faults, reproduce obsolete schematics, and restore boards to a condition that often meets or exceeds their original specifications.

4.1 Case study: Reverse engineering Siemens Eurocard boards for CP trains

Reverse engineering of legacy boards represents one of the core applications of the GreenUpPCB laboratory. These boards, widely used in railway rolling stock and industrial automation systems, frequently suffer from component obsolescence and a lack of documentation. To address this challenge, the ABI BoardMaster/RevEng Schematic Learning System, shown in **Figure 3**, provides a structured method for recovering accurate schematics, ensuring that critical assets remain serviceable and aligned with circular economy principles.

The workflow for the reverse engineering of a PCB follows five main steps:

1. The manual workflow begins with the safe initialisation of both hardware and software. At this stage, the system is kept in disconnect mode until powered on, preventing short circuits or voltage mismatches. Once the hardware is switched on and the RevEng software is launched, the platform is ready for secure board connections. The next phase involves hardware and cable set-up, during which ribbon harnesses are connected to the channel ports of the RevEng unit. Specific cable assemblies are selected in the software. Clip connectors are then attached to integrated circuits, while the power rails are

Figure 3.
ABI reverse-engineering equipment in use.

configured. With these preparations complete, the system is fully prepared for scanning and accurate pin referencing.

2. Following this set-up, each component on the board is defined within the software to ensure correct identification during the scan. Components such as integrated circuits, resistors, capacitors, and diodes are registered with their reference designators, part numbers, packages, and values. When components are absent from the existing library, they can be replaced with similar devices of the same pin count or created manually, with subsequent corrections to the schematic symbols as required. To optimise the scanning process, some devices are included for mapping, while others, such as connectors or non-critical jumpers, are excluded. This careful preparation guarantees that the subsequent scanning phase captures the full connectivity of the relevant devices.

3. The scanning itself is conducted using hooks, clips, or probes, depending on the physical accessibility of the components, as shown in **Figure 4**. For example, resistors and diodes may be scanned using hook leads attached directly to their pins, whereas integrated circuits are often scanned with clips that contact all pins simultaneously. When components are too close together for hooks or clips, probes become the most reliable option, often used in combination with a footswitch for efficiency. Once the scanning command is executed, the RevEng system automatically detects all electrical connections, including power rails and signal nets, and stores the results in the netlist

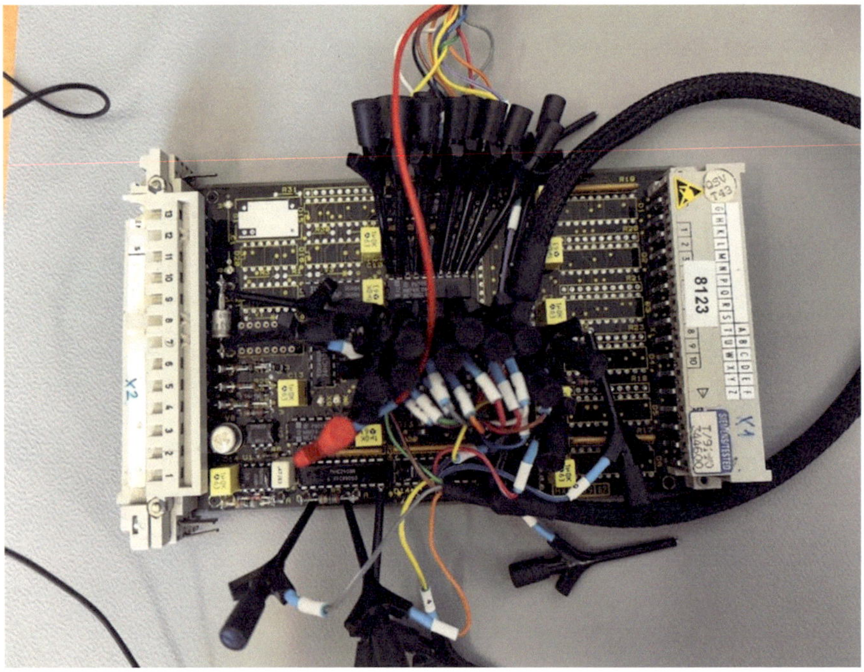

Figure 4.
ABI RevEng clips attached to a Eurocard board.

database. Coverage reaches 100% once every relevant connection has been identified and confirmed.

4. The next stage of the workflow is netlist generation. At this point, all detected connections are exported as a digital file, commonly in *.wrl or *.txt format. The structure of this file is divided into three main sections: components, values, and nodes. The components section lists all devices with their references, part numbers, and packages, while the values section defines the electrical values of resistors and capacitors. The nodes section contains the electrical connections, with pre-defined nets such as VCC and GND, as well as automatically generated labels for other signal paths. This netlist provides a comprehensive digital map of the board, bridging the gap between physical hardware and schematic representation.

5. Once exported, the netlist is imported into an EDA/CAD program to generate a complete schematic diagram. The user configures the page size and component clearance to match the board complexity, and then allows the software to automatically place all components in logical groups around a central reference device. This ensures that interconnections are displayed meaningfully. Autoconnect functions are then applied to wire the nodes, producing a complete schematic with all components placed and interconnected according to the imported data. The process results in a functional diagram that accurately represents the original Eurocard board and can be used for repair, redesign, or documentation purposes.

To provide a clearer overview of the methodology, **Figure 5** illustrates the full workflow adopted in the case study on reverse engineering Siemens Eurocard boards. The process begins with careful preparation and non-destructive netlist extraction, followed by logic function analysis, schematic reconstruction, validation, and testing. The workflow concludes with the renewal of obsolete components, documentation of the schematics, and reintegration of the reconditioned boards into the railway fleet.

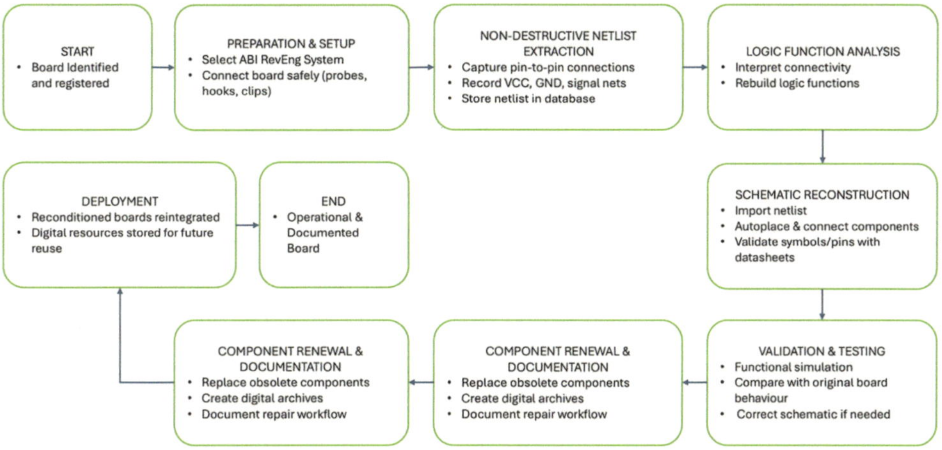

Figure 5.
Workflow for reverse engineering electronic boards using the ABI RevEng system, as applied in the CP case study.

Nonetheless, several practical considerations have emerged during the application of this workflow. Clearance values on the PCB must be adapted to the density of components for optimal readability. Selecting a central component with multiple interconnections is essential to achieve a logical arrangement of the schematic. Furthermore, library definitions must always be verified against official datasheets to ensure correct pin numbering, since inconsistencies may occur between systems. Ultimately, the reverse engineering process combines automated learning with manual oversight, producing schematics that are both technically accurate and practically usable.

In all repair and reverse engineering activities, GreenUpPCB applies strict safety and quality protocols aligned with the requirements of critical sectors such as rail transport and renewable energy. Electrostatic discharge control is ensured through the use of grounded workstations, antistatic equipment, and controlled environments. Every board undergoes functional testing after repair, retrofitting, or reverse engineering, guaranteeing that performance meets or exceeds original specifications. Full traceability is maintained by documenting diagnostic steps, replacement components, and test results, while attention to component provenance ensures that only certified and reliable parts are used.

5. Human capital and ecosystem development

A central pillar of GreenUpPCB is the development of human capital, which is essential for sustaining innovation and ensuring the long-term viability of circular electronics practices. The laboratory has adopted a multi-layered training strategy that combines micro-credentials, international exchanges, and open training initiatives. Within three years, the aim is to provide micro-credential courses that will certify at least 20 technicians with specialised skills in PCB diagnostics, rework, and reverse engineering. Erasmus + internships have already been established, welcoming students from institutions such as Üsküdar University [22], thereby fostering an international learning environment. In parallel, junior researcher placements are supported under national programmes [23], and training activities are opened to regional technicians, enabling them to upgrade their competencies and access advanced methodologies.

Beyond education, the laboratory plays a vital role in strengthening the local innovation ecosystem. Partnerships with regional companies ensure that SMEs benefit from diagnostic services and retrofitting expertise that would otherwise be difficult to access in low-density territories. GreenUpPCB collaborates with its partners not only by offering services but also by codeveloping research and development projects. In doing so, the laboratory acts as a bridge between academic research and industrial application, mentoring new ventures in the electronics sector and supporting the creation of start-ups that focus on repair, retrofitting, and reverse engineering.

This role is reinforced by the Polytechnic Institute of Castelo Branco's StartUp.CB initiative, which provides infrastructure, access to mentors, specialised training, and sustained guidance for start-up projects arising from or linked to the Polytechnic's research units [24]. Meanwhile, the Centro de Empresas Inovadoras (CEI, Castelo Branco) [25] expands this ecosystem by hosting FabLab CB [26], prototyping laboratories, and coworking spaces, and offering incubation and pre-incubation services. These facilities enable new electronics spin-offs and start-ups to rapidly move from concept to prototype and allow GreenUpPCB's repair and reverse engineering case studies to be tested, showcased, and scaled locally.

The laboratory also places significant emphasis on communication, awareness, and education directed at the wider community. Workshops are regularly organised in schools across the region. In May 2025, GreenUpPCB coorganised a thematic workshop in collaboration with Inova-Ria – Rede de Inovação em Aveiro, under the initiative Microelectronics for a Better Future [27]. The session focused on circularity in electronics, with discussions on eco-design, recovery, and materials recycling strategies. A photograph of the session is presented in **Figure 6**, illustrating the collaborative spirit of the initiative.

Participation in regional innovation fairs provides opportunities to showcase demonstrator projects, while online dissemination ensures that case studies and repair methodologies are shared with a broader audience. These outreach activities not only raise awareness of the importance of circularity in electronics but also inspire new generations of students to engage with science, technology, engineering, and mathematics (STEM) fields. Through this comprehensive approach to human capital development, GreenUpPCB cultivates a skilled workforce, empowers SMEs, and contributes to building a culture of technological resilience and sustainability. By linking training, research, and industrial practice, the laboratory demonstrates how capacity-building initiatives can transform regional ecosystems and support the wider European agenda for digital and green transitions.

Figure 6.
GreenUpPCB and Inova-Ria workshop on circularity, eco-design, and recycling (Castelo Branco, 2025).

6. The GreenUpPCB model: A framework for regional innovation hub on electronics

The GreenUpPCB hybrid model (**Figure** 7) synthesises the experience of the laboratory into a framework that can guide the creation of regional circular innovation hubs for electronics. Rather than viewing its activities as isolated technical services, the model integrates six mutually reinforcing dimensions: mission and objectives, infrastructure and technology, operational workflows, human capital development, ecosystem integration, and policy alignment. Together, these dimensions form a holistic and replicable blueprint for establishing electronics repair and retrofitting capacity in low-density regions.

6.1 Mission and objectives

At its core, the GreenUpPCB model seeks to extend the life cycle of electronic systems, reduce waste, and strengthen technological sovereignty. It does so by providing advanced diagnostic, repair, and reverse engineering services to sectors where electronics are critical – particularly renewable energy and rail transport. The mission is not only to solve immediate technical problems but also to anchor knowledge, skills, and infrastructure in regions that are typically underserved by high-tech innovation.

6.2 Infrastructure and technology

The model is built upon modular infrastructure and Industry 4.0 technologies. Diagnostic platforms, AI-assisted defect detection, CAD/CAM design tools, and precision rework stations enable the recovery and renewal of obsolete systems. Importantly, these technologies are scalable and portable, allowing the model to adapt to other regional contexts.

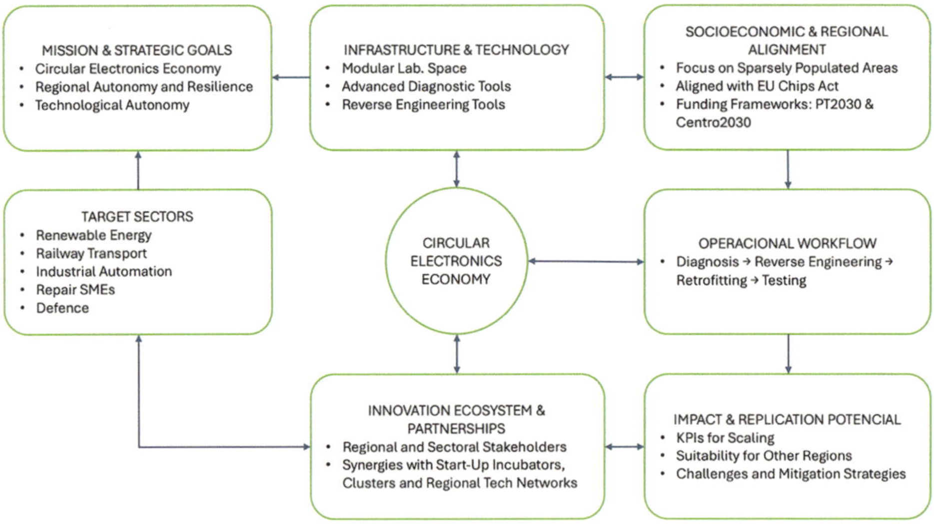

Figure 7.
Conceptual and logic-based model of the GreenUpPCB circular innovation hub.

6.3 Socio-economic and regional alignment

The model is designed to operate within, and contribute to, strategic policy agendas. At the regional level, it supports the Agenda for Microelectronics and Semiconductors [12]; at the national level, it complements the PNEC 2030 [11]; and at the European level, it aligns with the EU Chips Act [9] and the Green Deal [10]. This multi-level coherence ensures that local initiatives have visibility and support within broader transitions. A distinctive pillar of the model is its strong human-centred approach. Micro-credentials, Erasmus + exchanges, researcher placements, and open training opportunities create a skilled workforce embedded in the region. Training is not treated as an ancillary activity but as a driver of sustainability, ensuring that specialised knowledge is retained locally.

6.4 Operational workflows

Technical services follow structured processes that combine traditional repair with reverse engineering and rapid prototyping. By documenting and disseminating workflows, such as the reverse engineering of Siemens Eurocard boards for CP [4], GreenUpPCB ensures that tacit knowledge becomes transferable, strengthening both regional resilience and international learning.

6.5 Impact and replication potential

One important dimension of the GreenUpPCB model is its emphasis on measurable impact and replicability. From the outset, the hub has been designed not only as a service provider but also as a catalyst for systemic change in low-density regions. Impact is assessed across multiple dimensions, ranging from technical (eg, number of boards repaired, downtime reduced), educational (students and technicians trained), economic (SMEs supported, start-ups incubated), and environmental (waste avoided, equipment life cycles extended). By documenting case studies and disseminating open training resources, the model creates a knowledge base that can be transferred to other territories. Its modular infrastructure and portable methodologies make it adaptable to diverse regional contexts, while its alignment with European and national policy ensures institutional support. GreenUpPCB demonstrates a dual capacity: delivering immediate local benefits while offering a scalable framework for replication in other low-density regions facing similar challenges in renewable energy, transport, and industrial automation.

To make the impact of GreenUpPCB more tangible and to facilitate replication in other regions, a compact set of key performance indicators (KPIs) has been defined. These indicators capture not only the technical and operational efficiency of repair and retrofitting processes but also their educational, economic, and environmental contributions. **Table 1** summarises eight core KPIs that provide measurable benchmarks for assessing performance over a three-year horizon. They reflect industry-standard metrics such as repair yield and mean time-to-diagnosis, as well as broader sustainability goals, including waste and CO_2e avoided, technicians trained, and SMEs supported.

6.6 Innovation ecosystem and partnerships

The GreenUpPCB model demonstrates that a hub's impact depends on strong local and regional partnerships. At Castelo Branco, collaboration with CP [4], HFA

Domain	Indicator	Target
Technical quality	Repair yield (% of boards successfully restored)	≥ 85%
Technical efficiency	Mean time-to-diagnosis (average hours per PCB)	≤ 4 h
Operational impact	Downtime avoided (days of equipment unavailability saved in rail/energy)	150 + days
Life cycle extension	Average lifetime extension per PCB (years)	+5 years
Environmental impact	Electronic waste avoided (kg of PCBs not discarded)	500 kg
Environmental impact	CO_2e avoided (through reduced manufacturing of replacement boards)	20–30 t CO_2e
Human capital	Technicians trained/certified (micro-credentials, internships)	20
SME support	Regional SMEs supported (through services or joint R&D)	10
Ecosystem	Start-ups/spin-offs created	1
Knowledge transfer	Schools and awareness activities (STEM workshops)	10 schools

Table 1.
KPIs for the GreenUpPCB hub, designed to measure technical, economic, educational, and environmental impacts over a three-year horizon.

[21], StartUp.CB [24], CEI [25], and FabLab CB [26] has created an ecosystem where SMEs and start-ups can access advanced services while experimenting with new ideas. Workshops with networks such as Inova-Ria reinforce this collaborative dimension, embedding the hub within wider innovation systems.

7. Conclusion

GreenUpPCB exemplifies how a regional initiative can operationalise the principles of Industry 4.0 and the circular economy to create a sustainable and replicable model for electronics repair and retrofitting. By combining advanced diagnostic technologies, reverse engineering methods, and modular infrastructure, the laboratory has successfully addressed two pressing challenges: equipment obsolescence and resource scarcity in sectors that are strategically vital for Portugal, particularly renewable energy and rail transportation.

The experience of GreenUpPCB demonstrates that low-density territories can do more than passively benefit from innovation – they can actively lead it. By linking academic research with industrial needs, the hub provides concrete solutions to local companies while simultaneously offering training and capacity-building for technicians, students, and entrepreneurs. This dual role ensures that the laboratory's impact extends well beyond immediate technical services, fostering workforce development and strengthening the regional innovation ecosystem.

The journey, however, has not been without constraints. Recruiting highly specialised staff remains a challenge in a labour market where talent is often drawn to larger urban centres. Dependence on external suppliers for advanced diagnostic equipment can cause delays, while the continuity of funding is a recurring concern in research and development environments. Yet, these challenges have spurred adaptive strategies, such as modular infrastructure planning, the diversification of funding sources, and the establishment of strategic alliances with universities, SMEs, and regional innovation agencies. Together, these measures enhance the sustainability and resilience of the hub.

Equally important is the initiative's alignment with national and European policy frameworks. At the Portuguese level, it advances the Agenda for Microelectronics and Semiconductors [12] and the National Plan for Energy and Climate 2030 [11]; at the European level, it aligns with the EU Chips Act [9] and the Green Deal [10]. GreenUpPCB illustrates how ambitious policy strategies can be translated into practice through targeted interventions that create locally embedded hubs of expertise.

Ultimately, the case of GreenUpPCB demonstrates that electronics repair and retrofitting are not merely technical necessities but also drivers of socio-economic resilience. By extending the life cycle of PCBs, the laboratory reduces electronic waste, lowers costs, and ensures continuity in critical infrastructures. Just as importantly, it shows that innovation does not need to be confined to metropolitan centres: with the right vision and collaborative framework, smaller regions can become active participants in Europe's green and digital transitions.

In sum, the GreenUpPCB model provides a framework for sustainable smart manufacturing that is inclusive, regionally anchored, and policy-aligned. Its success offers a practical pathway towards greater technological sovereignty in Europe, while promoting a greener, more circular, and more resilient industrial future.

Acknowledgements

The authors acknowledge the cooperation of Eng. Hélder Martins, CEO of E-Novation, in establishing a protocol with IPCB that ensured the availability of ABI Electronics equipment at the GreenUpPCB laboratory.

Author details

Rogério Dionísio[1,2*] and Irem Ünal[3]

1 School of Technology, Polytechnic Institute of Castelo Branco, Castelo Branco, Portugal

2 CISeD – Research Center in Digital Services, Instituto Politécnico de Viseu, Viseu, Portugal

3 Üsküdar University, Istanbul, Turkey

*Address all correspondence to: rdionisio@ipcb.pt

References

[1] APREN. Boletim de Eletricidade Renovável – Junho 2025 [Internet]. Lisboa: Associação Portuguesa de Energias Renováveis (APREN); 2025. Available from: https://apren.pt [Accessed: 2025-September-09]

[2] INEGI. Portugal Parques Eólicos 2024 [Internet]. Porto: INEGI – Instituto de Ciência e Inovação em Engenharia Mecânica e Engenharia Industrial; 2024. Available from: https://www.inegi.pt [Accessed: 2025-September-09]

[3] APREN. Energia Renovável em Portugal – Brochura Institucional [Internet]. Lisboa: Associação Portuguesa de Energias Renováveis (APREN); 2023. Available from: https://apren.pt [Accessed: 2025-September-09]

[4] CP – Comboios de Portugal. Relatório de Sustentabilidade 2021–2023 [Internet]. Lisboa: CP – Comboios de Portugal; 2023. Available from: https://www.cp.pt [Accessed: 2025-September-09]

[5] bp L. Lightsource BP announces €900 million solar investment in Portugal – Solar projects pipeline. [Internet]. Lisbon: Lightsource bp; 2021. Available from: https://lightsourcebp.com/news/lightsource-bp-to-invest-e900million-in-portugal [Accessed: 2025-September-09]

[6] OECD. Regional Outlook 2023: Country Notes – Portugal. [Internet]. Paris: OECD Publishing; 2023. Available from: https://www.oecd.org/content/dam/oecd/en/publications/reports/2023-Regional-Outlook-Country-Notes-Portugal.pdf [Accessed: 2025-September-09]

[7] OECD. Rethinking Regional Attractiveness in the Centro Region of Portugal in the New Global Environment: Case Study. [Internet]. Paris: OECD Publishing; 2024. Available from: https://www.oecd.org/content/dam/oecd/en/publications/reports/2024/10/rethinking-regional-attractiveness-in-the-new-global-environment-case-studies_fbca2f00/rethinking-regional-attractiveness-in-the-centro-region-of-portugal_1352b712/5ca02c36-en.pdf [Accessed: 2025-September-09]

[8] Matos Silva M, Alves J, Ferreira A. Analytical model for the development strategy of a low-density rural region. Sustainability. 2022;**14**(7):4373. DOI: 10.3390/su14074373

[9] European Commission. Proposal for a regulation of the European Parliament and of the Council establishing a framework of measures for strengthening Europe's semiconductor ecosystem (Chips Act) [Internet]. Brussels: European Commission; 2022. Available from: https://eur-lex.europa.eu/legal-content/EN/TXT/?uri=CELEX%3A52022PC0046 [Accessed: 2025-September-09]

[10] European Commission. The European Green Deal. [Internet]. Brussels: European Commission; 2019. Available from: https://eur-lex.europa.eu/legal-content/EN/TXT/?uri=CELEX%3A52019DC0640 [Accessed: 2025-September-09]

[11] Agência Portuguesa do Ambiente (APA). Plano Nacional de Energia

e Clima 2030 (PNEC 2030) [Internet]. Lisboa: APA – Agência Portuguesa do Ambiente; 2020. Available from: https://apambiente.pt/clima/plano-nacional-de-energia-e-clima-pnec [Accessed: 2025-September-09]

[12] Portugal – Presidência do Conselho de Ministros. Estratégia Nacional para os Semicondutores. Resolução do Conselho de Ministros n.º 12/2024, de 8 de janeiro [Internet]. Lisboa: Diário da República; 2024. Available from: https://files.diariodarepublica.pt/1s/2024/01/00500/0012200138.pdf [Accessed: 2025-September-09]

[13] Amkor Technology Portugal (ATEP). Agenda da Microeletrónica – Ficha de Projeto [Internet]. Porto: Amkor Technology Portugal; 2024. Available from: https://micro-electronics.eu/wp-content/uploads/2024/01/Ficha-de-Projeto_Agenda-Microeletronica.pdf [Accessed: 2025-September-09]

[14] INOVC+. NOVC+ – Ecossistema de Inovação da Região Centro [Internet]. Coimbra: Universidade de Coimbra; 2021. Available from: https://inovc.uc.pt [Accessed: 2025-September-09].

[15] ABI Electronics. RevEng: Schematic Learning System – Product Overview [Internet]. Barnsley, UK: ABI Electronics Ltd.; 2023. Available from: https://www.abielectronics.co.uk/products/reveng-schematic-learning-system [Accessed: 2025-September-09]

[16] JBC Tools. Rework Stations – Precision Soldering Equipment [Internet]. Barcelona: JBC Tools; 2023. Available from: https://www.jbctools.com/rework-stations.html [Accessed: 2025-September-09]

[17] Challenger AI. PCB Defect Detection Dataset. [Internet]. Beijing: AI Challenger; 2018. Available from: https://challenger.ai/dataset/pcb [Accessed: 2025-September-09]

[18] Tang X, Zhai Y, Zhang S, Wu Z, Tang S DeepPCB: A Dataset for Printed Circuit Board Defect Detection and Classification. [Internet]. GitHub; 2019. Available from: https://github.com/tangsanli5201/DeepPCB [Accessed: 2025-September-09]

[19] Kaggle. PCB Datasets Collection [Internet]. San Francisco: Kaggle; 2020. Available from: https://www.kaggle.com/search?q=pcb+dataset [Accessed: 2025-September-09]

[20] ABI Electronics. BoardMaster 8000 Plus – Universal Diagnostic *System* [Internet]. Barnsley, UK: ABI Electronics Ltd.; 2023. Available from: https://www.abielectronics.co.uk/products/boardmaster-8000-plus-universal-diagnostic-system [Accessed: 2025-September-09]

[21] HFA. Henrique, Fernando & Alves, Lda. – Electronics Manufacturing Services [Internet]. Torres Vedras: HFA; 2023. Available from: https://www.hfa.pt [Accessed: 2025-September-09]

[22] Üsküdar University. Üsküdar University – Official Website [Internet]. Istanbul: Üsküdar University; 2023. Available from: https://uskudar.edu.tr/en [Accessed: 2025-September-09]

[23] Fundação para a Ciência e a Tecnologia (FCT). Bolsas de Investigação – Programas de Formação Avançada. [Internet]. Lisboa: FCT – Fundação para a Ciência e a Tecnologia; 2023. Available from: https://www.

fct.pt/apoios/bolsas [Accessed: 2025-September-09]

[24] StartUp.CB – Incubadora de Empresas do Instituto Politécnico de Castelo Branco. [Internet]. IPCB; 2024. Available from: https://www.ipcb.pt/investigar-e-inovar/inovacao/incubadora-de-empresas/ [Accessed: 2025-September-09]

[25] CEI – Centro de Empresas Inovadoras, Castelo Branco. [Internet]. Cei/ Cataa; 2024. Available from: https://www.cataa-cei.pt/ [Accessed: 2025-September-09]

[26] FabLab CB, CEI, Castelo Branco. [Internet]. FabLabs.io; 2024. Available from: https://www.fablabs.io/labs/fablabcastelobranco [Accessed: 2025-September-09]

[27] Inova-Ria. Inova-Ria – Rede de Inovação em Aveiro [Internet]. Aveiro: Inova-Ria; 2023. Available from: https://www.inova-ria.pt [Accessed: 2025-September-09]

Chapter 13

Applying Industry 4.0 Principles to Smart and Resilient Cities

Sergej Kozhevnikov, Vladimir Marik and Miroslav Svitek

Abstract

Urban structures initially appear to serve as environments for Industry 4.0, providing services to production units. However, a deeper analysis reveals that production units are understood as territorial entities with distinct needs for tailored solutions. It is clear that Industry 4.0 will lead to changes not only in the production processes themselves but also in the environment surrounding individual production companies, including energy, security, logistics, transport systems, and water or waste management systems. Due to current technological advancements, the fourth industrial revolution is permeating nearly all sectors, leading to significant changes and giving rise to Smart and Resilient Cities and Regions. This chapter demonstrates how the principles of Industry 4.0 can be applied across all areas of smart cities to enhance their sustainability – environmental, social, and economic – and its resilience. The latest AI trends in Smart Cities are mentioned.

Keywords: Industry 4.0, social-cyber-physical systems, smart cities, urban digital ecosystem, digital twin, resilient cities

1. Introduction

The fourth Industrial Revolution, commonly referred to as Industry 4.0, represents a transformation in the way industries operate and interact with digital technologies. Industry 4.0 is characterized by the integration of cyber-physical systems (CPS), internet of things (IoT), artificial intelligence (AI), big data analytics, cloud computing, and autonomous systems into industrial processes [1, 2]. These technologies enable real-time decision-making, predictive analytics, and decentralized control, fundamentally reshaping production lines, supply chains, and service delivery. And the principles of both digital twins and the system integration of physical manufacturing systems in the virtual space represent the core principles of Industry 4.0.

Originally rooted in manufacturing, Industry 4.0 principles have increasingly found relevance beyond factory walls. Urban systems and Smart Cities (SC) – composed of infrastructure, utilities, transportation, and governance – are now incorporating similar technologies to improve operational efficiency and service quality.

Smart cities are urban environments that utilize digital technologies to improve the quality, performance, and interactivity of urban services, while reducing resource consumption and increasing sustainability [3, 4]. SC concept integrates information and communication technologies (as well as Industry 4.0) across various domains, including energy, transportation, waste management, and governance, to create more livable and resilient communities [5].

Key components of smart cities include sensor networks, smart grids, urban mobility systems, e-governance platforms, and data-driven urban planning tools. Importantly, the goal is not only to make cities technologically advanced but also sustainable, inclusive, human, and adaptive to change, particularly in the face of global challenges such as climate change, rapid urbanization, and public health crises [6].

2. Industry 4.0 for urban innovations

2.1 Similarities between Industry 4.0 and smart cities

Despite emerging from different domains, both Industry 4.0 and Smart City paradigms share the same foundational principles. These similarities offer a powerful basis for smart city concepts and their implementation:

- *System integration:* Industry 4.0 principles support complete system integration of all the elements of the distributed manufacturing process. The integration is based on the digital twin/agent technology and strong virtualization of the decision-making processes.

- *Virtualization:* City systems are replicated in a twin city virtual model, which has the capability for simulation and prediction, linked to real-time data sources from sensors, and also explores the multi-agent systems (MAS).

- *Decentralization:* The complexity of a smart city structure necessitates decentralized management of all city systems and activities. This allows for independent decision-making, corrections, and the formation of alliances between intelligent components across systems to address critical situations or share resources.

- *Resilience and adaptability:* Both paradigms show the importance of being able to anticipate, absorb, and recover from disruptions. In Industry 4.0, resilience is achieved through flexible production, predictive maintenance, and supply chain adaptability [2], while in smart cities, resilience is expressed in climate adaptation, disaster preparedness, and infrastructure redundancy [7].

- *Sustainability:* Environmental, social, and economic sustainability is a cornerstone of both frameworks. Industry 4.0 leverages resource efficiency, circular economy principles, and low-carbon technologies [8], whereas smart cities prioritize green infrastructure, energy-efficient mobility, and smart grids [6].

- *Data-driven intelligence:* Both rely on real-time data acquisition, processing, and analytics. Industry 4.0 uses IoT, AI, and big data to optimize manufacturing and logistics [9], while smart cities apply the same tools to urban mobility, waste management, and public services [10, 11].

- *Interconnectedness and autonomy:* A central principle in both models is the creation of interoperable, autonomous, and cyber-physical systems. Factories in Industry 4.0 communicate across supply manufacturing and supply/chain facilities through CPS [1], while smart cities depend on IoT-enabled sensor networks and AI-driven decision-making platforms to coordinate transport, energy, and governance [12].

- *Interoperability*: The city is envisioned as a smart factory, where all components of its cyber-physical-social system are interconnected and communicate with one another within a secure network environment. This includes buildings, utilities, integrated transportation systems, street elements, and structures such as smart lighting, smart bins, smart stops, street cameras, hospitals, educational facilities, logistics centers, transportation hubs, business centers, retail and commercial facilities, energy centers, social networks, the IoT, and the internet of services.

- *Real-time capability:* The ability to collect data from the physical world in real time enables control mechanisms and the operational management of city systems to respond with timely solutions.

- *Service orientation:* The interaction among city systems, various neighborhoods, citizens, businesses, visitors, and others is interconnected through the internet of services, facilitating service delivery to a much wider audience.

- *Modularity:* This ensures the opportunity for system reconfiguration in the case of Industry 4.0 and flexibility for cities to adapt to immediate changes requested by clients or to address malfunctioning services in the smart city domain. Changes can be made within individual modules, or new modules with improved features, capabilities, or functionalities can be added.

These similarities demonstrate that the technological DNA of Industry 4.0 and smart cities is closely aligned. Both paradigms aim to integrate cyber-physical infrastructures, intelligent analytics, and enable self-regulating and even self-reconfiguring systems, which makes their merge into the smart resilient city a natural progression step [3, 13].

While the integration of Industry 4.0 technologies into industrial contexts has been widely studied [14], their systematic application to urban resilience and smart city design remains under-explored. Many smart city projects employ digital tools in isolated domains (e.g., traffic or energy) but lack a holistic framework that unites the full potential of Industry 4.0 principles [6].

2.2 Existing frameworks integrating Industry 4.0 and smart cities

The intersection of Industry 4.0 and smart city development has given rise to several conceptual and practical frameworks, although the field is still emerging and often fragmented. Most existing models present the integration of cyber-physical systems, IoT platforms, and real-time analytics into city infrastructures. These frameworks often draw on digital twin architectures, urban data platforms, and interoperable sensor networks to improve situational awareness and decision-making across urban domains [4, 10, 11].

The Smart Industry Readiness Index (SIRI), for example, developed by Singapore's Economic Development Board, bridges industrial digitalization with broader smart city planning by assessing both technological maturity and integration capabilities at various scales – from factories to urban infrastructure [15]. Meanwhile, the European Innovation Partnership on Smart Cities and Communities (EIP-SCC) proposes an integrative framework that aligns smart city initiatives with industrial innovation strategies, providing sustainability, open data, and cross-sector collaboration [6].

Another framework is the "City-as–a-Platform" model, which conceptualizes cities as integrated ecosystems where data flows from industry, infrastructure, and citizens merge in unified platforms for policy and operations [5]. This model supports scalability and adaptation, making it highly relevant for combining Industry 4.0 technologies with urban resilience efforts.

Despite these advances, there is a lack of standardized, cross-domain frameworks that fully operationalize Industry 4.0 principles in the smart city context, especially for resilience applications. This gap highlights the need for hybrid models that not only adopt advanced technologies but also account for urban governance, inclusivity, and ethical data use [14].

Our primary research employs a qualitative methodology, using literature synthesis and case study analysis. The literature review aggregates findings from peer-reviewed articles, policy documents, and technical reports to outline the main theoretical perspectives, challenges, and trends in the integration of Industry 4.0 technologies with smart urban systems [2, 10].

The case studies extract the literature synthesis by offering real-world illustrations of how smart cities adopt Industry 4.0 tools in practice. Case study methodology is particularly valuable here, as it allows for in-depth analysis of contextual complexity, policy frameworks, and technological integration across diverse urban settings. This aligns with a constructivist research philosophy, where meaning is drawn from practical, situated experiences rather than controlled experimentation [16].

2.3 Application of Industry 4.0 technologies in urban systems

The merging of urban management with advanced technologies from Industry 4.0 transforms traditional city functions into cyber-physical ecosystems. Below, we outline four key technologies and their integration into urban contexts:

2.3.1 Iot for real-time monitoring

The IoT enables continuous sensing, data collection, and communication among urban assets – from streetlights and traffic systems to waste bins and flood sensors.

Through interconnected sensor networks, cities can monitor energy usage, environmental quality, water levels, and structural health in real time [17].

For example, in flood-prone areas, IoT sensors can track rising water levels and alert emergency responders automatically. In energy systems, smart grids use IoT devices to dynamically allocate electricity based on usage patterns, reducing peak demand and outages [9].

2.3.2 AI for traffic optimization

Artificial Intelligence plays a leading role in urban mobility management, enabling adaptive traffic control, incident detection, and demand forecasting. AI algorithms process traffic sensor data, GPS inputs, and camera feeds to identify congestion patterns and dynamically adjust traffic lights, reroute flows, and reduce emissions [18].

Cities like Hangzhou, China, have implemented AI-powered traffic systems, resulting in up to a 15% reduction in congestion and more efficient emergency vehicle routing [19]. Such systems also improve public transportation scheduling and last-mile delivery logistics, contributing to both sustainability and economic productivity.

2.3.3 Digital twins for city planning

A digital twin is a virtual re-print of a physical urban system that can simulate behavior, predict outcomes, and optimize planning. By integrating real-time IoT data, AI, GIS, and 3D models, digital twins support scenario testing, infrastructure resilience planning, and asset lifecycle management [3].

City digital twin (CDT) integrates the physical, cyber, and social aspects of the city [20]:

- Data, information, knowledge: Utilizes sensors, cameras, mobile phones, Floating Car Data (FCD), and Cooperative Intelligent Transport Systems (C-ITS), including Vehicle-to-Vehicle (V2V), Vehicle-to-Infrastructure (V2I), and Vehicle-to-Everything (V2X).

- Modeling and simulation: Employs tools such as SUMO, VISSIM, and PALM for predictions and forecasting.

- Negotiation and decision-making: Incorporating multi-agent technologies (MAT) and AI.

- Visualization: Utilizes virtual and augmented reality (VR/AR).

Cities like Singapore and Helsinki are using digital twins for land-use simulations, transportation modeling, and climate adaptation planning. These tools allow urban planners to test policy interventions in silico before real-world implementation – increasing effectiveness and reducing risk [21].

Let us remark that once the digital twin is equipped with the capabilities of communicating and negotiating with other elements in the urban ecosystem, we have to talk rather about agents (actors). The agents, equipped with communication and negotiation capabilities, can behave – in comparison to digital twins –

pro-actively. The capabilities of the agents' behavior represent the main foundation for exploring MAS technology for urban system integration purposes.

2.3.4 Blockchain for governance and transparency

Blockchain introduces decentralized, immutable registries that can transform public sector operations. In smart cities, it can be used for land registry, public procurement, citizen identity verification, and energy trading [22]. Smart contracts also automate processes while ensuring transparency and reducing fraud.

For example, Estonia's e-governance system uses blockchain for secure data handling and public service delivery. Cities like Dubai are piloting blockchain-based urban governance platforms to manage transportation and utility billing with increased accountability. [23]

Together, these technologies form the technological foundation of smart cities. Their integration enables cities to operate as adaptive, data-centric ecosystems, capable of real-time response, long-term planning, and participatory governance.

2.4 Qualitative approach: Case studies

This section presents case studies of cities implementing Industry 4.0 technologies in urban infrastructure, governance, and resilience efforts. The selection highlights geographical diversity, technological integration, and lessons learned. Cases were selected based on their documented use of multiple Industry 4.0 technologies (e.g., IoT + AI + digital twins), evidence of improved urban resilience or service delivery, and the availability of open data or academic analysis [4, 11].

2.4.1 Singapore: A Smart nation built on Industry 4.0 foundations

Singapore's Smart Nation initiative demonstrates effective integration of Industry 4.0 tools across city operations. Its strategies include ubiquitous IoT, AI-driven systems, and city-scale digital twins. Through the Virtual Singapore project, the city has developed a high-fidelity, real-time digital replica used for planning, disaster response, and infrastructure monitoring [24].

The Smart Mobility 2030 program uses AI to analyze real-time traffic data, enabling dynamic traffic control and predictive congestion management. In housing, smart sensors regulate energy use and monitor environmental conditions. Meanwhile, GovTech Singapore employs blockchain in trial programs for secure digital identity and public record verification [25].

Main technologies:

- Digital twin (Virtual Singapore).
- AI in traffic and public services.
- Blockchain in e-governance.
- Cross-ministerial data-sharing platforms.

2.4.2 Amsterdam: Data-driven urban innovation

Amsterdam's Smart City Platform enables open collaboration among government, companies, and citizens. The city employs IoT, open data portals, and AI-driven analytics to address mobility, energy, and waste challenges. Its AI-based adaptive street lighting system, for example, reduces energy use by 30% by adjusting brightness based on pedestrian movement [26].

The City Data Platform aggregates data from various sources – transportation, pollution, energy grids – to feed into predictive analytics tools and urban dashboards. Additionally, blockchain is being piloted for decentralized energy trading and transparent subsidy management [27].

Main technologies:

- IoT-enabled public lighting.
- AI-powered traffic and energy management.
- Blockchain in peer-to-peer energy trading.
- Community participation and open innovation.

2.4.3 Curitiba: Citizen-centric smart resilience

Curitiba, Brazil, known for its innovative urban planning, is now implementing Industry 4.0 technologies to support social resilience and urban sustainability. The city uses mobile apps and IoT sensors to manage solid waste and public transportation, while AI tools support flood prediction and emergency response [28].

Through the Curitiba App, citizens receive personalized transport recommendations, report local issues, and track sustainability metrics. Integration with GIS and AI models has helped the city respond to urban flooding more effectively – a critical issue given the increase in extreme weather events [12].

Main technologies:

- AI + GIS for climate risk management.
- Mobile tech for citizen engagement.
- IoT in waste collection and transit.
- Strong focus on equity and inclusion.

3. Urban digital ecosystem

3.1 From digital twins to digital ecosystems

One of the most promising future trends shared by both Industry 4.0 and Smart City paradigms is the evolution from digital twins toward digital ecosystems:

(1) *Digital twins* – At present, both domains use digital twin technology as a key enabler. In Industry 4.0, digital twins replicate machines, factories, and supply chains for predictive optimization [3, 21]. In smart cities, digital twins simulate urban infrastructure, mobility, and environmental systems for planning and resilience [20, 21].

(2) *Twin-of-twins* – Instead of relying on a single central twin, each specialization manages its own digital twin that serves as a detailed digital representation of assets and services. Importantly, these twins are interconnected, forming a federation of twins that communicate through a standardized federation bus. This structure, in city cases, allows each local twin to interact with others and with Europe-wide data infrastructures, including Destination Earth (DestinE) or the climate/environment data hub (CEADS). In summary, the twin-of-twins approach [29] utilizes each capability, allowing them to harness collective intelligence [30].

(3) *Isolated digital ecosystems* – The next stage of development involves creating domain-specific digital ecosystems, where Industry 4.0 digital platforms interconnect factories and supply chains, and smart city platforms unify urban services and governance systems. [31]

(4) *Unified digital ecosystems* – Over time, the boundary between industrial and urban ecosystems will blur. The integration of data flows, platforms, and governance models may lead to one comprehensive digital ecosystem that simultaneously manages production, mobility, energy, logistics, governance, and citizen services. In this vision, the distinction between "Smart city" and "Smart industry" disappears, replaced by a seamlessly integrated, adaptive, and resilient digital ecosystem [32].

This transition highlights the co-evolutionary path of both paradigms: from parallel digitalization to full integration into a smart, resilient, and intelligent society.

3.2 Unified digital ecosystems

The transition from isolated digital twins to unified digital ecosystems will depend on the maturation and merging of two foundational technological domains:

1. **Data and knowledge management technologies**

   At the heart of a digital ecosystem lies the ability to organize, interpret, and operationalize data as actionable knowledge. Current digital twins primarily rely on data integration and simulation. To scale toward digital ecosystems, cities and industries require semantic interoperability, where information can be exchanged and understood consistently across heterogeneous systems.

   - Ontologies provide a key mechanism here, enabling a shared conceptual framework that aligns definitions of assets, processes, and relationships across multiple domains [33, 34]. Ontology-based systems allow machines and software agents to reason about data, ensuring that industrial, environmental, and governance data can be harmonized into one knowledge-driven ecosystem.

- This capability supports advanced tasks such as automated decision-making, policy simulation, and cross-sector optimization, where industrial production schedules can adapt dynamically to urban mobility flows or energy availability.

2. **Actor-based representation and simulation technologies**

   A second part is the ability to represent and coordinate diverse stakeholders and entities – citizens, machines, organizations, and infrastructures – within the ecosystem as autonomous, interacting actors.

   - Multi-agent systems (MAS) are particularly suitable for this task, as they allow the modeling of distributed, autonomous actors that interact, cooperate, or compete within a digital environment [35–37].

   - By applying principles of natural ecosystems – where actors adapt to changing conditions and maintain balance – MAS can simulate and coordinate the behavior of factories, transportation fleets, energy grids, and even citizens, treating them all as agents in one interconnected system.

   - This actor-centric view enables emergent intelligence, where local decisions and interactions scale up to system-wide resilience and adaptability.

Together, ontology-based knowledge management and multi-agent actor representation provide the cognitive and behavioral foundations for digital ecosystems. While digital twins replicate systems in isolation, digital ecosystems will require:

- Knowledge coherence across domains (ontology-driven interoperability).

- Dynamic coordination of heterogeneous entities (multi-agent interaction).

This dual foundation enables the vision of a unified, adaptive digital ecosystem, where industries and cities are no longer separate but operate as parts of a single, evolving socio-technical organism [33–36, 38].

4. The concept of the smart sustainable and resilient city

The Smart Sustainable and Resilient City (SSRC) aims to effectively utilize Industry 4.0 principles to create synergistic effects among various subsystems, such as transportation, logistics, safety and security, energy, and building management, all with a focus on improving the quality of life for its citizens. The SSRC concept represents a city as a digital platform and an ecosystem of smart services. In this framework, individuals, objects, documents, robots, and other entities can negotiate directly with one another based on resource-demand principles [39], resulting in appropriate, sustainable, and resilient solutions.

A smart, sustainable, and resilient city is a forward-thinking urban model that combines technological intelligence with the capacity to adapt, absorb, and recover from various disturbances, whether environmental, infrastructural, or socio-economic [39]. It brings together three contemporary urban paradigms:

- Smart cities – which prioritize efficiency, data-driven governance, and citizen-centric services using ICTs.
- Sustainable cities – designed to balance environmental protection, social equity, and economic development while ensuring a high quality of life.
- Resilient cities – which focus on robustness, redundancy, and adaptability to unforeseen challenges like climate change, pandemics, and cyber threats.

By integrating Industry 4.0 technologies into the urban fabric, cities can move from being reactive to proactive, using real-time sensing, predictive analytics, and automation to manage both daily operations and disruptions. By incorporating indicators such as air quality, energy demand, mobility efficiency, and quality of life, sustainable cities aim to provide long-term prosperity without compromising the needs of future generations. This hybrid model supports systemic urban resilience – not just in critical infrastructure, but also in socio-economic networks, urban planning, and participatory governance [13].

We propose a smart city universal platform designed as a digital ecosystem of services, inspired by complex systems science. Within this ecosystem, elements can negotiate, cooperate, or compete to achieve both local and global KPIs. At its core, the framework is resilient by design – a property of complex adaptive systems that enables them to react intelligently to unforeseen events and threats. Unlike pre-programmed systems, a resilient-by-design platform evolves under all conditions, adapting to disruptions as naturally as ecosystems like bee swarms or ant colonies.

A smart, sustainable, and resilient city is, therefore, one that uses knowledge-based systems and AI algorithms to ensure efficiency, service quality, and competitiveness, while meeting present and future needs across economic, environmental, social, health, and cultural dimensions. Events like the COVID-19 pandemic demonstrated the need for systems that can support solidarity, creativity, and rapid innovation beyond pre-defined scenarios.

Our proposed Smart City 5.0 framework incorporates several features:

- Recognition, reaction, prediction, and learning as core functions.
- Semantic interoperability to connect all services into one organism.
- Just-in-time reactions to threats, modeled after natural ecosystems.
- "What-if" analysis for predictive scenario testing.
- Multiple operational modes (normal, predictive, critical, post-critical) with clear transitions.
- Dynamic KPIs that adapt in real-time to changing conditions.

- Holistic decision-making that accounts for interdependencies among city actors.

The framework is implemented in two stages:

(1) Strategic modeling platform: For planning, assessment, and simulation of urban systems [40].

(2) Operational platform: A resilient-by-design service layer for real-time decision-making and adaptive smart services.

This dual structure allows both long-term planning and real-time adaptation, ensuring that smart cities can evolve sustainably and resiliently under all conditions.

4.1 Smart sustainable and resilient city architecture

The proposed smart city operational platform is built around several key components that, together, enable intelligent, adaptive, and resilient urban management.

(1) Multi-agent engine (ME) and multi-agent technologies (MAT).
At the core of the platform is the multi-agent engine, which deploys intelligent agents to represent both physical and virtual city entities. These agents ensure system connectivity, negotiate with each other, and cooperate to solve complex problems in real time under user-defined constraints and objectives. Beyond optimization, they can also conduct "what-if" analyses, simulating alternative scenarios to support predictive planning and adaptive decision-making.

(2) Smart city database (SCDB).
The SCDB serves as a secure repository for data collected from diverse city services. It ensures structured access to information according to predefined security levels.

(3) Smart city knowledge base (SCKB).
Developed as a city-wide ontology, the SCKB transforms raw data into knowledge following semantic web principles. It defines the behavioral logic for agents independently of hardcoded rules, which is crucial for adaptability. During critical modes, the SCKB becomes the backbone of system resiliency, enabling knowledge-driven responses.

(4) Intelligent Enterprise Service Bus (iESB).
The iESB connects all city services, regulating access to the SCDB and SCKB according to established rules. It also functions as the communication backbone for inter-agent messaging and service orchestration.

(5) Supplementary modules.
Additional components such as data mining tools, reasoning engines, VR/AR interfaces, and digital twin modules improves data interpretation, simulation, and visualization, further improving service quality.

The overall SSRC platform architecture is illustrated in **Figure 1**.

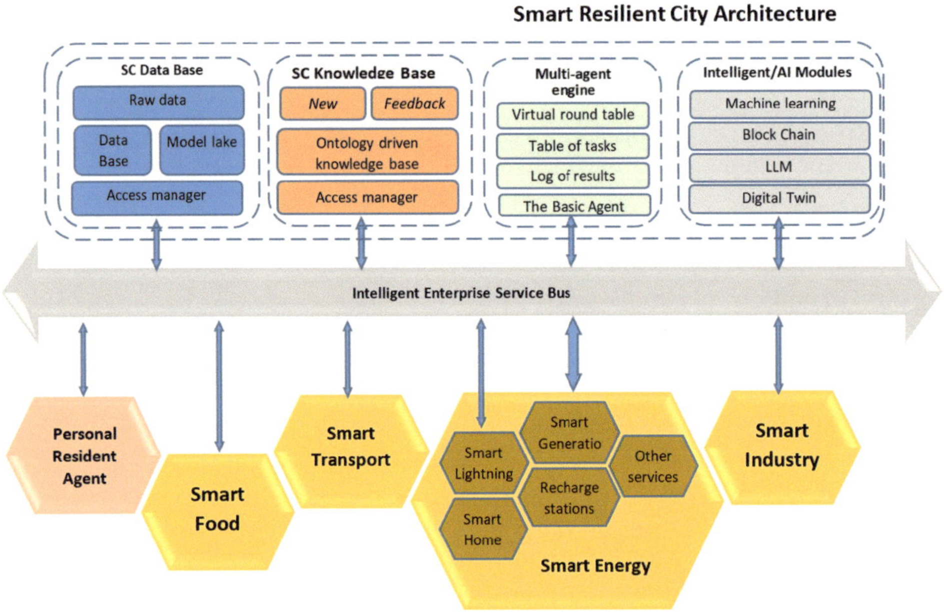

Figure 1.
SSRC platform architecture.

To ensure transparency, the SSRC platform requires clearly defined roles and responsibilities for its core components:

1) Agents and the multi-agent engine are maintained by the Municipal ICT team in collaboration with system integrators and certified vendors. They are responsible for deploying and updating intelligent agents, validating agent behavior, ensuring interoperability between services, and monitoring for biases, errors, or malfunctioning agents.

2) Data and knowledge bases are governed by the city data governance authority (Municipal IT & Innovation Office), ensuring data quality, security, interoperability, and compliance with privacy regulations (e.g., GDPR). The knowledge base is co-maintained with academic and industry partners to integrate ontologies and update knowledge models.

3) Intelligent Enterprise Service Bus is managed by the Municipal IT department, with oversight from a city-wide digital governance board. It is responsible for regulating data exchange, controlling access rights, and ensuring secure and seamless communication between agents and services.

4) The key performance indicators are set up and owned by city governance bodies (mayor's office, planning departments, transport, and energy authorities). They are reviewed by joint governance boards, including policymakers, urban planners, citizen representatives, and independent auditors. The main responsibility lies in defining, monitoring, and periodically revising KPIs to reflect changing priorities (e.g., shifting from mobility efficiency to

public health during pandemics). KPIs act as triggers for system modes and must be regularly evaluated to ensure alignment with sustainability and resilience goals.

5) The supplementary modules (AI, digital twins, VR/AR, etc.) can be maintained by certified specialized innovation labs and private-sector technology partners, with responsibilities such as updating algorithms, ensuring model accuracy, and integrating new features into the platform while respecting interoperability standards.

4.2 City strategic simulator for critical situations

In the first phase, the city strategic simulator is used for normal modes with the following features:

- Each city service maintains its own local KPIs.
- The platform monitors global KPIs, which may conflict with local service objectives.
- The system analyzes data dependencies between services.
- Each service is represented by an intelligent agent.
- A general ontology creates a unified semantic knowledge space.
- A city equalizer module visualizes KPI interdependencies and suggests interventions to improve both local and global performance.

The by-design resiliency approach demonstrates the ability of a smart city platform to react dynamically to unexpected events rather than relying solely on predefined scenarios. The system operates across the same four modes (normal, predictive, critical, post-critical), but in cases of accidents, catastrophes, outages, crises, or other emergencies, the knowledge base–driven engine enables rapid identification of appropriate solutions.

Illustrative examples of adaptive responses

- Transport repurposing: Taxi fleets and private vehicles can be redirected to serve as emergency delivery services or medical transport, with drivers volunteering freely or working for compensation during crises.
- Mass evacuation: Public transport routes can be reconfigured into evacuation services in the event of large-scale threats.
- Mobilizing human resources: Citizens with relevant expertise – such as pensioners, trained employees from adjacent sectors, or people with medical education – can be engaged as volunteers during health emergencies.

- Infrastructure reallocation: Underused facilities can be repurposed, e.g., vacant hotels functioning as temporary hospitals or stadiums serving as shelters for refugees.
- Communication and awareness: Public advertising billboards can be reprogrammed to display emergency instructions and action plans for citizens.
- Resource redistribution: In the case of power, water, or internet outages, the system can designate alternative access points (restaurants, residential buildings, or public facilities with owner consent).
- Individualized guidance: Every citizen can receive a personalized action or evacuation plan on their mobile device, including instructions on how to assist others.

While city governments typically maintain sets of regulations and emergency protocols, no framework can predefine every possible scenario. A by-design resilient system functions as an expert system, dynamically synthesizing data, knowledge, and contextual factors to propose effective responses. This allows decision-makers to act with greater confidence, speed, and adaptability when facing unpredictable threats or combinations of crises.

4.3 Replication and scalability for smart resilient cities

To support the replication and scaling of smart resilient city platforms, we propose a minimal, transferable framework:

1) Adopt open standards and models:
   - Use semantic ontologies for interoperability.
   - Leverage open data formats for data exchange.
   - Align with international standards (ISO 37120 on city indicators, ISO/IEC 30182 on smart city ontology).
2) Minimal viable architecture:
   - Data layer: City database + knowledge base with semantic interoperability.
   - Agent layer: Multi-agent engine for service representation and coordination.
   - Integration layer: Enterprise service bus for secure interconnection.
   - Application layer: Dashboards, digital twins, and predictive analytics.
3) Governance and privacy basics:

- Define a data governance authority with clear responsibilities for ownership, access, and audit.
- Implement privacy-preserving mechanisms (anonymization, federated learning, GDPR compliance).
- Ensure multi-stakeholder participation in KPI definition and review.

4) Pilot-to-scale path:

- Phase 1: Start with a single-domain pilot (e.g., mobility or energy).
- Phase 2: Demonstrate cross-domain integration via a shared knowledge base.
- Phase 3: Expand toward a multi-domain ecosystem and integrate with national/regional platforms (e.g., DestinE, CEADS).
- Phase 4: Institutionalize governance, funding, and citizen participation to sustain scaling.

5. Smart city digital testbed

In this section, we include a custom case and experimental setup based on the authors' own research focus. The Czech capital, Prague, has initiated a pilot deployment of a smart city digital Testbed focused on water, gas, energy, and mobility domains in the area of Evropská Avenue (www.gatetoeurope.eu). The capital, Taipei, in Taiwan collaborates with the authors on an urban traffic management study using Multiagent algorithms.

5.1 City simulation software for smart Evropská Avenue

As part of a sub-project from the National Center for Cybernetics and Artificial Intelligence, an interdisciplinary team of experts was formed, focusing on different areas such as transportation, energy, spatial planning, smart buildings, and environmental modeling. This collaboration led to the concept of the Smart Evropská Avenue[1], a digital smart city polygon in Prague. It also encompasses the surrounding area of Evropská Avenue, as shown in **Figure 2**, where various development projects are planned and where applications of the Industry 4.0 approach can be effectively tested, particularly in the construction industry.

So far, City Simulation Software (CSS) has been used on Evropská Avenue in Prague [41]. The image on the left in **Figure 3** shows the sociological data, and on the right, the environmental parameters displayed in the VR map. Thanks to CSS, it is possible to verify in a virtual environment whether these new buildings will cause future problems in the given urban area. Unfortunately, after the approved building

1 Home – Gate To Europe

Figure 2.
Map of Evropská Avenue in Prague.

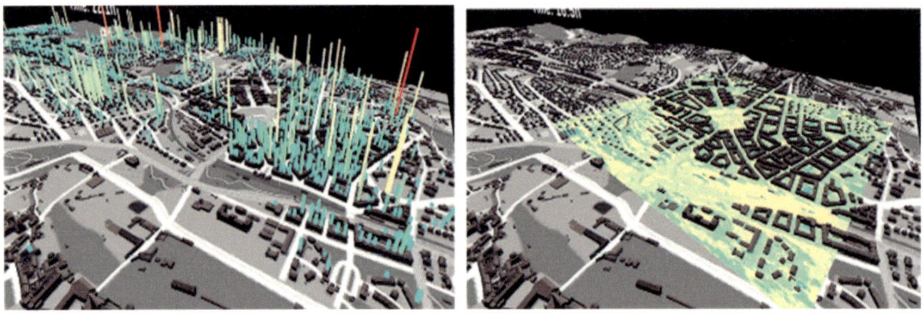

Figure 3.
The graphical outputs of City Simulation Software (CSS) to a virtual reality (VR) environment.

implementation, it will no longer be possible to change it, and the economic and societal consequences could be enormous.

The measurement was also made during the COVID-19 pandemic, allowing for a comparison of the mobility and environmental situation both during and after the pandemic, when normal life returned to Prague.

The core mobility KPIs – average travel time, vehicle speed, mean density, and congestion levels – served as primary measures of transport network efficiency. A baseline model was first constructed by integrating infrastructure geometry, traffic control parameters (signal plans, public transport priorities, pedestrian crossings), and urban planning objectives.

Thermal comfort was assessed using the Universal Thermal Climate Index (UTCI), which distinguishes six degrees of cold and heat stress. In parallel, the air quality KPI was evaluated following the European Air Quality Index (EAQI), a standardized method for assessing pollutant concentrations across time and space. Concentrations of $PM_{2.5}$, PM_{10}, NO_2, and O_3 were included, and the maximum concentration among these pollutants was selected as the representative KPI.

To interpret the outputs, post-processing tools were introduced. The first tool allowed for the combination of all KPIs into heatmaps, highlighting the frequency of threshold exceedances in time and space. The second tool averaged threshold

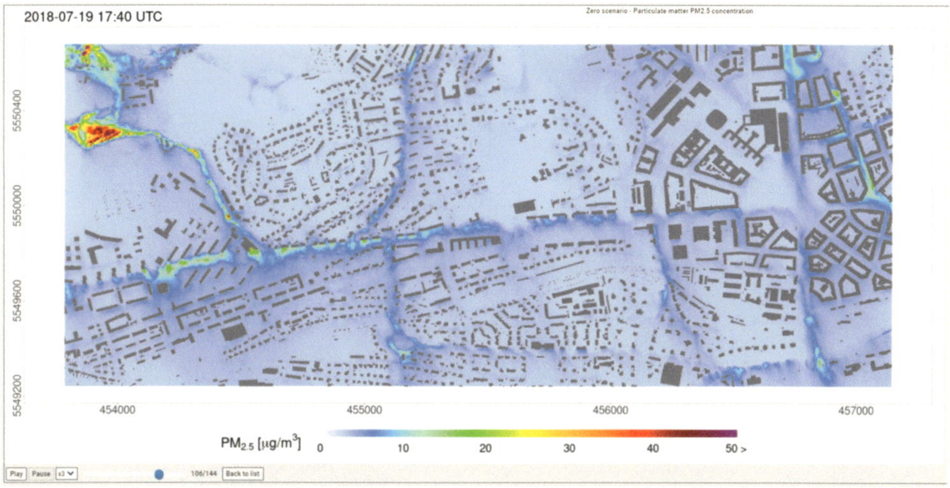

Figure 4.
Results of environmental analysis before COVID-19.

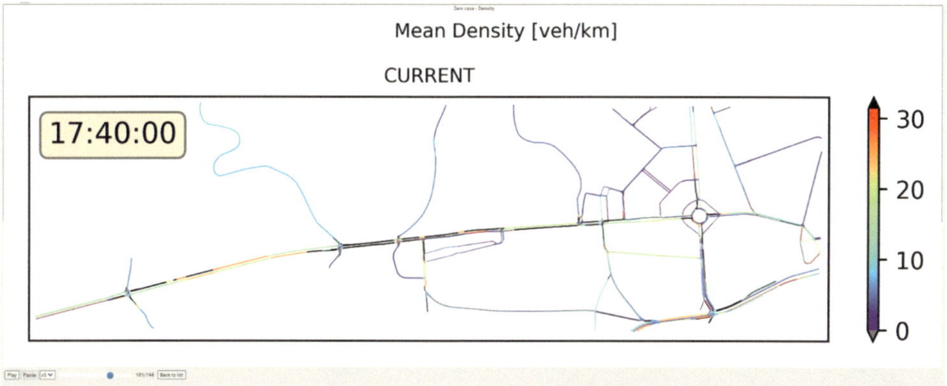

Figure 5.
Results of traffic analysis before COVID-19.

exceedances across mobility and environmental KPIs, resulting in a composite indicator that offered a holistic perspective of the system's impacts.

The results in **Figure 4** and **Figure 5** show the heatmaps of traffic (mean density) and environmental (PM 2.5) parameters before the COVID-19 pandemic; **Figure 6** and **Figure 7** show the heatmaps during the COVID-19 pandemic. The CSS results obtained can contribute to advanced city management by using data, models, and mainly extracted knowledge.

For practical implementation, it is necessary to establish a city data platform that offers open data in a machine-readable format and is usable by third parties. In Prague, this platform is known as Golemio[2], which serves as a fundamental part of the Smart Prague solution, managed by the Operator ICT[3]. This municipal

[2] Golemio: English page
[3] We are creating a new Prague. | oict

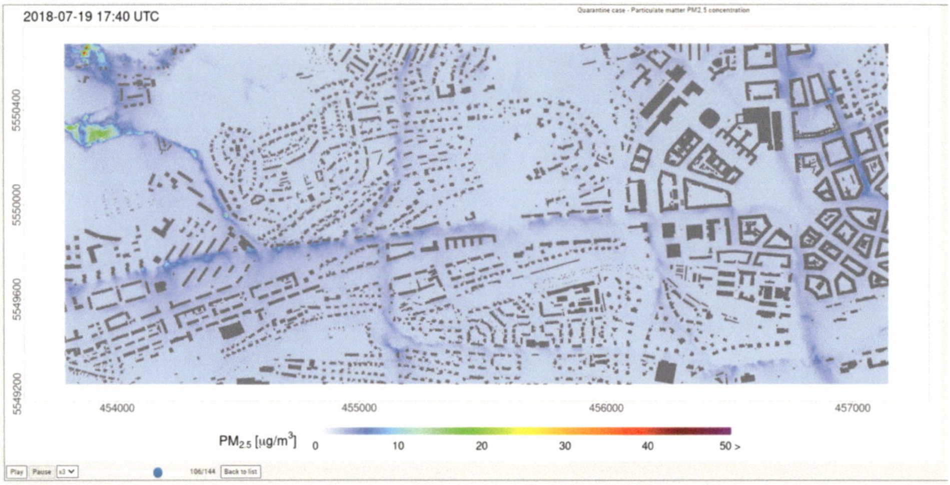

Figure 6.
Results of environmental analysis during COVID-19.

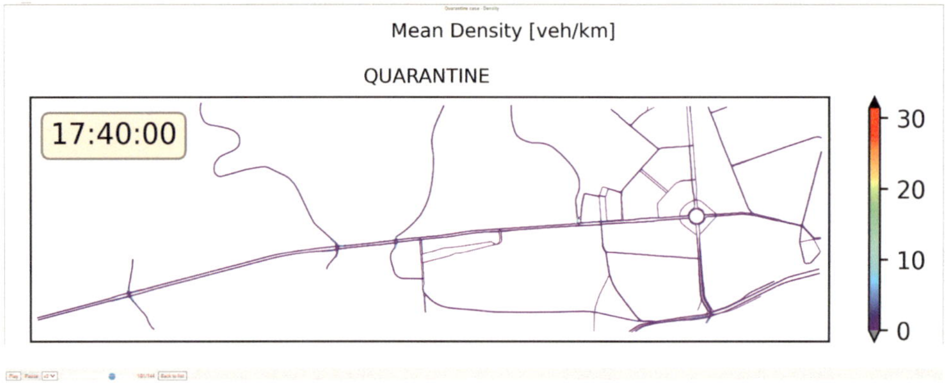

Figure 7.
Results of traffic analysis during COVID-19.

organization is fully responsible for the collection and management of selected datasets relevant to smart city solutions. These datasets are made accessible to the public and, under equal conditions, to third parties who may utilize them for developing applications and services for end users. The validation of the results of CSS is based on the data from the Golemio platform, as well as sensor-based measurements.

5.2 Enhancing urban traffic management in Taipei

The study applied Q-learning and Multiagent algorithms within the SUMO (Simulation of Urban MObility) environment to optimize traffic signal control at two major intersections in Taipei (Gongguan and Keelung-Xinhai). Real traffic data – including vehicle types, flow volumes, and traffic light phases – was collected on-site and used to calibrate the simulation. Traffic lights were modeled as reinforcement learning (RL) agents capable of adjusting signal timing dynamically.

A weighted reward system was designed to balance multiple objectives: minimizing waiting times, reducing total trip durations, and maximizing throughput, while also considering vehicle types and lane density.

Findings of the simulation (presented in **Figure 8**):

- Non-peak hours: Reduced trip durations by 29–39% and median waiting times by up to 44%, while increasing throughput by as much as 7.9%.

- Peak hours: Improvements in median trip duration ranged from 12–33%, with reductions in average trip times of up to 25%.

- Environmental impact: The system significantly reduced CO_2 emissions, especially for common vehicle categories, such as Euro-4 petrol cars.

The results of the comparison of RL and MAS algorithms (presented in **Figure 9**):

- The RL algorithm currently operates faster than the multi-agent algorithm. However, the execution time for each simulated second remains below 30000 microseconds (less than one real-time second), making both algorithms feasible for real-world deployment.

- The RL algorithm tends to balance the maximum waiting time across all vehicles, resulting in a more uniform distribution of delays.

- The multi-agent algorithm performs comparably to the RL algorithm only when the minimum traffic light duration is set to 20 seconds. This is primarily because the multi-agent approach does not currently account for lane-specific congestion levels.

- The multi-agent algorithm adjusts both the total waiting time per vehicle and the overall traffic light capacity, optimizing intersection throughput.

The analysis of algorithm behaviors shows that both algorithms demonstrate sufficient processing speed for real-time operation within the simulated environment. However, their fundamental approaches differ:

- The multi-agent algorithm follows an event-driven recalculation model, where each newly introduced object (e.g., a vehicle) can trigger a chain reaction that modifies the traffic light schedule. However, this approach lacks predictive capabilities, as it responds only to immediate conditions.

- The RL algorithm, on the other hand, operates based on data accumulation. While recalculations are rapid, the model incorporates historical data, which sometimes leads to decisions that do not fully reflect current road conditions.

The Taipei case illustrates the by-design resiliency principle: the ability of urban systems to respond intelligently to unpredictable and dynamic conditions.

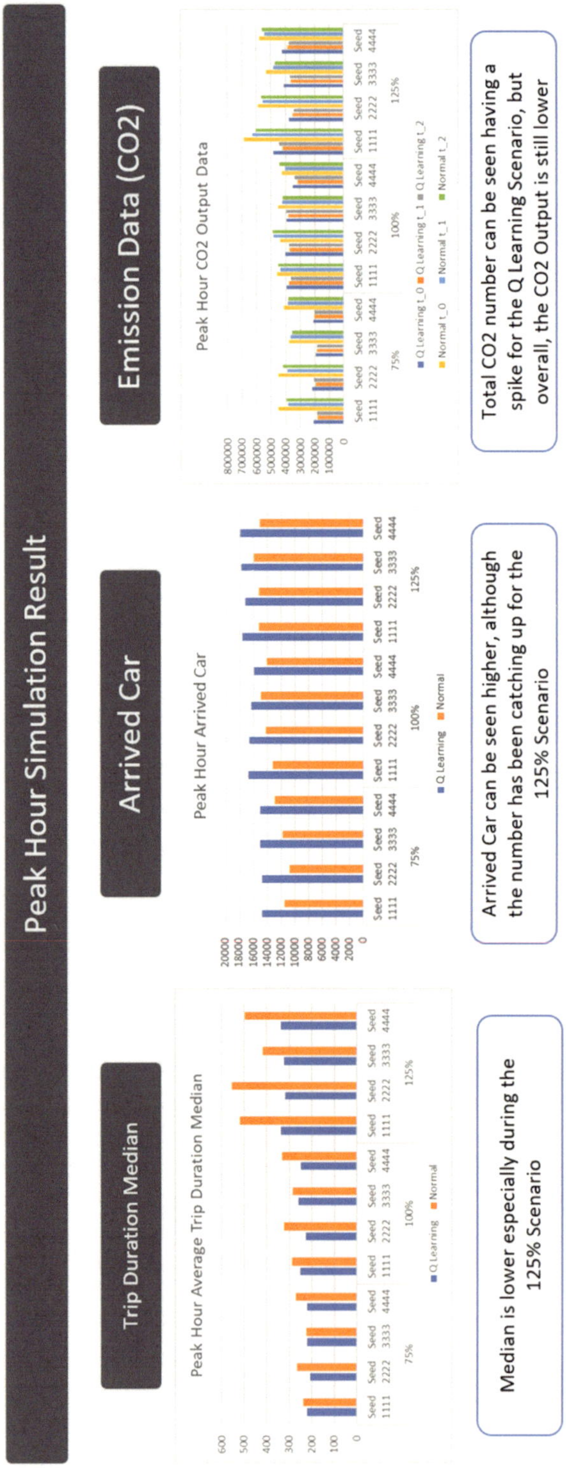

Figure 8.
Results of traffic optimization during peak hours.

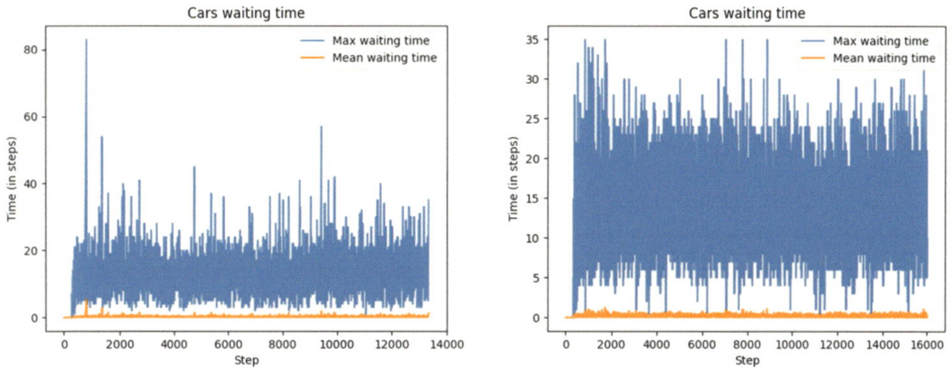

Figure 9.
The results of the comparison of MA (left) and RL (right) algorithms by "car waiting time" parameter.

Rather than relying on static schedules, the RL-based traffic management adapts to real-time changes, reduces inefficiencies, and contributes to sustainability targets.

6. Discussion

6.1 Toward digital ecosystems: Smart cities meet industry of the future

The combination of Industry 4.0 technologies and smart city infrastructure signals the appearance of digitally unified ecosystems, where cities and industries co-evolve through shared platforms, real-time data, and autonomous systems [13, 14]. As cities become increasingly data-centric, industries integrated within urban zones must align with these infrastructures to achieve mutual sustainability and resilience goals.

Emerging concepts like the "Industry of the Future" envision industrial processes that are not isolated but embedded within smart urban systems – sharing cloud platforms, energy networks, logistics data, and AI services [2]. This transformation requires horizontal integration across sectors and vertical integration from physical devices (e.g., IoT sensors) to intelligent decision-making layers (e.g., LLMs and AI agents). The MAS technology has been dominantly used for this purpose until now, but the Generative AI (LLM, SLM, RAG approaches) will definitely play a key role in the future.

6.2 Generative AI enters the scene

Large language models (LLMs), such as ChatGPT or other generative AI systems, and self-managing logistics (SML) platforms are increasingly seen as tools to optimize decision-making in smart cities by enabling natural language interfaces, automated policy simulations, and dynamic logistics coordination [42]. These technologies reduce friction in complex urban environments and improve situational awareness across city departments and industrial actors.

Both the integration efficiency and the level of intelligence of the smart city systems in future ecosystems are expected to be achieved by exploring the RAG (Retrieval-Augmented Generation) technology. This enables the engagement of individual subsystem data (architecture descriptions, negotiation protocols, financial documents, historical data, etc.) in the generation of more precisely targeted prompts for the corresponding foundation language models. The RAG technology enables the development of adequate reactions in a fast manner, based on very detailed, focused, and customized information.

It is expected that new language models will be designed and developed as smart city counterparts to LPM (large process models) [43] and LAM (large action models) [44]. The smart city LPM (SC-LPM) might be used for the complex planning/scheduling of a series of actions to achieve better behavior of the urban system from the point of view of, e.g., resilience or transport efficiency. The SC-LPM will generate digital twins of the complex processes to be accomplished. The SC-LAM systems might be used for the scheduling of individual immediate actions as elements of the complex plan. They are expected to generate digital twins of individual actions/operations.

The capabilities to generate digital twins of both the processes and the actions would strongly enrich the complex decision-making at the urban ecosystem level.

6.3 Flexible production: A key component in resilient cities

A core innovation from Industry 4.0 is flexible production – systems that can rapidly adjust manufacturing outputs in response to market demands, supply disruptions, or emergency needs [2, 45]. Within urban environments, this concept enables distributed micro-factories, on-demand 3D printing, and reconfigurable supply chains.

During the COVID-19 pandemic, cities that adopted localized flexible manufacturing – including 3D printing of medical equipment and personal protective gear – exhibited greater agility and resilience in response to supply chain shocks [46]. As climate change, geopolitical shifts, and energy crises increasingly disrupt global production systems, flexible production within cities will become a solution for urban resilience.

Additionally, flexible production aligns with the principles of the circular economy, where materials and products are reused, recycled, and regenerated within city boundaries – minimizing waste and dependency on global supply chains [8].

6.4 Cities and industries as co-evolving ecosystems

To move beyond isolated innovation, both smart cities and advanced industries must transition toward collaborative, adaptive ecosystems. This shift involves:

- Shared digital infrastructure (e.g., city-wide IoT platforms and edge-cloud computing).
- Joint governance and regulation through public-private-people partnerships.
- Semantic interoperability and standardization of urban-industrial data.
- Ethical AI frameworks ensure accountability, transparency, and inclusivity [47] [38].

DOI: http://dx.doi.org/10.5772/intechopen.1013144

These ecosystems foster collective intelligence, where insights from industrial AI systems improve urban planning, and real-time city data informs product design, logistics, and service delivery.

In this vision, cities no longer just host industry – they become active, intelligent partners in industrial systems, creating feedback loops between urban life, digital innovation, and economic production. The result is a new paradigm: the smart, resilient city as a living, learning industrial ecosystem.

7. Conclusion

The concept of smart, resilient cities is an interdisciplinary issue that presents a new way of managing the city with the help of available knowledge and technologies that were unimaginable until recently. In this concept, it is possible to use models of different situations that can occur in the city, analyze them, and ensure the implementation of the most optimal response with respect to given criteria.

While digital technologies such as digital twins, multi-agent systems, and reinforcement learning provide technical capabilities, their value lies in measurable contributions to resilience, sustainability, and equity:

- Digital twins simulate infrastructure and urban dynamics in real time. DTs improve resilience through predictive scenario testing (e.g., flood response planning). A measurable indicator is the reduction in recovery time after disruption events.

- Multi-agent systems enable autonomous negotiation among heterogeneous actors (citizens, services, assets), supporting sustainability by optimizing resource allocation (e.g., balancing electricity demand across microgrids). A relevant KPI is the percentage reduction in peak energy load or the increase in renewable energy share.

- Reinforcement learning improves system adaptivity by learning optimal responses from data streams, enhancing equity through dynamic traffic or service allocation (e.g., reducing average waiting time for public transport across neighborhoods). A concrete KPI is the variance reduction in service accessibility between districts.

Together, these technologies bridge the gap between technical optimization and public value creation, making outcomes measurable, comparable, and actionable across cities. It is clear that the concepts of Industry 4.0 will be permanently motivating for researchers in the field of smart cities, and the exploration of AI, especially generative AI, will strongly influence the future progress.

Acknowledgement

This work was supported by the European Union under the project Robotics and advanced industrial production (No. CZ.02.01.01/00/22_008/0004590).

List of abbreviations

AI	Artificial Intelligence
AR	Augmented Reality
C-ITS	Cooperative Intelligent Transport Systems
CDT	City Digital Twin
CEADS	Carbon Emission Accounts and Datasets
CPS	Cyber-physical System
CSS	City Simulation Software
DestinE	Destination Earth
DT	Digital Twin
EAQI	European Air Quality Index
EIP-SCC	European Innovation Partnership on Smart Cities and Communities
ESG	Environment, Social and Governance
FCD	Floating Car Data
GIS	Geographic Information System
GPS	Global Positioning System
iESB	Intelligent Enterprise Service Bus
IoT	Internet of Things
KPI	Key Performance Indicator
LAM	Large Action Models
LLM	Large Language Models
LPM	Large Process Models
MAS	Multi-agent Systems
MAT	Multi-Agent Technologies
ME	Multi-Agent Engine
PALM	High-resolution atmospheric and environmental simulation model
QoL	Quality of Life
RAG	Retrieval-augmented Generation
RL	Reinforcement Learning
SC	Smart Cities
SC-LAM	Smart City Large Action Models
SC-LPM	Smart City Large Process Models
SCDB	Smart City Database
SIRI	Smart Industry Readiness Index
SLM	Small Language Models
SML	Self-Managing Logistics
SSRC	Smart Sustainable and Resilient City
SUMO	Software for Simulation of Urban MObility
UTCI	Universal Thermal Climate Index
V2I	Vehicle-to-Infrastructure
V2V	Vehicle-to-Vehicle
V2X	Vehicle-to-Everything
VISIM	Microscopic traffic simulation software
VR	Virtual Reality

DOI: http://dx.doi.org/10.5772/intechopen.1013144

Author details

Sergej Kozhevnikov, Vladimir Marik* and Miroslav Svitek
Czech Institute of Informatics, Robotics, and Cybernetics, Czech Technical University in Prague, Prague, Czech Republic

*Address all correspondence to: marik@cvut.cz

References

[1] Kagermann H, Wahlster W, Helbig J, Recommendations for Implementing the Strategic Initiative Industrie 4.0. Final Report of the Industrie 4.0 Working Group; 2013

[2] Liao Y, Deschamps F, Loures EDFR, Ramos LFP. Past, present and future of Industry 4.0 – A systematic literature review and research agenda proposal. International Journal of Production Research. 2017;**55**(12):3609–3629. DOI: 10.1080/00207543.2017.1308576

[3] Batty M. Digital twins. Environment and Planning B: Urban Analytics and City Science. 2018;**45**(5):817–820. DOI: 10.1177/2399808318796416

[4] Batty M, Axhausen KW, Giannotti F, et al. Smart cities of the future. European Physical Journal Special Topics. 2012;**214**:481–518

[5] Nam T, Pardo T. Conceptualizing smart city with dimensions of technology, people, and institutions. In: Proceedings of the 12th Annual International Digital Government Research Conference: Digital Government Innovation in Challenging Times New York; New York: Association for Computing Machinery; 2011. p. 282–291. DOI: 10.1145/2037556.2037602

[6] Ahvenniemi H, Huovila A, Pinto-Seppä I, Airaksinen M. What are the differences between sustainable and smart cities? Cities. 2017;**60**:234–245. DOI: 10.1016/j.cities.2016.09.009

[7] Sharifi A, Yamagata Y. Resilient urban planning: Major principles and criteria. Energy Procedia. 2014;**61**: 1491–1495. DOI: 10.1016/j.egypro.2014.12.154

[8] Geissdoerfer M, Savaget P, Bocken NMP, Hultink EJ. The Circular Economy: A new sustainability paradigm?. Journal of Cleaner Production. 2017;**143**:757–768. DOI: 10.1016/j.jclepro.2016.12.048

[9] Gubbi J, Buyya R, Marusic S. Palaniswami M.Internet of Things (IoT): A vision, architectural elements, and future directions. Future Generation Computer Systems. 2013;**29** (7):1645–1660. DOI: 10.1016/j.future.2013.01.010

[10] Bibri SE. The IoT for smart sustainable cities of the future: An analytical framework for sensor-based big data applications for environmental sustainability. Sustainable Cities and Society. 2018;**38**:230–253. DOI: 10.1016/j.scs.2017.12.034

[11] Gil-Garcia JR, Pardo TA, Nam T. What makes a city smart? Identifying core components and proposing an integrative and comprehensive conceptualization. Information Polity. 2015;**20**(1):61–87. DOI: 10.3233/IP-150

[12] Peixoto JPJ, Costa DG, Portugal P, Vasques F. Flood-Resilient Smart Cities: A Data-Driven Risk Assessment Approach Based on Geographical Risks and Emergency Response Infrastructure. Smart Cities. 2024;**7** (1):662–679. DOI: 10.3390/smartcities7010027

[13] Bibri SE, Krogstie J. Smart sustainable cities of the future: An extensive interdisciplinary literature review. Sustainable Cities and Society. 2017;**31**:183–212. DOI: 10.1016/j.scs.2017.02.016

[14] Xu X, Xu LD, Li L. Industry 4.0: State of the art and future trends. International Journal of Production Research. 2018;**56**(8):2941–2962. DOI: 10.1080/00207543.2018.1444806

[15] Economic Development Board Singapore. Smart Industry Readiness Index [Online]. 2020. Available from:www.edb.gov.sg/content/dam/edb-en/about-edb/media-releases/news/the-smart-industry-readiness-index/the-sg-smart-industry-readiness-index-whitepaper%20(1).pdf [Accessed: 2026-January-23]

[16] Yin RK. Case Study Research and Applications: Design and Methods. 6th ed. Thousand Oaks, CA: SAGE Publications; 2018

[17] Zanella A, Bui N, Castellani A, Vangelista L. Zorzi M. Internet of Things for smart cities. IEEE Internet of Things Journal. 2014;**1**(1):22–32. DOI: 10.1109/JIOT.2014.2306328

[18] Saxena R. Artificial Intelligence in Traffic Systems. 2024. DOI: 10.48550/arXiv.2412.12046

[19] City Brain in Hangzhou: AI makes urban life smarter. 2018. [Online]. Available from: www.ehangzhou.gov.cn/2020-10/23/c_275749.htm [Accessed: 2026-January-23]

[20] ISO/IEC 30182. Smart city concept model — Guidance for establishing a model for data interoperability. 2017 [Accessed 2026-January-23]

[21] Ketzler B, Naserentin V, Latino F. Digital Twins for Cities: A State of the Art Review. Built Environment. 2020;**46**:547–573. DOI: 10.2148/benv.46.4.547

[22] Reyna A, Martín C, Chen J, Soler E, Díaz M. On blockchain and its integration with IoT: Challenges and opportunities. Future Generation Computer Systems. 2018;**88**:173–190. DOI: 10.1016/j.future.2018.05.046

[23] Khan S, Shael M, Majdalawieh M, Nizamuddin N, Nicho M. Blockchain for Governments: The Case of the Dubai Government. Sustainability. 2022;**14**:6576. DOI: 10.3390/su14116576

[24] Tan SY, Taeihagh A. Smart city governance in developing countries: A systematic literature review. Sustainability. 2020;**12**(3):899. DOI: 10.3390/su12030899

[25] Government Technology Agency of Singapore (GovTech Singapore). Our Smart City Projects & Solutions in Singapore. Available from: https://www.tech.gov.sg/about-us/what-we-do/our-capabilities/smart-city-technology/2025 [Accessed: 2026-January-23]

[26] City AS. Amsterdam Smart City Projects [Online]. 2015. Available from: https://amsterdamsmartcity.com [Accessed 2026-January-23]

[27] de Waal M, Dignum M. The citizen in the smart city: How the smart city could transform citizenship. IT—Information Technology. 2017;**59**(6):263–273. DOI: 10.1515/itit-2017-0012

[28] Macke J, Casagrande R, Sarate J, Silva K. Smart city and quality of life: Citizens' perception in a Brazilian case study. Journal of Cleaner Production. 2018;**182**:717–726. DOI: 10.1016/j.jclepro.2018.02.078

[29] van Dyck M, Lüttgens D, Piller FT, Brenk S. Interconnected digital twins and the future of digital manufacturing: Insights from a Delphi Study. Journal of Product Innovation Management.

2023;**40**(4):475–505. DOI: 10.1111/jpim.12685

[30] Yu T, Li Z, Sakaguchi K, Hashash O, Saad W, Debbah M. Internet of Federated Digital Twins: Connecting Twins Beyond Borders for Society 5.0. IEEE Internet of Things Magazine. 2024;**7**(50):64–71. DOI: 10.1109/IOTM.001.2300279

[31] Margherita EG, Braccini AM. Examining the development of a digital ecosystem in an Industry 4.0 context: a sociotechnical perspective. SN Business & Economics. 2021;**89**. DOI: 10.1007/s43546-021-00096-z

[32] Svítek M, Skobelev P, Kozhevnikov S. Smart city 5.0 as an urban ecosystem of smart services. Studies in Computational Intelligence. 2020;**853**:426–438. DOI: 10.1007/978-3-030-27477-1_33

[33] Corcho O, Fernández-López M, A. Methodologies G-P. tools and languages for building ontologies: Where is their meeting point?. Data & Knowledge Engineering. 2003;**46**(1):41–64. DOI: 10.1016/S0169-023X(02)00195-7

[34] TR G. Toward principles for the design of ontologies used for knowledge sharing. International Journal of Human-Computer Studies. 1995;**43**(5–6):907–928. DOI: 10.1006/ijhc.1995.1081

[35] Wooldridge M. An Introduction to MultiAgent Systems. 2nd ed. Hoboken, NJ: Wiley; 2009

[36] Weiss G, editor. Multiagent Systems. 2nd ed. Cambridge, MA, USA: MIT Press; 2016. p. 2–18

[37] Leitao P, Lai LL, Strasser T, Abrishambaf R, Barbosa J, Dai W, et al. 2660.1-2020 - IEEE Recommended Practice for Industrial Agents: Integration of Software Agents and Low-Level Automation Functions. 2021. p. 43. DOI: 10.1109/ieeestd.2021.9340089

[38] Floridi L, Cowls J, Beltrametti M, et al. AI4People—An ethical framework for a good AI society: Opportunities, risks, principles, and recommendations. Minds and Machines. 2018;**28**:689–707. DOI: 10.1007/s11023-018-9482-5

[39] Kozhevnikov S, Svitek M. From Smart City sustainable development to resiliency by-design. In: 2022 Smart Cities Symposium Prague, SCSP; 2022. DOI: 10.1109/SCSP54748.2022.9792566

[40] Svitek M, Kozhevnikov S, Tencar J, Bhattacharjee S, Benes V. Smart city 5.0 as the digital ecosystem of smart services: Practical applications, Book: Smart Cities and Digital Transformation: Empowering Communities. Limitless Innovation, Sustainable Development and the Next Generation. 2023;327–356. DOI: 10.1108/978-1-80455-994-920231016

[41] Daniel M, Dostál R, Kozhevnikov S, Matysková A, Moudrá K, Reteira AM, Přibyl O. City Simulation Software: Perspective of Mobility Modelling. In: 2021 Smart City Symposium Prague (SCSP), 2021. DOI: 10.1109/SCSP52043.2021.9447384

[42] Chui M, Harrysson M, Manyika J, et al. Notes from the AI Frontier: Applying AI for Social Good. McKinsey Global Institute. 2018. Available from: https://www.mckinsey.com/featured-insights/artificial-intelligence/applying-artificial-intelligence-for-social-good#/ [Accessed: 2026-January-23]

[43] Kampik T, et al. Large Process Models: A Vision for Business Process

Management in the Age of Generative AI. arXiv:2309.00900v3. 2025. [Accessed: 2025-January-17]

[44] Wang L, et al. Large Action Models: From Inception to Implementation. arXiv:2412.10047v2. 2025. [Accessed: 2025-January-13]

[45] Ivanov D, Dolgui A. Viability of intertwined supply networks: Extending the supply chain resilience angles towards survivability. International Journal of Production Research. 2020;**58**(10):2904–2915. DOI: 10.1080/00207543.2020.1750727

[46] Ivanov D, Das A. Coronavirus (COVID-19/SARS-CoV-2) and supply chain resilience: A research note. International Journal of Integrated Supply Management. 2020;**13**(1):90–102. DOI: 10.1504/IJISM.2020.107780

[47] William I, Kozhevnikov S, Sontheimer M, Chou S. Enhancing Urban Traffic Management in Taipei: A Reinforcement Learning Approach. In: 2024 Smart Cities Symposium Prague, SCSP 2024 – Proceedings; 2024. DOI: 10.1109/SCSP61506.2024.10552687

Chapter 14

Perspective Chapter: From Industry 4.0 to 5.0 – Human-Centered Digitalization for Sustainable Heritage Tourism

Krisztián Attila Bakon

Abstract

This chapter examines how Industry 4.0 technologies can be strategically repurposed beyond manufacturing to support human-centered, sustainable, and culturally embedded experiences in industrial-heritage tourism. While cyber-physical systems, artificial intelligence, and large-scale digitization have traditionally prioritized efficiency and automation, their potential to advance social, cultural, and environmental objectives remains underdeveloped. Building on recent scholarship in human-centered digitalization and the evolving paradigm of Industry 5.0, we propose the Human-Centered Heritage Experience Design Model – a conceptual framework that integrates augmented reality, virtual reality, mobile applications, and artificial intelligence-enhanced systems within heritage-driven industrial contexts. The model repositions technology as a facilitator of participatory engagement, digital storytelling, and stakeholder co-creation, emphasizing cultural authenticity, community agency, and socio-technical sustainability. Through illustrative application scenarios, we demonstrate how immersive technologies can enable inclusive learning, strengthen place-based identity, and reduce environmental impacts by minimizing reliance on physical infrastructure. At the same time, we address implementation challenges such as infrastructure disparities, digital ethics, institutional inertia, and cultural commodification. Ultimately, this work reframes Industry 4.0 tools as enablers of post-industrial regeneration, arguing that human-centered digitalization offers a viable pathway for aligning technological innovation with cultural preservation, community resilience, and the green transition in heritage tourism ecosystems.

Keywords: human-centered digitalization, industrial-heritage tourism, Industry 4.0 and Industry 5.0, augmented and virtual reality, sustainability in cultural tourism, socio-technical design frameworks

1. Introduction

Digital transformation is reshaping the landscape of cultural heritage tourism, presenting both unprecedented opportunities and complex challenges. As heritage institutions and communities grapple with the integration of advanced technologies – such as augmented reality (AR), virtual reality (VR), artificial intelligence (AI), and mobile platforms – the need for frameworks that align innovation with cultural integrity and sustainability becomes increasingly urgent. While digital tools can expand access, enhance engagement, and support conservation, their implementation often reflects technocentric biases that marginalize community voices, overlook socio-cultural complexities, and prioritize novelty over authenticity.

This chapter responds to these tensions by advancing a holistic framework for designing digital heritage experiences that are inclusive, participatory, and resilient. Drawing on interdisciplinary scholarship across digital heritage tourism, human-centered design, stakeholder theory, sustainability science, and emerging Industry 5.0 paradigms, it critically reviews current approaches and identifies theoretical and practical gaps in the literature. In particular, the chapter highlights persistent shortcomings in stakeholder integration, co-creation practices, and long-term sustainability planning, revealing the fragmented nature of existing models and interventions.

To address these limitations, the chapter introduces the Human-Centered Heritage Experience Design Model (HC-HEDM), a conceptual and operational framework that repositions technology as an enabler of community-authored, culturally situated, and ecologically responsible heritage experiences. Grounded in iterative, stakeholder-driven processes and informed by the triple bottom line ethos, the HC-HEDM offers a structured yet flexible approach for aligning digital innovation with human values and place-based knowledge systems.

Through a comprehensive literature review and synthesis of applied case studies, this chapter establishes the theoretical foundations and empirical relevance of the HC-HEDM. It demonstrates how the model responds to urgent calls for digital sovereignty, ethical innovation, and sustainable development in heritage contexts. By bridging disciplinary silos and embedding co-creation at its core, the HC-HEDM contributes a replicable and adaptable blueprint for reimagining heritage tourism in the era of Industry 5.0.

2. Literature review

This section establishes the theoretical foundations underpinning the HC-HEDM by examining the convergence of digital heritage tourism, human-centered design methodologies, Industry 5.0 paradigms, stakeholder engagement frameworks, and sustainability approaches. The review synthesizes existing scholarship to identify theoretical gaps and empirical limitations that justify the need for an integrated framework addressing contemporary challenges in heritage tourism development.

2.1 Methodological note

This chapter draws on interdisciplinary scholarship across digital heritage tourism, human-centered design, Industry 5.0 paradigms, stakeholder engagement, and sustainability frameworks to establish theoretical foundations for the proposed HC-HEDM. Section 4 synthesizes published case studies to illustrate practical applications; no new fieldwork was conducted for this chapter.

2.1.1 Digital heritage tourism: Evolution and current paradigms

Digital heritage tourism has evolved from basic multimedia guides to sophisticated immersive experiences incorporating AR, VR, and AI technologies. Early digitalization efforts focused primarily on enhancing visitor information delivery, but contemporary applications seek to transform the fundamental nature of heritage encounters through interactive storytelling and personalized content curation [1–3].

Recent studies demonstrate that AR technologies in heritage museums enhance visitor engagement by providing immersive audio-visual experiences that extend and complement physical artifacts. The integration of AR overlays enables visitors to access additional interpretive layers, effectively serving as virtual interpreters that provide contextual information unavailable through traditional static displays. Zhang and colleagues found that VR and AR technologies enrich historical narratives, fostering visitor immersion and satisfaction while promoting sustainable development goals. These technologies represent a shift from passive consumption to active participation in heritage interpretation [1, 4].

The COVID-19 pandemic accelerated digital transformation in heritage tourism, catalyzing the development of hybrid models that combine physical and virtual experiences. This forced adaptation revealed both the potential and limitations of digital heritage platforms, highlighting the need for more sophisticated approaches that maintain cultural authenticity while expanding accessibility. Karaman's research on Turkish heritage sites demonstrates how VR, AR, and AI can enhance social media marketing while supporting sustainable heritage preservation by reducing tourism's environmental impact [2, 3].

Recent research highlights the growing role of AI in heritage tourism as a driver of content personalization, visitor profiling, and adaptive user experience design. Asif et al. (2025) emphasize that AI can support sustainability goals by optimizing tourist flows, reducing site congestion, and tailoring cultural content based on real-time environmental and user data. However, they caution that, without inclusive design principles and proper governance, AI may reinforce socio-economic exclusion and undermine cultural authenticity, especially in underserved or marginalized communities [5].

Mobile applications designed for heritage engagement increasingly use locative media and participatory features to deepen visitor connections and community awareness. Han et al. (2014) presented the "Lost State" mobile app, which integrated geo-tagged heritage content contributed by local residents, offering visitors interactive navigation through historical narratives. The study emphasized the value of involving communities in co-creating digital content to foster place-based learning and emotional attachment, particularly among youth and marginalized users. The authors found that participatory mobile platforms could democratize heritage interpretation and serve as tools for informal cultural education [6].

Recent research also explores the intersection of place attachment and gamified mobile applications in promoting proximate visitor engagement. Xia (2024) examined how location-based storytelling and mobile app design enhance users' sense of belonging to local heritage spaces. The study highlights the role of digital narratives in reimagining community identity and supporting regenerative tourism practices. By blending emotional engagement with digital interaction, heritage apps can bridge generational knowledge gaps and enable visitors to experience intangible cultural assets beyond static exhibits [7].

However, current digital heritage approaches often prioritize technological sophistication over cultural sensitivity and community engagement. The emphasis on immersive technologies can overshadow the fundamental purpose of heritage preservation and transmission, potentially leading to what Buragohain describes as the risk of detachment from authentic cultural narratives. This technological determinism underscores the need for frameworks that position technology as a facilitator rather than a driver of heritage experiences [3].

2.1.2 Human-centered design in cultural heritage contexts

Human-centered design has gained prominence in cultural heritage as institutions recognize the limitations of technology-driven approaches that fail to address diverse stakeholder needs and cultural protocols. Radice's foundational work on participatory design within cultural heritage emphasizes that contemporary audiences expect active engagement rather than passive consumption, necessitating collaborative approaches to experience design. The research demonstrates that participatory methodologies can enhance visitor experiences while ensuring that heritage narratives remain community-controlled and culturally appropriate [8].

Recent systematic reviews reveal a growing adoption of participatory design and co-design approaches in digital cultural heritage projects, yet implementation remains inconsistent across diverse contexts. The analysis of 40 publications spanning 2014–2024 indicates a recurring tension between advanced technological solutions and the need for authentic, community-driven design. Many projects follow generic participatory design approaches that predominantly involve users during evaluation stages rather than throughout the design process, limiting genuine co-creation opportunities [9].

Lupo's post-pandemic analysis of participatory approaches highlights the need to reframe cultural heritage engagement within a "participatory continuum of phygital proximity" that acknowledges the interconnection of digital and physical experiences. This perspective recognizes that heritage institutions must evolve from content providers to facilitators of experiences, supported by digital technologies that enable novel practices of audience engagement while preserving cultural integrity [10].

The human-centered design literature in heritage contexts reveals significant gaps in addressing power dynamics between institutions, communities, and visitors. While participatory methodologies are well-established theoretically, practical implementation often struggles to balance professional expertise with community knowledge, highlighting the need for frameworks that explicitly address these tensions through structured stakeholder engagement processes.

Recent work by Maye and Claisse (2023) provides a multilayered analysis of co-design practices in cultural heritage, illustrating how participatory strategies foster both community ownership and long-term engagement. Their study proposes that successful co-design depends on inclusive mechanisms that bridge intergenerational and intercommunity divides, and that digital tools must be tailored to support the socio-cultural dynamics of each heritage context. Importantly, they emphasize that co-design should be viewed not merely as a method but as a continuous dialogic process, evolving alongside shifting community priorities and institutional capabilities [11].

Similarly, Pisoni et al. (2021) argue for the integration of human-centered artificial intelligence (HCAI) in cultural heritage design, advocating for systems that respect community values, accessibility needs, and participatory decision-making. Their research outlines a framework in which HCAI technologies – such as adaptive interfaces and on-site interactive systems – enhance inclusivity by enabling users to personalize their engagement without compromising cultural protocols. The authors stress the importance of co-developing AI-driven tools with stakeholders to avoid digital colonialism and ensure that innovation aligns with community-defined heritage goals [12].

2.1.3 Industry 5.0 and the human-centric paradigm shift

Industry 5.0 represents an evolution beyond the digitization focus of Industry 4.0, emphasizing human-centricity, sustainability, and resilience alongside technological advancement. This paradigm shift is particularly relevant for heritage tourism, which is fundamentally a human-centric service industry that must balance technological innovation with authentic cultural transmission and community empowerment [13, 14].

Recent literature highlights the increasing role of Industry 5.0 principles – particularly human-centric design, environmental sustainability, and social resilience – in shaping innovation strategies across various sectors, including tourism. Fraga-Lamas and Fernandez-Carames (2024) argue that transitioning to Industry 5.0 requires a rethinking of earlier automation paradigms to center human agency, inclusive innovation, and cultural responsibility. Their work emphasizes the integration of blockchain and digital technologies to create ecosystems that are not only smart but also ethically accountable and socially inclusive – a critical orientation for heritage tourism, where stakeholder diversity and cultural authenticity are paramount [15].

Ghobakhloo et al. (2024) similarly position Industry 5.0 as a move toward a "digital society" rather than just digital manufacturing. They propose a roadmap where digital technologies are leveraged not for efficiency alone but for building resilient, human-centered experiences. This reconceptualization has direct relevance for heritage tourism, where the visitor experience must harmonize immersive technology with the preservation of intangible and tangible cultural heritage [16].

Research in built cultural heritage conservation reveals that sustainability emerges as the most widely discussed Industry 5.0 principle, while resilience and human-centrism remain less developed in both theory and practice. Jiménez Rios and colleagues found that conservation professionals recognize the potential benefits of Industry 5.0 principles, including enhanced resource efficiency and adaptability, but face significant barriers, including lack of knowledge, inadequate training, and resource constraints [13, 17].

The tourism industry's transition toward Industry 5.0 accelerated during the COVID-19 pandemic, which catalyzed rapid digitalization while highlighting the importance of human-centric approaches. This evolution is characterized by a shift from merely digitalized tourism experiences to "sustainable smart tourism," where technological advancements serve human needs, promote environmental stewardship, and foster social equity. The convergence of technological sophistication with purposeful innovation creates more resilient, sustainable, and human-centered tourism ecosystems [14].

However, the application of Industry 5.0 principles to heritage tourism remains largely theoretical, with limited empirical evidence of successful implementation. The literature reveals a need for practical frameworks that operationalize Industry 5.0 values – especially inclusivity, co-creation, and sustainable innovation – within cultural contexts. More research is needed to translate these concepts into actionable strategies for heritage institutions, technology partners, and communities alike.

2.1.4 Stakeholder theory and community engagement in heritage tourism

Stakeholder theory provides essential frameworks for understanding the complex relationships between diverse actors in heritage tourism development. Aas et al.'s (2005) examination of collaborative approaches in Luang Prabang, Laos, demonstrates that effective heritage management requires systematic stakeholder engagement that addresses power dynamics, communication channels, and benefit distribution. The research identifies five critical aspects: communication between heritage and tourism groups, income generation for conservation, community involvement in decision-making, participation in tourism activities, and collaborative management effectiveness [18].

Community-based heritage tourism models emphasize the central role of local communities as both beneficiaries and stewards of heritage resources. Wijayanti's research reveals that community participation is driven by perceptions across multiple dimensions, including heritage diversity awareness, economic impact understanding, stakeholder engagement quality, conservation importance, tourism experiences, and development strategies. The study demonstrates that perception significantly influences participation, but heritage tourism management, stakeholder engagement, and tourism experiences require strengthened performance to motivate community involvement [19].

Recent studies highlight the importance of addressing power imbalances between different stakeholder groups. Alhadad's research on community involvement in heritage management identifies significant challenges in achieving balance among stakeholder goals, noting that "power-holders," such as governmental organizations, remain reluctant to distribute authority to local community members. This finding underscores the need for frameworks that explicitly address governance structures and decision-making processes [20].

Additional studies further underscore the importance of integrated stakeholder collaboration and community-based models in sustaining heritage tourism. Li and Hunter (2015) proposed a conceptual model where sustainable tourism development hinges on community empowerment, transparent stakeholder relations, and long-term local capacity building. They emphasize that stakeholder engagement must go beyond superficial consultation to involve communities in decision-making and benefit-sharing processes [21]. Similarly, Esichaikul and Chansawang (2022) documented how multistakeholder collaboration in Thailand's Sukhothai Historical Park, facilitated by the Designated Areas for Sustainable Tourism Administration (DASTA), led to the successful implementation of a community-based tourism framework. This model not only improved communication between government and local actors but also fostered local entrepreneurship and cultural stewardship, demonstrating the operational potential of stakeholder-centric governance in heritage site management [22].

The stakeholder literature reveals recurring themes around the need for long-term engagement, capacity building, and benefit-sharing mechanisms. However, most studies focus on individual stakeholder relationships rather than ecosystem-wide approaches that address interdependencies and feedback loops among multiple actor groups simultaneously [18, 19].

2.1.5 Sustainability paradigms in digital heritage tourism

Sustainability integration in heritage tourism has advanced from isolated environmental concerns to encompassing social, economic, and environmental dimensions within triple bottom line approaches [23]. This paradigm challenges traditional profit-centric tourism models by incorporating people and the planet alongside financial performance.

In heritage tourism contexts, sustainability integration addresses multiple interconnected challenges. Environmental sustainability focuses on heritage site conservation, carbon footprint reduction, and ecosystem protection through technologies that minimize physical impact while maintaining educational value [2, 13]. Social sustainability emphasizes community empowerment, cultural continuity, and equitable access to heritage resources and economic benefits. Economic sustainability ensures that tourism development generates lasting benefits for local communities while supporting heritage preservation activities [14, 19].

Digital technologies present both opportunities and challenges for sustainable heritage tourism development. VR and AR applications can reduce physical wear on fragile sites while expanding global access to cultural heritage [2, 3]. However, the digital divide creates new forms of exclusion, with socio-economic status, cultural relevance, and trust in digital systems shaping access to digitally mediated heritage experiences. Research indicates that existing digital tourism platforms can exacerbate inequalities between global digital elites and local, less-connected communities, potentially increasing economic disparity within destinations [24].

The sustainability literature in heritage tourism reveals significant gaps in understanding the long-term impacts of digital interventions. While short-term benefits are well-documented, empirical evidence of sustained community empowerment, cultural preservation, and environmental protection through digital heritage tourism remains limited.

Recent empirical research reinforces the potential of digital heritage technologies in advancing sustainability goals when implemented with cultural sensitivity and local collaboration. Yeniasir and Gökbulut (2022) found that digital storytelling and visualization tools used in Northern Cyprus significantly enhanced cultural tourism while fostering local identity and reducing environmental degradation associated with mass tourism [25]. Similarly, Gonçalves et al. (2022) demonstrated how community-driven digital initiatives in Tavira, Portugal supported the preservation of intangible cultural heritage by promoting sustainable tourism practices through inclusive social media strategies [26]. These studies underscore the importance of aligning digital innovation with participatory governance and place-based authenticity to achieve long-term sustainability in heritage tourism.

To ground human-centered digitalization within industrial-heritage practice, we note complementary strands: (i) industrial site reuse and industrial-heritage tourism foreground issues of interpretation, risk, and landscape valorization [27, 28]; (ii) ecomuseum scholarship stresses community custodianship and place-based co-governance [29]; and (iii) the tourism literature on authenticity and commodification offers critical lenses for avoiding reductive narrative packaging

[30, 31]. These strands align with HC-HEDM's emphasis on community agency, cultural fidelity, and multiactor governance.

2.1.6 Synthesis and research gaps

The preceding review highlights meaningful advances across five interrelated domains – digital heritage technologies, human-centered design, Industry 5.0 paradigms, stakeholder engagement, and sustainability frameworks. While each area has contributed valuable theoretical insights and practical tools, the literature remains notably fragmented. Few studies bridge these domains in a systematic and integrated manner, resulting in heritage tourism interventions that are often technologically sophisticated but socially or culturally incomplete.

Three overarching research gaps emerge from this analysis. First, much of the existing work in digital heritage tourism is driven by technological determinism, emphasizing novelty and immersion at the expense of cultural authenticity and community agency. This imbalance risks reducing heritage to entertainment content, undermining its social and epistemological value. Second, while participatory and human-centered designs are increasingly advocated, they are rarely operationalized in ways that address entrenched power asymmetries between institutions, communities, and visitors. Many co-design efforts remain superficial, limiting local stakeholders to consultative roles rather than enabling meaningful co-ownership. Third, although the triple bottom line is widely referenced, sustainability is often approached as a rhetorical add-on rather than as a guiding principle. There is a lack of operational frameworks that translate environmental, social, and economic sustainability into actionable metrics within the digital heritage tourism life cycle.

Together, these gaps underscore a critical need for a unified framework that reconceptualizes digital heritage tourism as a human-centered and ethically grounded practice. Such a model must integrate technology not as an end in itself, but as a facilitator of inclusive, culturally embedded, and environmentally responsible heritage experiences. It must also embed mechanisms for community governance, adaptive design, and ongoing impact evaluation to ensure long-term viability.

The HC-HEDM proposed in this work directly addresses these shortcomings. By merging participatory design methods, stakeholder theory, and sustainability principles within a dynamic, iterative structure, the HC-HEDM offers a pragmatic roadmap for aligning technological innovation with cultural stewardship. It advances the discourse by moving beyond theoretical advocacy to provide a structured, scalable, and context-sensitive approach to designing and managing digital heritage tourism experiences that are both meaningful and sustainable.

3. Conceptual framework: HC-HEDM

This section introduces the HC-HEDM, an integrative framework grounded in human-centered design, stakeholder theory, and sustainability science. Rather than treating digital tools as standalone innovations, the HC-HEDM repositions technology as an enabler of community-authored, culturally situated, and ecologically responsible heritage experiences. Through this model, the design of digital heritage tourism transitions from being technology-driven to stakeholder-led and impact-oriented.

3.1 Stakeholder ecosystem mapping

Effective heritage experience design begins with a nuanced understanding of the ecosystem of actors whose values and goals shape both development and outcomes. Heritage visitors today demand individualized engagement, seeking immersive encounters that resonate with personal interests and learning preferences. Local communities serve as custodians of intangible traditions and must be engaged not merely as informants but as co-creators, ensuring that digital narratives reflect community memory and cultural protocols. Heritage managers – including museum curators and site administrators – are evolving from gatekeepers to facilitators, orchestrating collaborative governance of content and access. Technology partners contribute essential expertise in AR/VR, mobile development, AI, and IoT, but must align innovations with cultural sensitivities, accessibility needs, and local priorities. As illustrated in **Figure 1**, these four primary stakeholder groups – heritage visitors,

Figure 1.
The HC-HEDM: stakeholder-driven design process integrating technology, human values, and sustainability principles across iterative phases.

local communities, heritage managers, and technology partners – form the core of the HC-HEDM ecosystem. This structure reflects the interdependence of experiential, cultural, managerial, and technical domains. Drawing from stakeholder salience theory [32], the HC-HEDM promotes adaptive engagement strategies based on power, legitimacy, and urgency, recognizing that stakeholder influence fluctuates dynamically across the design life cycle. Across these groups, power and interest fluctuate throughout the design process. Employing a dynamic power-interest matrix enables practitioners to calibrate engagement strategies: high-power, high-interest actors co-lead decision-making, whereas those with high interest but less influence contribute through targeted consultation. This approach fosters shared ownership, builds trust, and mitigates conflict as design progresses from conception to deployment.

3.2 Design process: Explore–express–evaluate

The HC-HEDM follows a cyclical, co-creative design process – explore, express, and evaluate – adapted from human-centered design frameworks [33]. This loop, visually represented in the inner ring of **Figure 1**, guides interdisciplinary teams through iterative stages of contextual research, co-design, and impact assessment. During the explore phase, interdisciplinary teams conduct ethnographic fieldwork, user interviews, and participatory mapping workshops to uncover community values, site constraints, and visitor expectations. This groundwork illuminates both tangible and intangible assets – from architectural features to oral histories – while identifying technological opportunities and infrastructural limits. Building on these insights, the express phase convenes co-creative prototyping sessions in which community members, visitors, and technologists collaborate to shape narrative frameworks and interaction models. Through low-fidelity storyboards, AR mockups, and mobile app wireframes, stakeholders negotiate the balance between immersive storytelling and cultural authenticity, ensuring that digital overlays and AI-driven personalization reinforce, rather than overshadow, local meanings. Sustainability considerations are woven into every design decision, prompting choices of materials, energy sources, and content that minimize ecological footprints and support equitable benefit-sharing. Finally, the evaluate phase employs mixed-methods assessment – combining visitor surveys, cultural impact studies, and sustainability metrics – to measure effectiveness across multiple dimensions. Continuous feedback loops allow rapid refinement: if an AR experience misrepresents community narratives or a mobile interface excludes older users, the design team revisits earlier phases to adjust content, interaction flows, and accessibility features. This cyclical process embeds resilience, ensuring that the heritage experience evolves responsively to stakeholder needs and environmental changes. Following the evaluate phase, the execute phase represents the full deployment and scaling of the co-designed heritage experience, incorporating lessons learned from the pilot implementations and ensuring sustainable, long-term operation.

3.3 Technology layer

Within the HC-HEDM, digital technologies form a supportive layer that operationalizes design outcomes. AR interfaces supplement on-site exploration with culturally appropriate overlays – such as multilingual captions and 3D reconstructions – while respecting sacred protocols and ensuring universal access. Immersive VR scenarios extend reach to remote users or fragile sites, offering empathetic re-enactments of historical events without physical impact. Complementing immersive media, AI algorithms tailor content recommendations based on visitor profiles, leveraging natural language processing to deliver accessible, context-sensitive narratives. Mobile applications equipped with offline modes and low-bandwidth options ensure that users in areas with limited connectivity can still participate fully. Finally, IoT sensors monitor environmental conditions – such as humidity and foot traffic – and feed real-time data to both community stewards and heritage managers, enabling proactive conservation interventions. These digital solutions must be designed in alignment with stakeholder roles. For instance, AR applications enhance visitor interpretation; AI supports personalized learning paths; IoT empowers heritage managers through real-time monitoring. European immersive heritage initiatives illustrate how co-developed AR/VR platforms can foster cultural empathy and operational sustainability when community actors guide content and design choices [34, 35].

3.4 Human dimension

Surrounding technology, the human dimension safeguards cultural integrity and fosters community agency. Central to this layer is the principle of digital sovereignty – the right of communities to control how their cultural narratives are represented, distributed, and interpreted through digital means. Community-led storytelling platforms not only preserve intangible heritage but also act as vehicles for intergenerational knowledge transfer, cultural resilience, and identity negotiation.

3.5 Sustainability integration

Encapsulating all layers, sustainability integration applies a triple bottom line ethos – encompassing environmental integrity, social cohesion, and economic resilience [23]. As shown in the outer ring of **Figure 1**, this layer reflects a circular and regenerative approach to heritage tourism, guiding both operational decisions and long-term strategy. Strategies include virtual visitation to manage ecological carrying capacity, localized training to build socio-cultural resilience, and community-centered value chains to ensure that economic returns stay within the host region. Participatory dashboards – co-managed by stakeholders – support adaptive governance by revealing patterns in visitor flows, cultural use, and environmental pressure points. When risks are identified, such as overuse or misrepresentation, stakeholders co-create timely corrective actions. In sum, the HC-HEDM offers a holistic, theory-informed model with layered interdependencies and dynamic feedback mechanisms. By centering human values and systemic sustainability, the model provides a practical blueprint for designing, implementing, and evaluating digital heritage experiences. Its modular structure enables adaptability across geographic and cultural contexts, offering potential for broader replication, comparative evaluation, and future empirical validation.

3.6 Operationalizing the HC-HEDM: From concept to practice

Having established the theoretical foundations and conceptual architecture of the HC-HEDM, this section transforms the framework into actionable guidance for practitioners and site managers. **Table 1** operationalizes each phase of the design process by mapping specific stakeholder roles, concrete activities, practical instruments, and measurable indicators, enabling systematic deployment across diverse heritage contexts.

As illustrated in **Table 1**, each phase involves distinct but overlapping stakeholder groups, ensuring continuous community engagement while leveraging specialized expertise. The progression from exploration through execution reflects the iterative, human-centered design principles embedded in the HC-HEDM, with indicative metrics providing concrete measures for evaluating cultural fidelity, accessibility, and sustainability impacts. The execute phase addition responds to practical implementation needs identified in the literature review, ensuring that co-designed heritage experiences achieve long-term viability through proper deployment protocols, staff training, and ongoing monitoring systems. This four-phase structure provides practitioners with acomprehensive roadmap from initial community engagement through to sustainable operation.

4. Practical application scenarios and case studies

This section illustrates how the HC-HEDM can be operationalized across diverse heritage contexts through the application of emerging digital technologies. The selected case studies demonstrate the model's relevance in addressing real-world

Phase	Primary Roles	Sample Activities/Instruments	Indicative Metrics (examples)
Explore	Communities, visitors, site managers	Ethnographic interviews; participatory mapping; accessibility audits (WCAG); stakeholder salience/power–interest matrix; inventory of tangible/intangible assets.	Stakeholder diversity index; % of underserved groups represented; accessibility conformance checklist; number of community-sourced narratives.
Express	Communities, tech partners, site managers	Co-design workshops; storyboarding; low-fi AR mockups; VR scene scripts; mobile wireframes; content governance charter.	Cultural fidelity score (community review: 1–5); prototype SUS score; % content co-authored by the community; estimated energy use/device requirements.
Evaluate	All actors	Mixed-methods assessment; on-site/off-site A/B tests; sustainability dashboard; governance retrospectives.	Learning outcomes (pre/post); congestion reduction; virtual/physical substitution ratio; benefit-sharing equity indicator; carbon savings estimate.
Execute	Site managers, tech partners, governance bodies.	Full system deployment; staff training programs; maintenance protocols; revenue-sharing agreements; long-term monitoring systems; scalability planning.	System uptime %; visitor satisfaction scores; revenue distribution equity; environmental impact reduction; staff competency levels; scalability success rate.

Table 1.
HC-HEDM as aready-to-use guide for practitioners: phases, stakeholder roles, activities, and indicative metrics.

challenges and opportunities, particularly through the integration of human-centered design, stakeholder participation, and sustainability objectives.

4.1 Local site revitalization: AR-enhanced community storytelling

AR technologies have shown considerable promise in revitalizing local heritage sites by enabling layered, interactive storytelling. In a study on AR travel apps for heritage tourism in India, Singh et al. (2024) [36] found that service quality, personal innovativeness, and perceived complexity significantly influenced users' behavioral intention to adopt AR applications. The app, which overlaid 3D architectural reconstructions onto heritage sites such as the Taj Mahal, was accessible via smartphones and designed for eco-friendly deployment. Structural equation modeling revealed that personal innovativeness was the strongest predictor of adoption, while perceived complexity negatively impacted use. Notably, sustainability concerns amplified the positive effects of service quality and innovativeness, underscoring the potential of AR not only to enhance visitor engagement but also to promote environmentally conscious design. These findings align with the HC-HEDM's explore–express–evaluate cycle by demonstrating how user preferences and contextual constraints shape both design and deployment strategies.

4.2 Immersive VR for cultural heritage education and access

VR technologies are increasingly being leveraged to support inclusive education and site conservation in heritage tourism. Buragohain et al. (2024) [3] examined a set of VR initiatives featuring photorealistic 3D reconstructions of culturally significant sites. These experiences were designed for both remote and on-site use, enabling broader participation among geographically distant users and mobility-impaired audiences. Importantly, VR use reduced physical interaction with fragile environments, thereby contributing to long-term conservation objectives. The study highlights how VR can function as a digital surrogate, offering immersive educational experiences while minimizing ecological impact. This reflects the HC-HEDM's sustainability integration layer, where technological solutions are aligned with environmental stewardship and access equity.

4.3 Urban cultural trails: Participatory mobile applications

In urban heritage contexts, mobile applications serve as platforms for participatory storytelling and localized engagement. In Barcelona's El Raval district, Murphy et al. (2022) [37] reported on the co-creation of a universal-design mobile application developed in collaboration with local artists, residents, and community organizations. The app allowed users to filter walking trails by interest categories (e.g., architecture, urban art, gastronomy) and functioned offline to accommodate connectivity limitations. Geotagged, community-contributed multimedia content enabled underrepresented narratives to surface, reinforcing a sense of cultural ownership among users. This participatory model directly supports the HC-HEDM's human dimension and stakeholder engagement strategies, emphasizing co-creation, local memory, and inclusive access.

4.4 Remote heritage experiences: Comparing VR and AR impacts

Comparative evaluations of immersive technologies provide critical insights for strategic deployment. Guo et al. (2024) [38] conducted an experimental study to compare the effects of AR and VR on user engagement in heritage tourism. Using the stimulus–organism–response framework, they found that VR's visual vividness enhanced perceived authenticity, while AR's interactivity was more effective at fostering presence. Both modalities influenced users' behavioral intentions positively, though no significant difference emerged regarding overall satisfaction or future visitation intent. These findings suggest that technology selection should be context-sensitive, depending on whether emotional immersion or real-time interaction is prioritized – an insight that supports the HC-HEDM's technology layer, where tools are tailored to situational and stakeholder needs.

Collectively, these case studies demonstrate the applied relevance of the HC-HEDM in designing heritage tourism experiences that balance technological innovation with cultural sensitivity and sustainable development. Each example reflects one or more dimensions of the model – whether through stakeholder co-design, context-aware technological implementation, or operational sustainability practices. By aligning digital tools with the socio-cultural realities of heritage environments, these applications exemplify how human-centered digitalization can translate theory into meaningful, context-responsive practice.

5. Implementation challenges, limitations, and strategic responses

While the HC-HEDM provides a theoretically grounded and operationally robust framework, its implementation across diverse heritage contexts is not without obstacles. This section critically examines the technical, social, institutional, and ethical challenges inherent in deploying the model, while proposing strategic responses aligned with its core principles of inclusivity, sustainability, and stakeholder co-governance.

A primary technical challenge stems from the persistent digital divide between urban centers and rural or economically disadvantaged regions. In areas with limited Internet infrastructure, low device penetration, or unreliable electricity, immersive experiences enabled by AR, VR, or AI may remain aspirational rather than feasible. This asymmetry risks undermining the HC-HEDM's human-centered ethos by excluding marginalized communities from meaningful participation in design and engagement.

To address this, design teams should adopt universal design and graceful degradation strategies that enable accessibility under constrained conditions. For example, mobile applications with pre-cached content, marker-based AR interactions, and lightweight multimedia formats can deliver narrative richness while minimizing bandwidth requirements. Integrating offline functionality and energy-efficient platforms ensures that the model remains viable across infrastructural contexts without compromising on experience quality or cultural fidelity.

As digital heritage systems increasingly rely on user data – ranging from geolocation and behavioral patterns to biometric feedback – new ethical and legal challenges emerge. Without rigorous data protection frameworks, the collection and processing of sensitive information can infringe upon user privacy and community sovereignty, particularly in contexts where cultural knowledge is considered sacred or non-transferable.

To safeguard against these risks, co-developed data governance protocols must be embedded in project life cycles. These should include consent-driven data architectures, the use of edge computing for on-device processing, anonymization standards, and community-led oversight mechanisms. Such safeguards align with the HC-HEDM's commitment to digital sovereignty, ensuring that technological personalization and analytics serve local values rather than extractive interests.

Social resistance poses another major challenge, particularly in communities where oral traditions, embodied practices, or analog forms of knowledge transmission dominate. Skepticism toward digital mediation is often grounded in past experiences of extractive research or exclusion from decision-making processes. Additionally, disparities in digital literacy or generational familiarity with technology may hinder meaningful community engagement.

Strategically, the HC-HEDM responds by advocating for long-term capacity-building and embedded co-design. Trust must be cultivated through transparent partnerships, repeated engagements, and shared authorship of content. Embedding local cultural practitioners within project teams not only enhances representational integrity but also counters epistemic asymmetries. Demonstrating tangible benefits – such as community-controlled tourism revenue or localized employment opportunities – reinforces the social legitimacy of digital interventions.

Heritage institutions themselves often face internal limitations – budget constraints, lack of digital expertise, and rigid bureaucratic structures – that can inhibit the adoption of human-centered and technologically enriched approaches. In many cases, preservation mandates are prioritized over innovation, and risk-averse governance cultures discourage experimentation.

Addressing these barriers requires inter-institutional alliances and hybrid governance models. Partnerships with academic institutions, startups, and technology incubators can provide access to funding, expertise, and research infrastructure. Fellowship programs and interdisciplinary knowledge exchanges can strengthen institutional capacity. Moreover, by positioning digital heritage tools as complements to, rather than replacements for, traditional conservation practices, resistance can be transformed into constructive engagement. Mature institutions can act as incubators, mentoring less-resourced peers through collaborative consortia.

A persistent risk in heritage tourism is the commodification of culture – where narratives are simplified, romanticized, or exoticized to appeal to commercial markets. Such distortions compromise cultural authenticity, erode community trust, and contribute to the replication of colonial patterns of representation.

The HC-HEDM addresses this risk through its multivocal narrative model, which mandates inclusive content governance and critical historical framing. Storyline audits conducted by community advisory boards, participatory editorial processes, and mechanisms for revising narratives in response to community feedback are all essential components. These ensure that content not only represents diverse perspectives but also evolves dynamically in accordance with community priorities and cultural sensitivities.

Importantly, the challenges outlined above do not occur in isolation – they are interdependent and often reinforce one another. For example, institutional resistance may exacerbate technological exclusion, while ethical lapses may amplify social resistance. As such, addressing these challenges requires not only discrete technical fixes but also the cultivation of resilience at the system level.

This calls for adaptive governance mechanisms, iterative evaluation protocols, and anticipatory design practices that embed flexibility and reflexivity into the development process. By treating the HC-HEDM as a living framework – responsive to shifting contexts and emerging risks – practitioners can move beyond prescriptive models toward resilient heritage ecosystems capable of sustaining cultural value in an age of digital transformation.

In sum, while technological, social, and institutional barriers remain formidable, they are not insurmountable. When addressed holistically – through inclusive co-design, ethical governance, strategic partnerships, and system-level adaptability – the HC-HEDM can be effectively deployed to advance digital heritage tourism that is not only innovative but also just, sustainable, and community-empowering.

6. Future directions and open questions

As the field of human-centered digital heritage tourism matures within the emergent Industry 5.0 paradigm, it confronts a rapidly shifting terrain defined by technological acceleration, evolving visitor expectations, and the imperative of sustainability. The HC-HEDM offers a structured foundation for navigating these complexities, yet several frontier domains demand deeper inquiry and strategic development. This section outlines key trajectories for future research and practice, organized around five critical dimensions: emerging technologies, digital ethics, resilience, interdisciplinarity, and evaluation methodologies.

The convergence of AI, extended reality (XR), and blockchain technologies with cultural heritage introduces novel affordances for enhancing interpretive depth, participatory design, and operational scalability. AI-driven personalization is poised to transform heritage experiences by dynamically adapting content to individual preferences, learning styles, and accessibility profiles. Through real-time data analytics, these systems can also support adaptive crowd management, environmental monitoring, and content optimization – critical functions for sustainable tourism ecosystems.

Meanwhile, XR technologies – encompassing augmented, virtual, and mixed reality – enable immersive storytelling that transcends spatial and physical constraints. These tools offer unprecedented opportunities to engage remote or mobility-impaired audiences, simulate fragile or vanished heritage environments, and experiment with multisensory narrative formats. As hardware becomes more affordable and authoring tools more intuitive, XR integration within community-led design processes will become not only feasible but essential.

Blockchain and distributed ledger systems also merit deeper exploration. Their potential for transparent provenance tracking, fair compensation, and decentralized governance could redefine cultural ownership and intellectual property in digital heritage systems – particularly for Indigenous or historically marginalized communities. However, empirical evidence on real-world applications remains limited and is urgently needed.

As digital interventions become more sophisticated, so too do their ethical implications. Concerns over algorithmic bias, surveillance, data privacy, and the commodification of culture must be addressed not only at the technical level but also through participatory and anticipatory governance frameworks. There is a pressing need for models of digital ethics-by-design that align with community values and legal protections, particularly in contexts where sacred knowledge or oral traditions are at risk of misappropriation.

Participatory governance – already embedded within the HC-HEDM – must evolve to include oversight mechanisms for algorithmic transparency, consent-based data practices, and inclusive access. Special attention should be given to the digital marginalization of older adults, persons with disabilities, and low-income users, for whom current platforms may be inaccessible or culturally alienating.

The COVID-19 pandemic revealed the systemic vulnerabilities of heritage tourism and catalyzed interest in hybrid and virtual modalities. Future models must incorporate resilience as a core design value, not an emergency retrofit. This includes developing adaptive governance structures, investing in decentralized community-based infrastructure, and embedding scenario planning into design processes.

Long-term resilience also entails environmental adaptability. Climate change poses significant risks to heritage sites globally, from sea-level rise to extreme weather. Digital surrogates and remote access platforms – when governed equitably – can reduce physical strain on sites while maintaining interpretive richness. However, future work must consider not only ecological sustainability but also the carbon footprint of digital systems themselves, including server loads, data storage, and device obsolescence.

The complex interplay of technology, culture, ecology, and economics in digital heritage tourism necessitates robust interdisciplinary collaboration. Yet, beyond academic interdisciplinarity, there is a growing demand for transdisciplinary models that integrate not only researchers but also community stakeholders, practitioners, and policy-makers as co-producers of knowledge.

Future frameworks should actively incorporate methodologies from anthropology, environmental science, design research, critical heritage studies, and human-computer interaction. This epistemological hybridity is essential to producing both culturally sensitive and technically feasible solutions that can scale across diverse geographies and heritage types.

Finally, the field must advance beyond anecdotal or short-term assessments toward systematic, mixed-methods evaluation frameworks capable of capturing the long-term social, cultural, economic, and environmental impacts of digital heritage interventions. This includes developing standardized, yet context-sensitive metrics aligned with the HC-HEDM's layered dimensions: stakeholder engagement quality, cultural fidelity, technological usability, and sustainability performance.

Longitudinal studies are particularly needed to understand how digital experiences influence community well-being, intergenerational knowledge transfer, and equitable benefit-sharing over time. Impact frameworks should also account for unintended consequences, such as technology abandonment, cultural fatigue, or platform misuse.

The future of human-centered digital heritage tourism lies not in the unchecked proliferation of immersive technologies, but in the deliberate alignment of innovation with cultural responsibility, ethical governance, and sustainability goals. The HC-HEDM provides a foundational blueprint for this alignment, but its continued relevance will depend on how it evolves in response to emerging technologies, shifting power relations, and global uncertainties.

Key imperatives for future research and practice include the following:

- Operationalizing digital ethics and inclusive governance within co-design workflows;
- Exploring the sustainability implications of digital infrastructures;
- Building resilience into both content and infrastructure;
- Deepening cross-sectoral, transdisciplinary collaboration; and
- Developing robust metrics for evaluating cultural, social, and ecological impacts.

Only through such efforts can digital heritage tourism fulfill its promise?: not simply to digitize the past, but to co-create futures where cultural memory, technological agency, and human dignity converge.

7. Conclusions and future research directions

This chapter has introduced the HC-HEDM as an integrative framework that repositions digital heritage tourism through the convergence of human-centered design, stakeholder theory, and sustainability science. By placing heritage visitors, local communities, site managers, and technology partners within a dynamic explore–express–evaluate process, HC-HEDM enables the design of culturally grounded, socially inclusive, and ecologically responsible heritage experiences.

The HC-HEDM advances the field by operationalizing abstract concepts – such as participatory design, stakeholder salience, and the triple bottom line – into a structured methodology applicable across diverse heritage contexts. It reconceptualizes technology not as an end in itself, but as a cultural mediator and facilitator of co-creation. The model also embeds ethics, sustainability, and power-awareness as foundational elements, addressing long-standing limitations of technocentric and top-down approaches in digital heritage research.

Case applications – from community-authored AR storytelling in Finland to VR-enabled conservation in Peru – demonstrate the model's scalability and contextual flexibility. These implementations validate HC-HEDM's capacity to foster shared authorship of narratives, reduce ecological burdens through digital surrogacy, and build local capacity in digital heritage content creation.

Despite its strengths, the implementation of HC-HEDM faces structural barriers. Digital divides, particularly in rural or marginalized contexts, continue to constrain access to immersive and personalized technologies. Cultural resistance, lack of digital literacy, and institutional conservatism can impede co-design processes and stakeholder alignment. Data privacy, algorithmic opacity, and the risk of cultural commodification further complicate the ethical deployment of AI and XR in sensitive heritage settings.

These challenges point to unresolved research gaps. There remains a need to:

- Operationalize HC-HEDM in low-resource environments, identifying design adaptations for minimal infrastructure;
- Investigate privacy-preserving AI and consent-aware personalization that respects cultural sovereignty;
- Develop evaluation frameworks that go beyond usability to assess long-term impacts on cultural vitality, social equity, and environmental performance; and
- Refine governance models that allow for community-led decision-making and adaptive design.

Emerging technologies such as blockchain, AI-generated storytelling, and metaverse-based cultural immersion present exciting but complex opportunities. Integrating these into HC-HEDM will demand new theoretical anchoring in areas such as digital ethics, post-colonial critique, and co-production theory. Moreover, ensuring that these tools serve local knowledge systems and do not replicate extractive logics will be central to future research.

Equally important is the cultivation of cross-sectoral consortia that bring together academic researchers, community organizations, heritage professionals, and tech developers. These alliances can serve as innovation ecosystems for testing and iterating human-centered digital heritage strategies under real-world conditions.

Ultimately, the HC-HEDM offers a replicable yet adaptable blueprint for bridging the gap between technological innovation and cultural stewardship. It contributes both theoretically and practically to the evolving discourse on Industry 5.0, digital heritage, and sustainable tourism. By embedding participation, ethics, and resilience into the DNA of digital heritage design, HC-HEDM enables the creation of experiences that are not only technologically sophisticated but socially just, culturally authentic, and future-ready.

This chapter synthesizes secondary cases and conceptual literature; it does not report new fieldwork. While HC-HEDM is designed to be context-adaptable, validation to date is indicative and site-specific.

Two actionable steps are proposed for practitioners and site managers: (1) a small co-design pilot applying HC-HEDM at a selected industrial-heritage site (covering the full explore–express–evaluate–execute cycle with a community advisory group); and (2) a pre-/post-assessment using the indicators in **Table 1** to track cultural fidelity, accessibility, congestion reduction, and environmental gains (e.g., virtual/physical substitution ratio and carbon savings).

The task now is not merely to apply this model but to continuously evolve it – through empirical research, critical reflection, and pluralistic engagement – so that digital heritage tourism becomes a domain where innovation serves people, communities, and the planet alike.

Author details

Krisztián Attila Bakon
Department of Applied Informatics, Faculty of Information Technology, University Center for Circular Economy Nagykanizsa, University of Pannonia, Nagykanizsa, Hungary

*Address all correspondence to: bakon.krisztian@pen.uni-pannon.hu

References

[1] Chen Y, Wang X, Le B, Wang L. Why people use augmented reality in heritage museums: A socio-technical perspective. Heritage Science. 2024;**12**(1):108

[2] Karaman MK, Deniz Ö. The role of VR, AR, and Artificial Intelligence in the sustainable preservation of cultural heritage: Social media marketing of historical sites, new media education in Turkey, and their contributions to SDGs. Journal of Lifestyle and SDGs Review. 2025;**5**(3):e05260–e05260

[3] Buragohain D, Meng Y, Deng C, Li Q, Chaudhary S. Digitalizing cultural heritage through metaverse applications: Challenges, opportunities, and strategies. Heritage Science. 2024;**12**(1):295

[4] Zhang Y, Papp-Váry Á, Szabó Z. Digital engagement and visitor satisfaction at world heritage sites: A study on interaction, authenticity, and recommendations in coastal China. Administrative Sciences. 2025;**15**(3):110

[5] Asif F, Jamshed K, Shah SHA. Prospects and challenges for AI and technological innovation in the European travel and tourism sector: Impact, effect and sustainable development goals. Festival and Event Tourism and Hospitality Series. 2025;140–161

[6] Han K, Shih PC, Rosson MB, Carroll JM. Enhancing community awareness of and participation in local heritage with a mobile application. In: Proceedings of the 17th ACM Conference on Computer Supported Cooperative Work & Social Computing, CSCW 14, New York, NY, USA: Association for Computing Machinery; 2014. p. 1144–1155

[7] Xia S. Narratives of place: Blending place attachment and digital innovation to enhance proximate visitor engagement: A research project on reimagining local exploration in collaboration with be here then, 2024

[8] Radice S. Designing for participation within cultural heritage. Participatory practices and audience engagement in heritage experience processes. 2014

[9] Nikolakopoulou V, Koutsabasis P. The 'making' of participatory and co-design for digital experiences in cultural heritage: A review. CoDesign. 2025;**21**(4):876–902

[10] Lupo E. Design and the reframing of participatory approaches in cultural heritage and museums beyond pandemic crisis. In: Molestina DDS, Galluzzo L, Rizzo F, Spallazzo D, editors. IASDR 2023: Life-Changing Design. Milan, Italy: Design Research Society; 2023. p. 1–15

[11] Maye L, Claisse C. Co-design within and between communities in cultural heritage: Current and open questions. Multimodal Technologies and Interaction. 2023;**7**(1):1–10

[12] Pisoni G, Dıaz-Rodriguez N, Gijlers H, Tonolli L. Human-centered artificial intelligence for designing accessible cultural heritage. Applied Sciences. 2021;**11**(2):870

[13] Jiménez Rios A, Nogal M, Plevris V, Ramirez R, Petrou ML. Towards enhanced built cultural heritage

conservation practices: Perceptions on industry 5.0 principles and enabling technologies. The Historic Environment: Policy & Practice. 2024;**15**(4):466–492

[14] Tran K, Bui TK, Do BXC. Digital pathways to sustainability: Empirical evidence of tourism industry transformation in the industry 5.0 era. Journal of Management Changes in the Digital Era. 2025;**1**(1):108

[15] Fraga-Lamas P, Tiago M, Fernández-Caramés AM, da Cruz R, Ivan Lopes S. An overview of blockchain for Industry 5.0: Towards human-centric, sustainable and resilient applications. IEEE Access. 2024;**12**:116162–116201

[16] Ghobakhloo M, Mahdiraji HA, Iranmanesh M, Jafari-Sadeghi V. From industry 4.0 digital manufacturing to industry 5.0 digital society: A roadmap toward Human-centric, sustainable, and resilient production. Information Systems Frontiers. 2024;**26**:2593–2618

[17] Jiménez Rios A, Plevris V, Nogal M. Towards industry 5.0: A stakeholder analysis to understand the human role in the adoption of a heritage bridge human-centric digital twin framework. Structure and Infrastructure Engineering. 2025;**15** (4):1–15

[18] Aas C, Ladkin A, Fletcher J. Stakeholder collaboration and heritage management. Annals of Tourism Research. 2005;**32**(1):28–48

[19] Wijayanti A. Devi E, Putri H, Indriyanti I, Rahayu E, Ulumuddin IAhmad A. Community-based heritage tourism management model: Perception driving participation. African Journal of Hospitality, Tourism and Leisure. 2023;**12**(4):1630–1645

[20] Esraa Fathy Abdelaziz Alhadad. The Contribution of Community Involvement in Heritage Management: a Cultural Mapping Project. [PhD thesis]. Brandenburgische Technische Universität Cottbus-Senftenberg; 2023

[21] Yulong L, Hunter C. Community involvement for sustainable heritage tourism: A conceptual model. Journal of Cultural Heritage Management and Sustainable Development. 2015;**5** (3):248–262

[22] Esichaikul R, Chansawang R. Community participation in heritage tourism management of Sukhothai Historical Park. International Journal of Tourism Cities. 2022;**8**(4):897–911,01

[23] Elkington J. The triple bottom line. Environmental Management: Readings and Cases. 1997;**2**(1997):49–66

[24] Minghetti V, Buhalis D. Digital divide in tourism. Journal of Travel Research. 2010;**49**(3):267–281

[25] Yeniasır M, Gökbulut B. Effectiveness of usage of digital heritage in the sustainability of cultural tourism on islands: The case of northern Cyprus. Sustainability. 2022;**14**(6):3621

[26] Gonçalves AR, Dorsch LLP, Figueiredo M. Digital tourism: An alternative view on cultural intangible heritage and sustainability in Tavira, Portugal. Sustainability. 2022;**14** (5):2912

[27] Philip Feifan A. Industrial Heritage Tourism. Vol. **43**. Bristol, UK: Channel View Publications; 2015. p. 254

[28] Storm A. Post-industrial Landscape Scars. New York, NY, USA: Springer; 2014

[29] Peter Davis. Ecomuseums. 2011

[30] MacCannell D. The Tourist: A New Theory of the Leisure Class. Berkeley, CA, USA: University of California Press; 2013

[31] Cohen E. Authenticity and commoditization in tourism. Annals of Tourism Research. 1988;**15** (3):371–386

[32] Mitchell RK, Agle BR, Wood DJ. Toward a theory of stakeholder identification and salience: Defining the principle of who and what really counts. Academy of Management Review. 1997;**22**(4):853–886

[33] Brown T. Change by Design: How Design Thinking Transforms Organizations and Inspires Innovation. New York, NY: Harper Business; 2009. p. 70–93

[34] Kassahun Bekele M, Pierdicca R, Frontoni E, Malinverni ES, Gain J. A survey of augmented, virtual, and mixed reality for cultural heritage. Journal on Computing and Cultural Heritage. 2018;**11**(2):1–36

[35] Champion E. Virtual heritage: A guide. Digital appl. Archaeol. Cultural Heritage. 2021;**23**:e00188

[36] Singh P, Sharma M, Daim T. Envisaging AR travel revolution for visiting heritage sites: A mixed-method approach. Technology in Society. 2024;**76**:102439

[37] Murphy C, Carew PJ, Stapleton L. Towards a human-centred framework for smart digital immersion and control for cultural heritage applications. IFAC-PapersOnLine. 2022;**55**(39):30–35. 21st IFAC Conference on Technology, Culture and International Stability TECIS 2022

[38] Guo Y, Shizhu L, Shen M, Huang W, Xiaolie Y, Zhang J. Differences in heritage tourism experience between VR and AR: A comparative experimental study based on presence and authenticity. Journal on Computing and Cultural Heritage. 2024;**17**(2):1–21

Chapter 15

Artificial Intelligence and Eco-Museums: Climatic Challenges and Ethical Perspectives within European Cultural Sustainability

Mauro De Bari

Abstract

Eco-museums – process-based, place-bound, and community-governed – center intangible cultural heritage (ICH) and a symbiotic relationship with the landscape. As digitization becomes routine in the European GLAM sector (galleries, libraries, archives, and museums), artificial intelligence (AI) introduces opportunities for access and mediation alongside environmental and ethical risks. The author advances a guiding principle and a 5.0 orientation in which technology is a means proportionate to human and planetary objectives. A central concern is the preservation of silent knowledge – embodied, practice-based competencies transmitted intergenerationally – which underpins ICH. AI must, therefore, assist rather than replace living practices. Grounded in the European framework (Green Deal, Ethics Guidelines for Trustworthy AI, AI Act), the article derives operational criteria for eco-museums: editorial transparency and human-in-the-loop review; governance of cultural data (consent, licensing); minimization of compute through short, contextual texts, retrieval-over-generation, and reuse/caching; clear labeling of AI involvement; and prudential choices of infrastructure and providers. The environmental analysis highlights inference-driven energy and water footprints and rebound effects; the ethical analysis addresses language standardization, bias, disembodiment of practice, and organizational dependencies. Benefits are attainable when AI serves as an editorial and linguistic aid, providing multilingual/plain-language access, micro-itineraries, and concise narrative syntheses that connect material and intangible heritage, all while remaining locally reviewed and anchored in context. The article concludes that AI is acceptable in eco-museums only when it serves the mission and measures its effects – proportional, transparent, auditable, and locally governed.

Keywords: eco-museums, silent knowledge, climate change, artificial intelligence (AI), sustainability, responsible digital innovation

1. Introduction

In recent years, the European cultural sector has moved beyond the so-called phase of digital transformation (DT): platforms, databases, and online workflows are now routine, although an equivalent level of digital literacy has not matched the widespread, almost compulsive use of such tools [1, 2]. Against this backdrop, artificial intelligence (AI) has become the new disruptive factor, capable of profoundly affecting both managerial processes and cultural mediation [3, 4]. For example, one of the most widely discussed recent cases concerns the Albanian state's decision to "employ" an AI avatar, Diella, as a minister to combat corruption. This move rides the latest trends while cleverly framing them as a celebration of the country's cultural heritage, as signaled by the traditional attire in which the Albanian Prime Minister Rama has introduced the avatar [5].

The author argues that, as communities approach the DT era, the spread of AI across the GLAM field (galleries, libraries, archives, museums) and the cultural industries raises a twin set of issues: on the one hand, opportunities in terms of efficiency and broader access – particularly around community inclusion; on the other, environmental impacts and ethical dilemmas that require a rethinking of practices and new customs [6]. In particular, growing compute demand for training and using generative systems entails energy and water consumption, as well as hardware-related externalities, which cannot be ignored by institutions that profess a commitment to sustainability [7].

Eco-museums are among the institutions potentially most exposed to this tension. By history and mission, they are not containers of objects but museums-as-process, rooted in place, where communities and landscapes are integral to heritage [8, 9]. In less than 50 years, the eco-museum structure has become an operational pact between inhabitants, local institutions, and places [10], applying both to tangible assets and, above all, to intangible cultural heritage (ICH) [11]. The strong centrality of ICH requires that any digital innovation, AI included, be functional to the living transmission of knowledge, practices, and memories, without reducing them to *simulacra* or decontextualized repertories.

This nexus between eco-museums and ICH has clear foundations in international frameworks: the 2003 UNESCO Convention defines the intangible as a set of practices, expressions, and knowledge transmitted from generation to generation – living and dynamic [11]; the Faro Convention underscores the link between heritage, cultural rights, and democratic participation [12]. Operationally, this entails community ownership of content and safeguarding mechanisms that foster continuity and transformation, not crystallization. Bringing AI into this ecosystem, therefore, means asking not only what can be automated, but how and within what limits to do so, preserving the voice, context, and responsibility of the communities that generate living heritage.

At the European policy-institutional level, significant steps toward the legitimization and regulation of AI in cultural institutions include the Green Deal (climate neutrality by 2050) [13], the Ethics Guidelines for Trustworthy AI (transparency, human oversight, justice, social and environmental well-being) [14], and the regulatory evolution culminating in the AI Act (risk-based classification with corresponding obligations) [15]. These are not abstract frameworks but, potentially, concrete criteria for eco-museums: for selecting suppliers (disclosure on energy and water), for designing workflows (minimizing generation, prioritizing retrieval and re-use), for labeling generated content, and for local review (human-in-the-loop) before publication.

Within this framework, the author highlights the problematic over-reliance on ChatGPT as the representative case, given that it is one of the best-known and "user-friendly" AIs – a tool fully capable of underpinning user-facing services [16]. The point is not to imply the "automation of the museum," but to understand whether and how a large language model can assist with low-criticality tasks (short mediation texts, micro-educational trails, and linguistic simplification) while maintaining human control and minimizing computational demand: concise prompts, length limits, retrieval-over-generation, and systematic re-use of approved outputs. The focus is not solely on text quality but also on computational parsimony as an ethical dimension, in line with the community nature of the eco-museum. The balance between benefits (accessibility, clarity, inclusion) and costs/risks (consumption, bias, opacity) becomes the testing ground for responsible adoption.

The resulting position for eco-museums and the cultural sector more broadly is as follows: AI is acceptable when it serves the mission and when it measures its own effects. More specifically, it serves the mission if it broadens access without flattening modes of expression, if it supports the transmission of ICH without supplanting it, and if it renders content more readable for diverse audiences; it measures its effects when it makes computational demand visible (how many requests, how many tokens, how many re-generations), when it makes editorial pipelines explicit (what was generated and from which local sources one started), and when it institutionalizes community review. This human- and planet-centered stance realigns AI with responsible innovation in a European key and with the civic-educational profile of eco-museums, avoiding both uncritical enthusiasm and prejudicial rejection.

2. Origins, definitions, and principles

The eco-museum was conceived as a process-based museum and a community pact. This participatory, place-based institution sustains silent knowledge over time, preventing the loss of traditions and cultural heritage. As an alternative to the object-centered model, the earliest French formulations of the 1970s [17] shifted attention to an inseparable triad – territory, community, and heritage – linking memory, environment, and local development. Early cases are emblematic: at Le Creusot (1971), in a mining and metallurgical district, the eco-museum became a tool for culturally repurposing industrial heritage [18]; at Fresnes, marked by the detention of Jews during the Second World War, the eco-museum approach helped restore civic meaning to a site of memory through collective storytelling and regeneration [19]. From the outset, these trajectories showed that the "collection" extends beyond material objects to the practices, traces, and testimonies that bind places to the lives passing through them. The concept finds a clear synthesis in the Altkirch Declaration (2004): the eco-museum is a territory-based museum grounded in participation, integrating cultural, natural, and intangible heritage to foster social cohesion and sustainable development [20]. This is not a slogan but a codification of principles already in use, reaffirming that territory is not a backdrop but a constitutive element of the museum.

Accordingly, the eco-museum acts as a community instrument for caring for, interpreting, and enhancing cultural and environmental heritage for local well-being – effectively embodying the community. Emphasis falls on purpose (care and transmission) and on cultural ownership (the community decides what counts and how it is told). The result is a focal point that goes beyond exhibition as an end in itself: the eco-museum activates processes and enables collective choices and practices.

Operationally, eco-museums employ tools such as landscape charters, participatory mapping, oral-history archives, intergenerational workshops, thematic itineraries, and place-based storytelling. The goal is not accumulation but to keep alive the relationships among places, knowledge, and people, maintaining continuity while allowing transformation. This aligns with the UNESCO 2003 Convention on Intangible Cultural Heritage: intangible heritage is living, context-dependent, and requires safeguarding that preserves use and meaning for the communities that generate it.

The territory itself is the primary "object-document," living evidence of collective identity. It must be understood beyond administrative boundaries as a relational space in which safeguarding means transmission and reactivation rather than crystallization. Here, the eco-museum takes shape as a pact among community, institutions, and landscape: the community recognizes, selects, and narrates; institutions facilitate and steward; the landscape provides the weave within which practices and memories make sense [21].

This approach entails an ethics of participation: the community does not hand over its heritage to an institution – it governs it [20]. Hence, the shift of focus from digital products to living contexts, ensuring that digitization supports rather than disembodies practices [22].

3. Organizational forms and functions

In the shift from an "alternative territorial institution" to a dynamic, shared space, the eco-museum has consolidated a flexible organizational morphology: not a single building–container, but a form that interweaves places, practices, and people within a pact among community, institutions, and landscape. In this configuration, the eco-museum serves as a community instrument for caring for, interpreting, and enhancing its cultural and environmental heritage, in the service of local well-being and sustainable development [23].

Operationally, recurring tools give concrete shape to the "distributed museum": participatory mapping to identify meaningful places and practices; oral-history archives and intergenerational workshops to activate transmission and learning; place-based narratives (trails, heritage walks, events) that circulate stories and know-how anew. These devices are not designed to "fix" traditions but to accompany their vitality within contexts of practice, avoiding drifts toward the decontextualized musealization of ICH [24].

The educational function is structural: eco-museums operate as community laboratories and partners to schools, supporting forms of active, place-rooted learning (local history, ecological knowledge, craftsmanship). At the same time, they experiment with accessible mediation for diverse audiences, fostering inclusion and cultural citizenship. Where appropriate, they promote slow, identity-rooted tourism attentive to local ecologies and proximity economies, without slipping into a spectacle.

From an organizational standpoint, an eco-museum works when decisions are shared, roles are clear, and community participation is not reduced to symbolic consultation. This approach enables the integration of scarce resources (volunteerism, distributed skills) with institutional support and project funding, while safeguarding the cultural ownership of those who inhabit the places. Absent this architecture, the risks grow: standardized language and the outsourcing of narrative choices.

Typical functions can be summarized along four interlinked axes:

1. **Documentation and care of memory** – collecting testimonies, recordings, and materials that, once returned to the community, support processes of recognition and awareness; not a closed archive but a common good in evolution.

2. **Mediation and narration of place** – producing situated narratives (briefs, micro-itineraries, events, performative formats) that connect places, practices, and lived landscapes, while maintaining the link between the tangible and the intangible.

3. **Education and inclusion** – co-designing with schools, associations, and local groups pathways that foster experiential learning, intergenerational exchange, and participation by new or marginalized publics.

4. **Sustainable local development** – supporting economic and social practices consistent with cultural diversity and proximity economies (craft know-how, agro-pastoral value chains, festivals), while avoiding commodifications that distort community meanings.

These functions call for hybrid competencies: curatorial, educational, project design, and documentary.

At the system level, building networks of eco-museums – national and translocal – proves crucial for sharing criteria, tools, and minimum standards (participatory processes, consent models, formats for returning materials to the community) and for strengthening organizational capacity in fragile contexts. Networks also help navigate heterogeneous regulatory frameworks, fostering operational convergence and the exchange of good practices.

In sum, the eco-museum is a social and cultural infrastructure that succeeds when it braids light, distributed organizational forms with functions of documentation, mediation, education, and local development – all rooted in participation and community ownership. This framework – process museum, shared governance, participatory tools, and networks – forms the basis on which, in the sections that follow, we will assess whether and how AI tools can be aligned with the mission, without slipping into the substitution of practice or the homogenization of languages.

4. Essential international experiences

International experiences confirm that an eco-museum works when cultural ownership remains with communities, practices are enacted (not merely documented), and mediation keeps its bond with territory and landscape. The four cases cited here are not a part of an exhaustive inventory but directional indicators showing how the model adapts to diverse contexts without losing its founding traits.

- **Lai Chi Wo (Sinosphere)** [25, 26]. In a historic village marked by terraced agriculture and tight interrelations among settlement, forest, and water cycles, the eco-museum approach has operated as a device for socio-ecological regeneration. Care for the agricultural landscape, the revival of traditional techniques, and the intergenerational transmission of knowledge have been brought back to the center not as folklore, but as living practices with contemporary utility (environmental education, local resilience). The community's role has remained decisive in selecting themes, narrating places, and defining educational activities, with governance that unites local actors and technical expertise without dispossessing the former of their voice.

- **Écomusée du Fier Monde (Canada)** [27]. Born within an urban fabric shaped by layered histories of labor, migration, and social movements, this eco-museum has made co-curation with associations and neighborhoods its distinctive hallmark. The memory of work and struggles has been presented not as a repertoire of objects but as oral history, neighborhood routes, co-produced temporary exhibitions, and educational programs developed with schools and civic networks. Its public function is explicit: to give space to plural narratives, make social transformations legible, and foster participation and inclusion through accessible language without sacrificing rigor and accountability to the communities involved.

- **Eco-museum of Nordfjord (Norway)** [28, 29]. In a coastal and rural context organized as dispersed nodes and thematic centers, the eco-museum connects fishing, seamanship, craftsmanship, rural life, and oral storytelling. Its polycentric structure avoids the "headquarters-museum" risk and allows ties to be rewoven among trades, landscapes, and everyday practices. Mediation favors walks, demonstrations, workshops, and small situated displays, with communities playing a strong role in selecting knowledge and testimonies. The integration of the material (tools, boats, architectures) and the intangible (technical know-how, local vocabularies, life stories) is treated as a continuum: what matters is the relationship to places and to the people who inhabit them.

- **Ecomuseu de Itaipu (Brazil)** [30]. In a territory transformed by major infrastructure, the eco-museum approach has focused on reweaving memory and environmental education, holding together the effects of territorial transformations with the persistence of local knowledge and identities. Its programming puts ecosystems and communities into dialog, with workshops, visits, and narrative devices that show how innovation can be oriented toward environmental justice and active citizenship, avoiding the spectacle of places and keeping the focus on rights, responsibilities, and long-term horizons.

From these four cases emerge operational constants that are also useful for the chapter's discussion of AI: (a) the centrality of the community in deciding what is told and how; (b) distributed, polycentric forms of organization capable of holding together places, practices, and memories; (c) accessible mediation that privileges clear language, situated narratives, and experiential learning; (d) an integrated material/intangible vision in which objects make sense because they are connected to knowledge, relationships, and landscapes; and (e) an explicit relationship to sustainability and rights that prevents drifts toward appropriation or the digital musealization of practices. It is on this warp and weft that the scope and conditions of any potential use of AI will later be assessed: if it does not reinforce these elements, the innovation is off-target; if it supports them with proportionality and transparency, it can become a tool that serves the eco-museum's mission.

5. Proportion over volume: AI that serves cultural and environmental ends

As discussed above, eco-museums live in symbiosis with nature. That bond sets a *conditio sine qua non*: digital tools, however helpful, should never erode the eco-museum's mission. Hence, the following are theoretical reflections on a conscious, nonintrusive use of AI. Against this background, with digitization now embedded in heritage workflows, the critical question is not whether to use technologies but how to align them with cultural and environmental ends [1, 3]. The lexicon of Industry 4.0 – automation, interoperability, and data integration – has streamlined cataloging, management, and access [31]. At the same time, the use of a shared digital language, from digitization to digitalization, has increased the computational loads that accompany next-generation AI. The next step, for a deliberate use of this tool, is the 5.0 inflection: technology no longer as a neutral multiplier of output, but as a means proportionate to human and planetary objectives, evaluated by impacts rather than by performance alone [32].

This inflection entails a shift in the decision-making scale. The unit of account is not the algorithm in the abstract but the editorial deliverable: every use of AI must be situated in places and practices, traceable in its sources and local review, reversible without loss of meaning, and measurable in the marginal computational load it introduces. The European framework on climate, trustworthy AI, and risk-based regulation provides an external constraint. Still, the real threshold lies inside editorial workflows: here, one decides whether AI adds public value or merely volume.

In the eco-museum context, this means combining tools inherited from 4.0 – helpful in describing and connecting knowledge and lived landscapes – with 5.0 safeguards: community ownership, protection of ICH, transparency in choices, and local control over steps. Responsible innovation is not a checklist of principles but an operating method: anticipate effects (environmental and cultural) by setting ex-ante limits on use; include territorial actors in the editorial chain; monitor language and representations to avoid bias; and correct course in response to documented community feedback.

Within this profile, the question "Why AI in eco-museums?" must be posed rigorously. The issue is not whether to adopt AI, but when and under what conditions. From this follows the proportionality criterion that governs the whole chapter: rely on AI only where it yields a recognizable cultural gain, and in the minimum necessary amount. In practical terms, this points toward brief, contextual texts; retrieval-over-generation where validated local materials exist; systematic reuse of approved outputs; local human review before publication; clear labeling of AI use; targeted prompts and lengths within budget; at most one re-generation; and traceability to local snippets and a community glossary. If even one of these prerequisites is missing, it is preferable not to generate AI content; that said, there are concrete reasons why some uses of AI can align with the eco-museum mission:

- AI can serve as a linguistic and editorial aid to improve clarity and accessibility, useful in resource-constrained settings and with heterogeneous audiences (local visitors, schools, families, older adults). Generating short briefs, plain-language materials, and local translations can support educational work without supplanting community voices, provided the algorithm's ancillary role is clear and local human review occurs before publication [33].

- The centrality of ICH – practices, know-how, rituals, performing arts, and ecological knowledge – requires tools that help contextualize and interpret without "fixing" traditions. The 2003 UNESCO Convention defines ICH as living heritage, transmitted across generations and continually renewed. AI can help draft comprehensible explanations or thematic companion itineraries to accompany practices – not replace them – safeguarding communities' cultural ownership of content and representations, with editorial control and transparency about tool use [34].

- In territories with infrastructural gaps and limited professional staffing, AI – if used sparingly – can facilitate distributed curation: drafting panels, micro-educational itineraries, and summaries for intergenerational pathways. Yet, there are risks of "disembodying" the digital experience: poorly mediated technology can turn ICH into a simulacrum detached from context, encouraging passive consumption and spectacle. Adoption must, therefore, remain instrumental and supervised, with review practices, citation of local sources, and accountability of knowledge-bearers within the editorial process.

- Finally, the environmental question arises from AI's cumulative effects on energy, data, and computation. Responsible use in an eco-museum adopts a demand-before-supply principle: generate only what is needed, in short texts, limiting re-generations, and favoring retrieval and the reuse of validated community materials. This keeps the local "computational cost" auditable and consistent with Europe's sustainability horizon [35].

6. European framework: Sustainability, cultural rights, and trustworthy AI

In the European context, the digital transformation spread gives way to a phase in which innovation is assessed in terms of sustainability and responsibility [13]. In its most recent outlooks, the European Union ties the use of AI to climate and social objectives, on the assumption that rising computational loads affect the cultural sector's environmental footprint and call for corrective action [13, 32]. The reference framework is the European Green Deal, with climate neutrality on the horizon for 2050 and the idea that the transition is both a challenge and an opportunity for "a better future for all" [13]. In parallel, the Commission has advanced the Ethics Guidelines for Trustworthy AI, which set requirements for human oversight, robustness, privacy, fairness, and social and environmental well-being, steering adoption toward a balance between innovation and responsibility [15, 36].

Alongside this strategic dimension is an evolving regulatory framework for AI. Europe is moving toward a regulation that classifies systems by their risk to health, safety, fundamental rights, the environment, and democracy, with graduated obligations for transparency, risk management, and technical documentation [15]. For cultural institutions, this entails assessing use cases, tracing data flows, and making the use of generative systems in mediation and education transparent, with specific attention to the overall environmental impact of the tools adopted [36].

The climate–culture link is explicit: used without restraint, AI amplifies energy and water consumption and produces externalities along the hardware supply chain; governed responsibly, it can serve educational purposes, access, and inclusion in line with European values. An indiscriminate increase in computational demand (training, storage, inference) would have negative climate repercussions; for this reason, the GLAM sector should define metrics and mitigation practices. In this sense, the EU framework is not ornamental: it invites measurement and reduction of the footprint, tying AI adoption to choices of parsimony (reuse, short texts, retrieval) and to policies of accountability toward visitors and communities.

For eco-museums, the European framework translates into operational consequences:

- Editorial transparency: When an LLM is used in mediation, it is appropriate to inform the public about AI use and to cite the local sources employed in retrieval, while maintaining human-in-the-loop review and terminological coherence with community glossaries [36].

- Minimization of computational demand, consistent with the EU's environmental approach – produce short, situated texts; limit re-generations; prefer retrieval-over-generation; and reuse approved snippets.

- Governance of cultural data, with informed consent and clear licenses, in line with Faro and the centrality of ICH as living heritage (not "digitally musealized").

- Measurement: Pragmatism would align with the spirit of European policy: not offloading sustainability to generic pledges, but supporting the institutionalization of monitoring routines and continuous improvement proportionate to eco-museum resources [31].

In this way, the EU framework – Green Deal, trustworthy AI, and risk-based regulation – ceases to be mere "context" and becomes a working method capable of combining innovation, the protection of cultural rights, and footprint reduction.

The hypothesis is that, in this near-post-digital transformation European landscape, AI, and in particular generative language models, will enter heritage workflows, promising accessibility and editorial simplification, but at the cost of increasing processing demands (energy, water, hardware) and ethical–cultural risks (prejudice, rights, opacity).

In eco-museums, where heritage, place, and community coincide, the issue becomes more pressing. Every innovation must support the living transmission of

ICH and the community's ownership of content, avoiding the "digital musealization" of practices and knowledge, and governing the uses of cultural data.

7. Implications for AI in the eco-museum context

In light of the foregoing, the adoption of AI in eco-museums is not a question of *if* but of *conditions*: to align with the mission and with European frameworks, especially ecological ones, it must be tied to a principle of proportional use and to accountability mechanisms that hold together living heritage, community, and sustainability [13, 36].

- Centrality of the community and cultural ownership. In eco-museums, the community is not a target audience but the subject of the museum process; consequently, AI may be used only within editorial pipelines that preserve community voice, control, and the right to review or refuse content that concerns them. This entails clear procedures for informed consent, explicit licensing rules, and traceability of local sources used in mediation. This is consistent with the Faro perspective and with the participatory approach the draft identifies as a condition for the future: "reclaiming responsible cultural ownership" is the precondition for any innovation, digital or algorithmic.

- Avoiding "disembodiment" and the digital musealization of ICH [12, 34]. Algorithmic mediation must accompany practices, not replace them. Digital tools risk turning the intangible into a decontextualized simulacrum or spectacle; the same holds for AI: generated texts, translations, or summaries must remain anchored to use contexts, places, and relationships that give heritage its meaning. Misuse – for example, standardized productions or tech "showcases" – encourages passive consumption and weakens the experiential learning eco-museums promote.

- Sustainability: reduce computational demand. The escalation of AI workloads (training, storage, inference) has direct and indirect climate impacts, calling for corrective measures and a culture of responsible use. In eco-museums, this translates into parsimonious choices: generate only what is essential for mediation (plain language, context notes, micro-itineraries), limit re-generations and length, and prioritize reuse of already approved community materials and contextualization over mass production of new texts. The goal is clear: secure gains in access and intelligibility without pushing computational demand beyond what is proportionate to the mission.

- Transparency and human control along the chain. Whenever AI shapes texts and mediation apparatuses, this should be disclosed to the public in plain terms, with local human-in-the-loop review before publication. Transparency is not a box to tick; it makes choices and responsibilities auditable, aligns the eco-museum's actions with European guidelines on trustworthy AI, and protects trust with visitors.

- Avoiding technological shortcuts and managing divides. It is also essential to avoid showcase-style drift (e.g., low-interactivity 360° tours touted as innovation), which can exacerbate infrastructural divides affecting many inland areas. In such contexts, AI must be designed to fit the territory: restrained applications oriented toward mediation and education, interoperable, with sustainable organizational costs, and without shifting decision-making and control away from communities. Eco-museum networks can reduce inequalities by sharing models, tools, and baseline skills [37, 38].

8. Implications for AI in the eco-museum context

In order to operationalize a responsible and sustainable approach to AI adoption in eco-museums, this section outlines a set of proportionality criteria that should guide any decision to implement generative technologies. These criteria are derived from the ethical, environmental, and cultural imperatives (**Table 1**) discussed in the previous sections and are aligned with the European framework for trustworthy and sustainable AI (Green Deal, Ethics Guidelines, AI Act).

The application of AI is deemed acceptable only when all of the following conditions are satisfied:

1. **Mission alignment:** The use of AI must demonstrably support the eco-museum's mission of safeguarding ICH, enhancing accessibility, and promoting sustainable local development. AI should function as an assistive instrument, not a substitute for embodied practices or community voice.

2. **Retrieval over generation:**Whenever possible, AI should draw upon validated, community-approved local content through retrieval mechanisms

Dimension	Indicator	Purpose
Accessibility benefit	Readability improvement ratio (e.g., % of texts simplified).	Measures how effectively AI improves clarity and comprehension.
Accessibility benefit	Multilingual reach (e.g., the number of supported local languages).	Tracks the ability to serve diverse audiences.
Efficiency	Output reuse rate (% of outputs reused instead of being regenerated).	Encourages content caching and editorial parsimony.
Computational cost	Average tokens per prompt (target: ≤200 tokens).	Controls text length and processing load.
Computational cost	Regeneration rate (max: 1 per output).	Limits unnecessary iterations and inference overhead.
Sustainability	Caching rate (outputs stored for reuse).	Supports lifecycle efficiency and reduces redundant computation.
Auditability	Number of AI queries per month (tracked locally).	Enables periodic review of volume and impact.

[1]*These KPIs are not prescriptive but indicative. They allow eco-museums to monitor the trade-offs between cultural benefits and computational demands, aligning usage with proportionality and sustainability principles.*

Table 1.
Key performance indicators (KPIs) for responsible AI use in eco-museums.[1]

rather than generating new content from scratch. This reduces both cultural risks (such as decontextualization and misrepresentation) and environmental costs.

3. **Human-in-the-loop editorial review:** All outputs must undergo local human review before public dissemination. Editorial control ensures cultural accuracy, guards against bias, and maintains the legitimacy of the eco-museum's interpretive authority.

4. **Transparent labeling and source attribution:** Content generated with AI must be clearly labeled, and any local sources used must be explicitly cited. This transparency is crucial for both editorial accountability and public trust.

5. **Compute minimization:** The computational footprint should be minimized by enforcing concise prompts, limiting generations (e.g., to a single regeneration if strictly necessary), and avoiding unnecessary model calls. This includes the reuse of validated outputs and the avoidance of content overproduction.

6. **Consent and licensing:** Any cultural data used in AI workflows must be governed by informed consent protocols and clear licensing agreements that reflect the community's rights and expectations.

7. **Auditability and measurability:** Institutions should adopt proxy indicators – such as the number of AI queries, average tokens per output, reuse ratios, and regeneration rates – to assess and document the environmental and editorial impact of AI use over time.

These criteria are not intended as mere recommendations but as thresholds for ethical and sustainable deployment. Where one or more criteria cannot be fulfilled, the use of AI should be avoided. This proportional approach (**Figure 1**) ensures that innovation remains subordinate to the eco-museum's core values: cultural continuity, environmental responsibility, and community ownership.

9. Environmental risks

Among the various issues linked to the unreflective use of AI, one of the primary risks concerns the environment. The escalation of computational demand – training, storage, and inference – cumulatively increases electricity consumption and pressure on water resources, with externalities across the entire hardware supply chain. On current projections, global data center electricity demand could more than double by 2030 ($\approx$945 TWh), with AI as the primary driver [39]. The growth of computing loads, especially since the advent of "large" models, makes a nonnegligible contribution to the ICT sector's climate-altering emissions, with direct and indirect effects that call for corrective measures and an explicit accountability framework in GLAM.

Training generative models is energy-intensive in itself; however, in the day-to-day practice of eco-museums, inference weighs more because it multiplies into countless requests and re-generations. Peer-reviewed accounting for BLOOM-176B shows approximately 25–50 tCO_2e for the final training run (method-dependent)

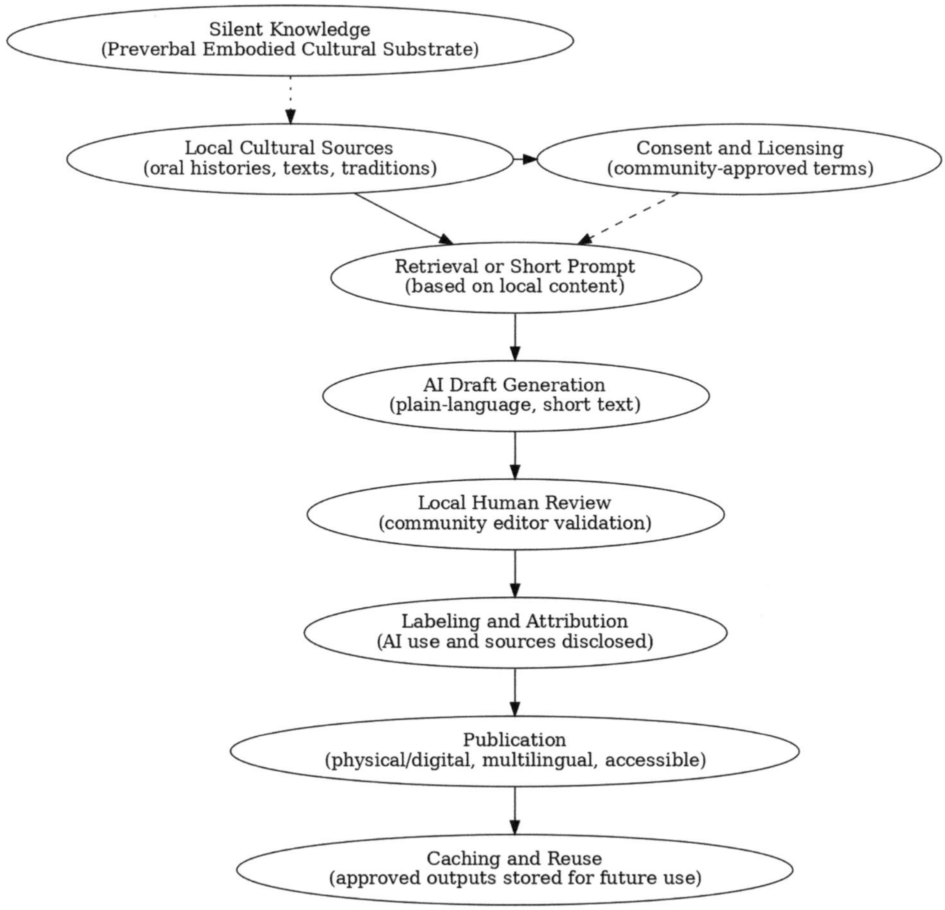

Figure 1.
From silent knowledge to sustainable AI-supported mediation in eco-museums.

and highlights that, once deployed, inference can become a major share of lifecycle impact [40]. The technological trajectory of the past decade shows a rapid rise in computing power used for AI; recent metering and estimates suggest a typical general-purpose LLM query today is on the order of approximately 0.3 Wh (model- and load-dependent) [41]; operationally translated: even seemingly harmless but repeated uses – re-generating the same draft multiple times, producing texts that are too long, serializing translations – inflate computational demand beyond what is proportionate to the mission.

The practical rule is simple: fewer calls, higher quality. Work with short texts, prefer adaptation from local materials (retrieval-over-generation), limit re-generation to one motivated attempt, and plan for the reuse of approved outputs. In this way, AI remains a useful tool without turning into a multiplier of consumption.

For the same amount of energy, where and how inference is run matters: data centers with fossil-heavy energy mixes and water-intensive cooling systems worsen the impact. Evidence shows nontrivial water use both on-site (cooling) and upstream (power generation); for example, training GPT-3 at a U.S. site was estimated to directly evaporate approximately 700,000 liters of freshwater, and day-to-day use can amount to roughly 10–30 milliliters per prompt (~0.5 liters over 20–50 prompts), depending on the site/technology [42, 43]. This figure should not be "absolutized," but it is relevant because it shifts attention from emissions alone to water resources, which are particularly critical in many European regions. Lifecycle studies also indicate that advanced liquid cooling (cold-plates/immersion) can reduce GHG by approximately 15–21%, energy by 15–20%, and blue-water by 31–52% compared to air cooling [43].

A single request to a general-purpose LLM consumes nonzero energy, even when answers look "small" (~0.3 Wh typical), and aggregate query volumes are the main driver of operational impact [39, 41]. Even if these are estimates, the operational message is clear: volumes of use and text length are material drivers of impact. For small, distributed institutions like eco-museums, controlling these two factors is the first lever of sustainability.

Beyond compute, the material chain of devices (servers, networks, endpoints) entails mineral extraction, manufacturing, transport, and electronic waste. Semiconductor fabs use high-GWP fluorinated gases (e.g., PFCs, NF_3, N_2O) and large amounts of energy and water; regulators and industry recognize this as a priority for abatement [44, 45]. Impacts are not exhausted by the kWh consumed: expanding installed capacity and shortening life cycles generate environmental degradation and pressure on value chains, especially in the absence of refurbishment and recovery strategies. Even if eco-museums do not run data centers, they participate as users: the demand they generate still feeds the supply chain.

Another risk is opacity: providers often do not report, in a granular fashion, energy mix, water consumption, and per-service/per-region metrics, making comparable assessment difficult for cultural institutions. Where available, prefer providers offering credible disclosures and, when configurable, regions with lower grid-carbon and lower water stress; note that public initiatives (e.g., IEA's Energy and AI Observatory) are improving transparency [46].

The ease of generating texts and translations introduces a rebound: the easier it is to produce content, the more content gets produced. In GLAM, this can induce "over-production" of labels, briefs, and micro-narratives, multiplying model calls without a corresponding increase in value for visitors.

Uncritical adoption of high-intensity, general-purpose models can be disproportionate for short editorial tasks. The environmental risk here is organizational: choosing tools more "powerful" than necessary creates structural computational waste. The chapter urges calibrating the model to the use case (reducing complexity, retrieval-over-generation, reuse) and limiting lengths and versions to avoid the model's authorial drift.

Implications for Eco-museums:

1. AI's environmental footprint is not abstract but the result of everyday editorial choices (how many requests, how long, how many re-generated). Where teams set thresholds and practice systematic reuse, impact decreases without sacrificing clarity and accessibility.

2. Transparency toward the public and community concerns not only the ethics of mediation but also environmental ethics: disclose when and why AI is used, and shows that practices of computational parsimony are in place.

3. Provider selection and compute-region choices (when configurable) affect the marginal impacts on energy and water; in the absence of complete data, it is reasonable to work with user-side proxies and prudential criteria.

4. Guard against rebound with editorial rules and retrospective cycles on actual tool use (how many re-generations were truly needed? Which texts can be cached and republished?).

In sum, the environmental risks of AI for eco-museums stem less from "grand projects" and more from diffuse, unguided use of generative tools: energy and water for inference, hardware-chain externalities, informational opacity, and editorial rebound. Impact can be reduced only if demand is measured (with simple proxies), production is contained (short texts, reuse, retrieval), the tool is proportionate to the task, and transparency is required of providers, in line with the European Green Deal horizon and the notion of trustworthy AI [47].

10. Ethical implications

The ethical and cultural risks tied to the use of AI in eco-museums can be read along four interwoven axes: (a) representation and bias; (b) cultural rights, consent, and licensing; (c) editorial transparency and human control; and (d) dependencies and divides – with a transversal risk of disembodying experience and digitally musealizing ICH when algorithmic mediation separates practices from the contexts that keep them alive [34, 48].

Language models tend to flatten registers and vocabularies: without editorial constraints and community glossaries, generative output can level out local differences (dialects, technical terms, variants) and induce stereotypes or oversimplifications, especially around minority cultures or nonstandardized practices. In the eco-museum sphere, where ICH is living heritage, the damage is twofold, as linguistic specificity is lost and situated transmission is weakened [48, 49]. AI governance must be consistent with communities' cultural ownership, with procedures for informed consent, explicit licensing, and clear labeling of generated content, together with a right of review or refusal for knowledge-bearers; where these steps are missing, the risks rise for appropriation and improper reuse of recorded intangible materials, contradicting Faro's participatory approach and the processual nature of ICH.

Transparency requires a human-in-the-loop pipeline and explicit disclosure whenever AI intervenes in mediation, so that authorship and sources remain auditable and editorial responsibility remains intact. Even for low-criticality tasks, data minimization and the exclusion of personal or sensitive data are necessary, and proportionality (limits on length and versions) should guide use. Closed or costly infrastructures and connectivity gaps can shift control away from territories, while showcase technologies with low interactivity (e.g., certain 360° tours) risk sterility and distance people from practices, creating organizational and linguistic lock-in. The strongest ethical risk is disembodiment: when AI (or the digital) replaces living practice with its representational double, ICH is uprooted from its vital matrices – body, relationship, place – and slides toward esthetic consumption; mediation should therefore accompany practices, not replicate them, with generated content pointing back to contexts and uses, citing local sources, and being corrected with the community. Operational mitigations follow: proportionality of use (only what is essential, short texts, retrieval-over-generation); reuse and caching (approved snippets, limited re-generations, versioning, and batching); community glossaries and co-curation; transparency by design (labels, source citation, a brief transparency note, prepublication checklist); governance of cultural data (consent, licenses consistent with community wishes, rights to withdraw/correct); and attention to divides (restrained platform choices, portable formats, networks that share consent models, glossaries, and quality rubrics).

AI use is ethically acceptable in eco-museums when (1) it strengthens accessibility and understanding without replacing practice; (2) it respects cultural ownership and community voice; (3) it is transparent and locally reviewed; (4) it limits computational demand and avoids overproduction; and (5) it reduces dependencies and divides through proportionate technical choices. Otherwise, risks of simulacrum, standardization, and appropriation prevail – at odds with the eco-museum's civic mission and the living nature of ICH.

11. The ChatGPT case

This section presents ChatGPT not as a technical protocol to be implemented but as an illustrative case that helps ground the broader arguments of the chapter. Its purpose is to explore, through a widely known and accessible tool, how generative AI may, or may not, align with the specific missions and constraints of eco-museums. The analysis is deliberately framed from a critical operational standpoint, not a computational one. Within this approach, ChatGPT serves as a proposal to reflect on the conditions under which generative AI can support accessibility and mediation in ways that remain consistent with the values of community ownership, contextual fidelity, and environmental responsibility. Its selection reflects both its pervasiveness in the cultural sector and its symbolic role in current AI discourse. On the benefits side, it can translate, contextualize, and interpret cultural narratives in support of digital and hybrid experiences – repositories, online exhibitions, installations – overcoming linguistic and geographic barriers, facilitating intercultural dialog, and sustaining educational outreach. The generation of short texts, light-touch curatorship, and collaboration tools can help renew educational and mediation practices, provided the community's voice and ownership remain clear and the algorithm stays an editorial aid under human control. Alongside these outcomes, however, there is an environmental flip side and necessary ethical cautions.

On the former, currently available orders of magnitude suggest that a single query can require 10 to 100 times the energy of an email, and that flows of hundreds of millions of queries could match the daily electricity consumption of tens of thousands of U.S. households; on water, processing 5–50 queries can require up to half a liter for cooling, with seasonal variation and dependence on data-center location, while recent water-use increases attributable to AI applications (on the order of + 34% and + 30% between 2021 and 2022 for major providers) signal a systemic risk if use is not measured and contained [50].

These figures should not be taken as absolute – because they depend on infrastructure and energy mix – but they are sufficient to justify parsimony and to discourage ornamental or redundant uses. Ethically and culturally, risks include language standardization, bias, and supply-chain opacity (models, data, footprint), which call for participatory frameworks and transparency: whenever ChatGPT intervenes in mediation, institutions should disclose this to the public, cite local sources, document editorial choices, and maintain local human-in-the-loop review.

What follows is a disciplined approach grounded in proportionality and computational parsimony: generate only what is needed for legibility and access, set thresholds on length and versions to avoid unnecessary re-generations, prefer retrieval-over-generation where validated community materials exist, and reuse approved snippets via caching. This is combined with an editorial pipeline that recognizes community ownership (outputs as drafts to be reviewed with knowledge bearers, shared glossaries, rights of review or refusal, labeling of AI involvement, and attribution of local sources), minimal metrics and auditability via user-side proxies (number of requests, average tokens per request, reuse/cache rate, re-generations) to enable periodic retrospectives and guide containment. Infrastructure and localization choices are oriented, where configurable, toward renewable mixes and more efficient cooling with provider disclosure on energy and water, or, where that is not possible, toward prudential criteria (shorter texts, batching, caching) made explicit to audiences. Finally, there is an emphasis on avoiding showcase uses that fuel sterile overproduction and decontextualized simulacra of ICH, reaffirming that AI is acceptable only when it strengthens transmission and in-person experience rather than replacing them.

Within this frame, the opportunity set (multilingual and plain-language access; assembling micro-narratives and maps across material and intangible heritage; editorial support for small teams; expanded educational outreach when integrated into experiential pathways) must always be weighed against costs and risks (energy and water for inference scaling with volumes and length; bias and register standardization; supply-chain opacity; editorial rebound as production becomes easier), which can be mitigated through parsimony, retrieval, reuse, local review, labeling, and monitoring proxies. The implication for the chapter is that the ChatGPT case, read through eco-museum constraints, does not deny the usefulness of LLMs; it specifies the conditions under which that usefulness can be converted into public value without offloading environmental and cultural costs. In line with the European framework (Green Deal, trustworthy AI) and with the 5.0 approach, adoption makes sense when it is proportionate, transparent, and auditable; otherwise, it yields simulacra and footprints disproportionate to the mission. The methodological section that follows translates these principles into a replicable test design – tasks, rubrics, proxies, and safeguards – to verify in practice how far ChatGPT can help without exceeding the bounds of parsimony.

12. Conclusion

Eco-museums, as process-based and place-bound institutions, safeguard ICH by sustaining silent knowledge – embodied, practice-based competencies transmitted across generations. In this context, AI is not an end in itself but a contingent means. Its acceptability depends on proportionality and accountability: technology must remain subordinate to the cultural mission and to ecological constraints.

The article argues that the decisive unit is the editorial deliverable, not the algorithm. AI use should be situated in places and practices, traceable to local sources, reversible without loss of meaning, and measured for its marginal computational load. Read through the European framework (Green Deal, Ethics Guidelines for Trustworthy AI, AI Act), this translates into a concise set of conditions: community ownership and the right of review; governance of cultural data (consent, licensing); minimization of compute (short, contextual texts; retrieval-over-generation; reuse/caching; limits on re-generations and length); labeling of AI involvement with human-in-the-loop review; and prudential choices regarding infrastructure and providers.

Under these conditions, AI can support access, mediation, and inclusion – multilingual plain-language materials, micro-itineraries, and succinct narrative links between material and intangible heritage – without supplanting practice or diluting local registers. Where conditions cannot be met, restraint is the responsible choice: unmeasured use risks disembodiment of ICH, standardization of language, rebound effects in production, and disproportionate environmental footprints.

The resulting stance is clear. For eco-museums and, by extension, for heritage institutions committed to participation and sustainability, AI is optional and conditional: adopt only when it serves transmission, protects community voice, and measures its effects. Otherwise, decline. This proportional approach preserves the primacy of place and practice, ensuring that digital means do not erode the silent knowledge on which ICH rests.

Author details

Mauro De Bari
Department of Research and Innovation for Humanities (DIRIUM), University of Bari Aldo Moro, Bari, Italy

*Address all correspondence to: mauro.debari@uniba.it

References

[1] Nikolaou P. Museums and the post-digital: Revisiting challenges in the digital transformation of museums. Heritage. 2024;**7**(3):1784–1800. DOI: 10.3390/heritage7030084

[2] CHARTER Alliance. Building an innovative and enterprising cultural heritage sector: CHARTER's 12 recommendations [Internet]. 2024. Available from: https://charter-alliance.eu/charter-recommendations-for-cultural-heritage/ [Accessed: 2025-November-24]

[3] European Parliamentary Research Service. Artificial intelligence in the context of cultural heritage and museums: Complex challenges and new opportunities [Internet]. 2023. Available from: https://www.europarl.europa.eu/RegData/etudes/BRIE/2023/747120/EPRS_BRI(2023)747120_EN.pdf [Accessed: 2025-November-24]

[4] Mazzanti P, Ferracani A, Bertini M, Principi F. Reshaping museum experiences with AI: The reinherit toolkit. Heritage. 2025;**8**(7):277. Available from: https://www.mdpi.com/2571-9408/8/7/277

[5] Reuters. Albania appoints AI bot as minister to tackle corruption [Internet]. 2025 September 11. Available from: https://www.reuters.com/technology/albania-appoints-ai-bot-minister-tackle-corruption-2025-09-11/ [Accessed: 2025-September-16]

[6] International Federation of Library Associations and Institutions (IFLA). Innovations are opportunities [Internet]. Notes ethical issues and environmental impact of AI in libraries/GLAM. Available from: https://www.ifla.org/news/innovations-are-opportunities/ [Accessed: 2023-January-23]

[7] Li P, Yang J, Islam MA, Ren S Making AI Less "Thirsty": Uncovering and Addressing the Secret Water Footprint of AI Models [Internet]. Available from: https://arxiv.org/abs/2304.03271 [Accessed: 2023-April-6]

[8] Rivière GH. The ecomuseum—an evolutive definition. Museum International. 1985;**37**(4):182–183

[9] Davis P. Ecomuseums: A Sense of Place. 2nd ed. London: Bloomsbury; 2011

[10] Davis P Ecomuseums and the representation of place. In: ICOM 24th General Conference, Milan; 2016

[11] UNESCO. Convention for the Safeguarding of the Intangible Cultural Heritage. 2003. Available from: https://ich.unesco.org/doc/src/01852-EN.pdf [Accessed: 2025-November-24]

[12] Council of Europe. Framework Convention on the Value of Cultural Heritage for Society (CETS No. 199) [Internet]. Available from: https://rm.coe.int/1680083746 [Accessed: 2005-October-27]

[13] European Commission. The European Green Deal. COM(2019) 640 final. Available from: https://eur-lex.europa.eu/resource.html?format=PDF&uri=cellar:b828d165-1c22-11ea-8c1f-01aa75ed71a1.0002.02/DOC_1 [Accessed: 2019-December-11]

[14] European Commission. Ethics guidelines for trustworthy AI

[Internet]. Luxembourg: Publications Office of the European Union; 2019. Available from: https://op.europa.eu/en/publication-detail/-/publication/d3988569-0434-11ea-8c1f-01aa75ed71a1/language-en [Accessed: 2025-November-24]

[15] European Union. Regulation (EU) 2024/1689… (Artificial Intelligence Act) [Internet]. Official Journal of the European Union. 2024 Jul 12. Available from: https://eur-lex.europa.eu/legal-content/EN/TXT/PDF/?uri=OJ:L_202401689 [Accessed: 2025-August-16]

[16] Reuters. OpenAI's weekly active users surpass 400 million [Internet]. Available from: https://www.reuters.com/technology/artificial-intelligence/openais-weekly-active-users-surpass-400-million-2025-02-20 [Accessed: 2025-February-20]

[17] Bruno MC (ed.). A History of Museology. Icofom/icom; 2019. p. 166–168 (nota sul termine "ecomuseum" proposto da H. de Varine nel 1971). Available from: https://icofom.mini.icom.museum/wp-content/uploads/sites/18/2023/03/2019_history_of_museology_bruno.pdf [Accessed: 2025-November-24]

[18] de Varine-bohan H. Un musée "éclaté": Le Musée de l'homme et de l'industrie, Le Creusot-Montceau-les-Mines. Museum International (Edition Francaise). 1973;**25**(4):242–249. Available from: https://unesdoc.unesco.org/ark:/48223/pf0000012428_fre

[19] Doulut A. La Déportation Des Juifs de France. Changement D'échelle. Paris: CNRS Éditions; 2025

[20] Corsane G. The ecomuseum ideal and issues related to its use. Museum and Society. 2006;**4**(1):1–5

[21] Šola T How Ecomuseums Changed our Job [Internet]. ECOVAST; 2018. p. 11. (Community agreement; roles of community/institutions/landscape). Available from: https://ecovast.hr/wp-content/uploads/2018/10/1-1_Tomislav_Sola.pdf [Accessed: 2025-November-24]

[22] Bakka E. Safeguarding of intangible cultural heritage—The spirit and the letter. Mots. 2015;**29**:135–169

[23] Canavese G, Gianotti F, de Varine H. Ecomuseums and geosites: Community and project building. International Journal of Geoheritage and Parks. 2018;**6**(1):43–62. DOI: 10.17149/ijg.j.issn.2210.3382.2018.01.004

[24] Wulf C. Unknowing and Silent Knowledge as a Challenge: Iconic, Performative, and Material Perspectives. In Unknowing and Silent Knowledge as a Challenge. SSRN Electronic Journal; 2017. DOI: 10.2139/SSRN.3647381

[25] The University of Hong Kong – Centre for Civil Society and Governance. Sustainable Lai Chi Wo Programme [Internet]. 2013–2017. Available from: https://ccsg.hku.hk/ruralsd/en/pages/sustainable-lai-chi-wo-programme-2013-2017/ [Accessed: 2025-November-24]

[26] Countryside Conservation Office (HKSAR). Lai Chi Wo [Internet]. Available from: https://cco.gov.hk/en/project-sites/lai-chi-wo [Accessed: 2025-November-24]

[27] Écomusée du fier monde. A museum anchored in its community/Musée d'histoire et musée citoyen (official site) [Internet]. Montréal. Available from: https://ecomusee.qc.ca/en/ [Accessed: 2025-November-24]

[28] Nordfjord Museum of Cultural History. Official site (MISF) — About/ Plan your visit [Internet]. Sandane: Musea i Sogn og Fjordane. Available from: https://misf.no/en/nordfjord-museum [Accessed: 2025-November-24]

[29] Nordfjord Museum of Cultural History. Plan your visit (location & opening hours) [Internet]. Sandane: Musea i Sogn og Fjordane. Available from: https://misf.no/en/nordfjord-museum/plan-your-visit [Accessed: 2025-November-24]

[30] Itaipu Binacional. Ecomuseu [Internet]. Foz do Iguaçu: Itaipu Binacional. Available from: https://www.itaipu.gov.br/sustentabilidade/social/esporte-e-cultura/ecomuseu/ [Accessed: 2025-August-16]

[31] EuropeanaTech. AI in relation to GLAMs – Task Force Report and Recommendations. The Hague: Europeana Network Association; 2023. Available from: https://pro.europeana.eu/files/Europeana_Professional/Europeana_Network/Europeana_Network_Task_Forces/Final_reports/AI%20in%20relation%20to%20GLAMs%20Task%20Force%20Report.pdf [Accessed: 2025-November-24]

[32] International Energy Agency. Energy and AI. Paris: IEA; 2025 Apr 10. Available from: https://www.iea.org/reports/energy-and-ai

[33] Münster S, et al. Artificial Intelligence for digital heritage innovation. Heritage (Basel). 2024;7(2):38. Available from. https://www.mdpi.com/2571-9408/7/2/38

[34] UNESCO. Safeguarding without freezing. Intangible Cultural Heritage portal; 2025 Apr 1 (last update). Available from: https://ich.unesco.org/en/safeguarding-00012 [Accessed: 2025-August-16]

[35] Li P, Yang J, Islam MA, Ren S. Making AI less "Thirsty": Uncovering and addressing the secret water footprint of AI models. arXiv:2304.03271. 2023. Available from: https://arxiv.org/pdf/2304.03271

[36] High-Level Expert Group on AI. Ethics Guidelines for Trustworthy AI [Internet]. Brussels: European Commission. Available from: https://digital-strategy.ec.europa.eu/en/library/ethics-guidelines-trustworthy-ai [Accessed: 2019-April-8]

[37] Europeana Network Association. AI in relation to GLAMs – Task Force Report and Recommendations [Internet]. 2023 Sep 11. Available from: https://pro.europeana.eu/project/ai-in-relation-to-glams

[38] Europeana Foundation. Europeana Impact Playbook [Internet]. Available from: https://pro.europeana.eu/page/europeana-impact-playbook [Accessed: 2025-November-24]

[39] International Energy Agency. Energy demand from AI [Internet]. Paris: IEA; 2025. "Global Electricity Consumption for Data Centres Is Projected to Double to ~945 TWh by 2030 in the Base Case." Available from: https://www.iea.org/reports/energy-and-ai/energy-demand-from-ai

[40] Luccioni AS, Viguier S, Ligozat A-L. Estimating the Carbon Footprint of BLOOM, a 176B Parameter Language Model. Journal of Machine Learning Research. 2023;**24**:1–56

[41] Epoch. How much energy does ChatGPT use? Gradient Updates

[Internet]. Available from: https://epoch.ai/gradient-updates/how-much-energy-does-chatgpt-use [Accessed: 2025-February-7]

[42] Elsworth C, et al. Measuring the environmental impact of delivering AI at Google scale [Internet]. Google. Available from: https://services.google.com/fh/files/misc/measuring_the_environmental_impact_of_delivering_ai_at_google_scale.pdf [Accessed: 2025-August-21]

[43] Li P, Yang J, Islam MA, Ren S. Making AI less "Thirsty": Uncovering and addressing the secret water footprint of AI models. arXiv. 2023. Available from: https://arxiv.org/abs/2304.03271

[44] Alissa H, Nick T, Raniwala A, et al. Using life cycle assessment to drive innovation for sustainable cool clouds. Nature. 2025;**641**:331–338

[45] Microsoft. Microsoft quantifies environmental impacts of datacenter cooling. [Internet]. Available from: https://news.microsoft.com/source/features/sustainability/microsoft-quantifies-environmental-impacts-of-datacenter-cooling-from-cradle-to-grave-in-new-nature-study/ [Accessed: 2025-May-1]

[46] US EPA. Semiconductor Industry [Internet]. ("Semiconductor manufacturing uses high-GWP fluorinated compounds.") Available from: https://www.epa.gov/eps-partnership/semiconductor-industry [Accessed: 2025-Mar-21]

[47] OECD. Measuring the Environmental Impacts of Artificial Intelligence Compute and Applications [Internet]. 2025. Available from: https://www.oecd.org/content/dam/oecd/en/publications/reports/2022/11/measuring-the-environmental-impacts-of-artificial-intelligence-compute-and-applications_3dddded5/7babf571-en.pdf

[48] Bender EM, Gebru T, McMillan-Major A, Shmitchell S On the Dangers of Stochastic Parrots: Can Language Models Be Too Big? In: Proceedings ACM FAccT; 2021

[49] Deas N, Ippolito D, Callison-Burch C. An Evaluation of African American Language Bias in Natural Language Generation. Singapore: EMNLP Findings; 2023

[50] O'Brien M, Fingherut H. 2023. "Artificial intelligence technology behind ChatGPT was built in Iowa — With a lot of water." AP. September 9. https://apnews.com/article/chatgpt-gpt4-iowa-ai-water-consumption-microsoft-f551fde98083d17a7e8d904f8be822c4 [Accessed: 2025-November-24]

Section 4

Governance, Ethics, Sustainability and Data

Chapter 16

Security and Privacy in Human-Centered Modern Technologies

Ahmed Abi Sen Adnan

Abstract

The world is witnessing rapid and significant advancements in modern technologies. After the Fourth Industrial Revolution, which focused on the Internet of Things (IoT), artificial intelligence (AI), automation, big data, and other digital technologies, the Fifth Industrial Revolution emerged, emphasizing personalization, human-centered design, sustainability, creating human value, and strengthening human–machine collaboration. However, this move toward personalization and creating human value has refocused attention on one of the biggest challenges facing technology: ensuring the security and privacy of user data. As modern devices increasingly function like continuous monitoring tools, vast amounts of data are collected about users. Meanwhile, attackers have more resources and tools with greater computational capabilities to analyze users' data and extract private and sensitive information. This chapter discusses this critical issue, outlining the main concerns and threats related to the industrial revolutions and their technologies. It also reviews current security and privacy solutions, highlighting their strengths and weaknesses, and finally proposes more adaptive solutions to address this issue in the new reality.

Keywords: Fourth Industrial Revolution, Fifth Industrial Revolution, privacy, security, protection, IoT, mobile computing, AI, metaverse, quantum

1. Introduction – The Fourth Industrial Revolution (Industry 4.0)

The Fourth Industrial Revolution has played a major role in advancing technology. It has brought together physical systems and digital systems, turning almost everything around us into something "smart." This revolution emerged in the second decade of the 21st century as a result of combining several advanced technologies – most notably the Internet of Things (IoT), artificial intelligence (AI), digital computing, digital twins, big data analytics, blockchain, and quantum computing [1, 2].

One of its key features is smart automation. Factories today are managed by intelligent machines capable of making semi-independent decisions using sensor data, machine learning algorithms, and contextual information. IoT tools have also made it possible to connect all elements of supply chains in real time, providing continuous data that improves production efficiency and supports better decision-making [3, 4].

IntechOpen

This transformation is not limited to industry alone; it has extended to healthcare, education, business, energy, agriculture, entertainment, and smart cities – bringing with it thousands of new and adaptive applications [5, 6].

However, modern technologies always come with challenges, and if left unaddressed, they can turn into risks. The biggest challenges are related to information security and privacy. Since data now flows continuously in real time, everything is being constantly monitored, creating massive amounts of data that are analyzed by smart algorithms – revealing sensitive and private information. For this reason, interest in cybersecurity has grown to protect new systems and their users from emerging threats. In the industrial domain specifically, this has given rise to what is known as cyber–physical systems (CPS) [7, 8].

To better understand the advances of the Fourth Industrial Revolution, we can look at some of its applications – both from an industrial perspective and from a consumer perspective.

1.1 Industrial applications of Industry 4.0

Industry 4.0 has introduced a paradigm shift in industrial environments by integrating digital technologies with physical production systems. These technologies enable smarter applications [9], which are:

- Smart factories: Production lines interconnected through IoT and 5G.
- Digital twins: Virtual simulations of factories or machines used to test and improve performance.
- Predictive maintenance: Using AI to predict equipment failures before they occur.
- Collaborative robots (Co-bots): Robots that work side by side with humans in factories.
- 3D printing: Rapid production of spare parts or industrial prototypes.
- Big data analytics: Optimizing production processes and supply chains.
- Autonomous logistics vehicles: Self-driving trucks and robots transporting goods within warehouses.

1.2 Consumer applications of Industry 4.0

Industry 4.0 technologies are increasingly embedded in everyday life, enabling smarter, personalized, and more connected consumer products and services [10], which are:

- Smart homes: IoT devices like Alexa and Google Home are used to control lighting, security, and air conditioning.
- Smart and autonomous cars: For example, Tesla's Autopilot.

- Wearable devices: Smartwatches that track fitness and health.
- Virtual and augmented reality (VR/AR): Devices such as Oculus and HoloLens are used for gaming and training.
- Smart e-commerce: Personalized recommendations on platforms like Amazon and Netflix.
- Interactive e-learning: Platforms such as Coursera and Duolingo.
- Digital health applications: Apps for tracking calories, sleep, and exercise.

2. Introduction – The Fifth Industrial Revolution (Industry 5.0)

Building on the technologies of the Fourth Industrial Revolution, Industry 5.0 emerged to address its limitations. It incorporates all the tools of Industry 4.0 but places stronger emphasis on human creativity [11].

Instead of people fully depending on technology, the focus is on humans working in harmony with technology. Rather than complete automation, Industry 5.0 brings the human element back to the center of industrial and technological processes. In other words, AI and robots are not meant to replace humans but to work alongside them [12].

Another core focus of Industry 5.0 is sustainability and human well-being. While Industry 4.0 prioritized efficiency and productivity, Industry 5.0 emphasizes responsible production, environmental protection, lowering carbon footprints, and improving quality of life. This includes the adoption of green technologies such as renewable energy, sustainable supply chains, and smart agriculture [13, 14].

Because humans are central to this revolution, issues of security, privacy, ethics, and innovation are critical. For example, blockchain can increase supply chain transparency, quantum computing can strengthen cybersecurity, and AI can help balance privacy with security while enhancing both [15].

2.1 Industrial applications of Industry 5.0

Industry 5.0 extends industrial transformation by emphasizing human-centric innovation, sustainability, and collaboration between advanced technologies and human expertise [16]. Here are some of the main applications:

- Human-centric manufacturing: Customizing products based on individual customers' needs.
- Sustainable manufacturing: Factories powered by renewable energy.
- Circular economy: Advanced recycling and reusing resources to minimize waste.
- Collaborative AI-powered robots: Robots that understand human language and gestures, and adapt accordingly.

- Industrial bioengineering: Producing bio-based materials, such as bioplastics.
- Smart materials: Materials that repair themselves or adapt to environmental changes.
- Distributed manufacturing: Producing goods closer to consumers, instead of relying on centralized factories.

2.2 Consumer applications of Industry 5.0

Industry 5.0 enhances consumer experiences by delivering personalized, ethical, and sustainable solutions that balance technological advancement with human well-being [17]. Here are some of the main applications:

- Next-gen wearables: Health-monitoring devices directly connected to doctors.
- Green apps: Tracking carbon footprints and energy consumption.
- Ethical digital assistants: AI systems designed to respect privacy and avoid bias.
- Extended reality for education (XR): Virtual labs and immersive learning for schools and universities.
- AI-personalized shopping: Products designed specifically for each individual consumer.
- Home medical devices: Smart tools for diagnosis and treatment at home.
- Virtual sustainable communities: Online platforms promoting environmental and social awareness.
- Sustainability-focused apps: Consumer apps to track and reduce carbon usage.

Briefly, Industry 4.0 was about automation, speed, and efficiency, while Industry 5.0 is about humans, sustainability, and personalization. The two revolutions complement each other, but Industry 5.0 shifts the focus from purely technological progress to responsible, human-centered innovation [18].

2.3 Comparison between the Fourth and Fifth Industrial Revolutions

The following table highlights the key differences between the Fourth and Fifth Industrial Revolutions in terms of technology, societal impact, and the nature of innovation [18, 19]:

Table 1 shows a comparison between the Fourth and Fifth industrial revolutions.

Factor	Fourth revolution	Fifth revolution
Core focus	Smart automation and digital transformation.	Human-centered, collaboration with AI.
Technology	AI, IoT, robotics, big data, digital twins, computing models, blockchain.	Same technologies of the Fourth Revolution + focus on sustainability and green tech.
Human vs. machine	Machines replace humans in most processes and jobs.	Humans collaborate with machines and lead them.
Objectives	Automation, efficiency, productivity, speed.	Human well-being, sustainability, mass personalization.
Challenges	Cybersecurity, privacy, job losses.	Balancing technology and humanity, AI ethics, digital justice, balancing security and privacy.

Table 1.
Comparison between the Fourth and Fifth industrial revolutions.

3. The growing tension for Industry 4.0 and Industry 5.0 with security and privacy risks

Industry 4.0 focused heavily on digital transformation and automation through technologies such as IoT, cloud computing, AI, and robotics. However, this progress also highlighted a major tension: the balance between innovation and efficiency on one hand, and security and privacy on the other [20].

To achieve smart automation and seamless connectivity, Industry 4.0 requires continuous flows of massive amounts of industrial, commercial, and even personal user data. This constant connectivity, however, also exposes systems to cyberattacks and hacking risks. A successful attack could disrupt supply chains, damage infrastructure, or leak sensitive information [21].

The constant sharing of data among service providers, factories, suppliers, and customers raises additional issues of trust, confidentiality, intellectual property protection, and personal privacy. A breach of an industrial control platform or the leakage of production data could cause huge financial losses to competitors. Likewise, the exposure of user data could harm all parties involved [22].

Thus, Industry 4.0 faces an ongoing challenge: finding the right balance between harnessing digital technologies and ensuring strong cybersecurity and privacy protection for both organizations and individuals.

In Industry 5.0, these concerns take on even greater importance because this revolution emphasizes human-centric goals. Cybersecurity and privacy are treated as essential pillars in the development and integration of modern technologies. The focus on personalization, human–machine collaboration, and sustainability naturally demands the ethical use of technology, ensuring human well-being and the protection of sensitive data [23].

However, personalization and integration also mean collecting and processing vast amounts of personal, behavioral, and biometric data – from wearable devices, customized manufacturing systems, or AI platforms that monitor health, preferences, and daily activities. While this human-centric approach promises empowerment, inclusiveness, and ethical use of technology, it is not enough on its own. Attackers are becoming more advanced, with diverse tools and greater resources, leading to more frequent and more sophisticated cyberattacks [24].

This raises serious concerns about privacy, especially when it comes to highly sensitive data from wearables, behavioral monitoring, or biometric records. Any data breach, unauthorized data exploitation, identity theft, or even bias by service providers could undermine trust in the very technologies that Industry 5.0 seeks to promote [25].

Therefore, Industry 5.0 must also strike a balance between innovation and governance. Strong and unified frameworks, strict cybersecurity measures, and robust privacy-preserving technologies tailored to each system or application are essential. Only then can the pursuit of human progress truly serve humanity – without turning into new risks for the very people it seeks to empower.

3.1 Key elements of protection in the age of industrial revolutions

- Ethical innovation [26]
  This means developing technologies and applications in ways that respect human values, such as transparency, fairness, and non-harm. Developers should always ask themselves: Is what I am building ethically sound? Does it avoid harming people or the environment? Does it serve the greater good? Examples include designing algorithms free from bias or building systems that reduce carbon footprints.

- Governance frameworks [27]
  This refers to laws, policies, and regulations that ensure the safe, transparent, and fair use of data to build trust between users and service providers. Governments and organizations need to set rules for AI use, data access, and the responsibilities of different stakeholders. A well-known example is the European GDPR (adopted in 2016, enforced in 2018), as well as U.S. and Saudi data protection laws.

- Security [28]
  Security focuses on protecting systems from hacking, denial-of-service attacks, and unauthorized access, while ensuring availability and safeguarding data against theft or tampering. It also covers protecting critical infrastructure and maintaining business continuity. Security is often summarized by the CIA triad:

  o Confidentiality: Keeping data hidden using encryption.

  o Integrity: Ensuring data is not altered through hashing and digital signatures.

  o Availability: Detecting and handling intrusions or attacks to ensure continuous service.

- Privacy [29]
  Privacy focuses on individuals' rights to protect their data from unauthorized use and to control how it is shared. It also addresses preventing misuse of data for political or commercial purposes and avoiding identity disclosure, profiling, or unethical tracking. Privacy is described by the IPT triad:

  o Identification: Preventing identity disclosure by using pseudonyms or anonymization.

- ○ Profiling: Preventing the collection of excessive user data that could be analyzed to reveal sensitive personal information.
- ○ Tracking: Preventing the monitoring of a user's location or behavior to stop unethical targeting.

- Trust [30]
  Trust is the ultimate outcome of applying protection measures. It reflects how confident users feel about relying on technologies and systems safely. Trust is built through transparency in algorithms, codes, and clear rules regarding how data is collected, used, and shared.

In reality, while governance frameworks and ethical policies are critical, they are not enough on their own. Attackers are constantly advancing their techniques, and some service providers may secretly bypass regulations. For this reason, many technical solutions for security and privacy have emerged. Before diving into these solutions, however, it is important to review the modern risks that accompany Industry 4.0 and Industry 5.0, along with some real-world use cases.

4. Security and privacy threats in Industry 4.0

Industry 4.0 has brought a radical transformation in the way goods and services are produced and delivered. It has merged digital and physical systems into one interconnected environment, relying heavily on big data, AI, advanced communications, IoT, digital twins, smart transportation, robotics, and more.

However, these emerging technologies also carry significant security, privacy, and trust challenges, mainly due to their dependence on data. Below are some examples of threats targeting the main pillars of Industry 4.0:

1. Cyberattacks on infrastructure [31]
   - ○ Continuous internet connectivity makes industrial systems vulnerable.
   - ○ The *Stuxnet* attack on SCADA systems, which disrupted factories and power plants using malicious software, was a significant event.
   - ○ Insider attacks or the planting of malware at physical sites can also cause destructive outcomes.
2. Industrial IoT (IIoT) breaches [32]
   - ○ Poorly secured sensors and robots can be exploited.
   - ○ Attackers may disrupt production lines, manipulate robots, or compromise product quality.
3. Big data security and privacy [33]
   - ○ Continuous collection of massive data exposes users and companies.

- Example: analyzing household energy use could reveal when no one is home; analyzing industrial data could expose trade secrets or sensitive designs.

4. AI and machine learning attacks [34]

    - AI can also be weaponized.
    - *Adversarial attacks* can trick robotic vision systems.
    - Manipulating supply chain algorithms can lead to incorrect decisions.
    - Ransomware (e.g., *Colonial Pipeline*) can halt operations by encrypting systems and demanding payment.

5. Blockchain vulnerabilities [35]

    - While blockchain provides higher security and transparency, it is not immune.
    - Attacks include smart contract manipulation or *51% attacks* to alter transaction records.
    - Attackers may even collaborate in "attack pools."

6. Cloud computing threats [36]

    - Heavy reliance on cloud storage increases risks.
    - *DDoS attacks* can cripple service providers.
    - Poor configurations can lead to major data leaks.

7. Digital twin exploits [37]

    - Virtual replicas of factories or cities can be hacked.
    - Manipulating a power plant's digital twin, for instance, could cause real blackouts.
    - Breaching smart city twins could expose traffic patterns, population data, and private behaviors.

8. 3D printing (additive manufacturing) risks [38]

    - Interference with digital design files could insert hidden flaws into medical, military, or aerospace parts.
    - Intellectual property theft and unauthorized copying are also significant risks.

5. Security and privacy threats in Industry 5.0

Unlike Industry 4.0, which emphasized automation, Industry 5.0 focuses on human–machine collaboration, sustainability, resilience, and ethical innovation. However, these pillars also bring new cybersecurity and privacy challenges:

1. Sustainable smart factories [39]
    - o Rely on AI, human creativity, and clean energy.
    - o Constant monitoring and sensors may leak sensitive data if compromised.
2. Collaborative robots (Co-bots) [40]
    - o Malicious software could turn co-bots into physical threats to workers.
    - o They could also collect sensitive employee data without consent.
3. Personalized products [41]
    - o Require detailed consumer data.
    - o If misused or leaked, this could lead to severe privacy violations.
4. Personalized medicine and genetic data [42]
    - o Genetic data is more sensitive than traditional medical data.
    - o Misuse could have devastating personal consequences.
5. Smart homes [43]
    - o Hacked IoT devices (locks, cameras, AC systems) could enable break-ins or spying.
    - o Devices like Alexa may record conversations and activities, creating massive privacy risks.
6. Personalized and adaptive education [44]
    - o Manipulated algorithms could push biased or harmful content.
    - o Some adaptive games and platforms have been linked to psychological harm or even suicide.
7. Over-reliance on AI [45]
    - o Automated decisions may be biased or opaque.

- o Attackers can exploit or manipulate AI models to cause injustice.

8. Immersive experiences (metaverse, XR) [46]

   - o Used in tourism, training, or entertainment.
   - o Tampering could cause physical harm or allow attackers to deeply profile user behavior.

9. Autonomous vehicles (cars and drones) [47]

   - o Increasingly common in transport, logistics, healthcare, and defense.
   - o Hacking could redirect or disable them, causing accidents or enabling surveillance of passengers.

10. Wearable devices [48]

    - o Collect sensitive health metrics (heart rate, oxygen, blood sugar).
    - o Vulnerable to hacking, interception, and insecure cloud storage.

11. Virtual sustainable communities [49]

    - o Platforms for environmental and social awareness.
    - o Risks include spreading false information, data misuse, and targeted cyberattacks on successful platforms.

Briefly, the Industry 4.0 threats focus more on infrastructure, automation systems, and industrial data, while the Industry 5.0 threats expand to include human-centric risks, such as privacy of personal, genetic, and behavioral data, ethical manipulation, and physical safety in collaborative environments.

6. Current security and privacy protection techniques

Security techniques can be grouped into several main categories, each containing many methods and approaches. In general, they all derive from the three classic security objectives: confidentiality, integrity, and availability. Below are some of the most common:

1. Encryption [50]

   - o The first and most important line of defense.
   - o Protects confidentiality by converting text, images, or other data into unreadable symbols.

  - Decryption is only possible with the correct key (symmetric or asymmetric).

2. Firewalls [51]

   - Monitor network traffic and block unauthorized access using IP or MAC addresses.

3. Antivirus systems [52]

   - Detect and remove malware, such as viruses, trojans, and worms, which are constantly evolving.

4. Authentication [53]

   - Verifies the user's identity before granting access.

   - Can use usernames/passwords, biometrics (fingerprints), or multi-factor authentication (e.g., SMS or email codes).

5. Intrusion detection systems [54]

   - Monitor systems for unusual activity or intrusion attempts.

   - May apply patches, detect repeated requests from one source, or use AI to identify suspicious behavior.

6. Identity and access management [55]

   - Define and enforce user permissions so that each user or device accesses only authorized resources.

7. Backups [56]

   - Keep multiple copies of data for recovery in case of loss or attack.

8. Digital Signatures [57]

   - Ensure data integrity and authenticity using hashing and cryptographic signatures.

9. Steganography [58]

   - Hide secret data inside other media (text, images, audio, video).

   - Often combined with encryption for highly sensitive communications.

10. Blockchain [59]

- o Distributed ledger system with encrypted, linked blocks that cannot be easily altered.
- o Provides transparency and strong protection but requires majority control (51% attack) to break.

7. Privacy protection techniques

Privacy focuses on preventing identity exposure, tracking, or profiling. Key methods include:

1. Encryption: can enhance privacy, but service providers (who hold the data) may themselves be a threat [25].
2. Anonymization/Pseudonyms: Replace real identities with codes or nicknames [60].
3. Mix zones: Regularly change identities over time to enhance privacy [61].
4. Dummy data: Send misleading or false data alongside real data to confuse attackers [62].
5. Obfuscation: Modify or add noise to data (e.g., slightly altering GPS coordinates) [63].
6. Data summarization: Send only aggregated summaries instead of full raw data [64].
7. Trusted third party: Hide user identity from service providers through an intermediary [65].
8. Cloak area: Multiple users share the same location data or queries to mask individuals [66].
9. Peer cooperation: Users share and relay requests to reduce direct exposure to providers [67].
10. Private information retrieval: Request large datasets so that providers cannot identify what the user really wants [68].

Some hybrid techniques combine multiple approaches for stronger protection, though often at the cost of performance. Unfortunately, all current methods face growing challenges as attackers gain new capabilities, especially with AI tools and quantum computing, which threaten to break traditional defenses.

8. Future recommendations for security and privacy

To meet new challenges, researchers propose modern approaches such as:

1. Dynamic protection and adaptive security [69]
   - No single method fits all apps and devices.
   - Systems should automatically analyze the nature of the application or service and apply the best protection using AI.
   - Depend on these phases (detect, assess, adapt, verify, and log).
2. Balancing security and privacy [70]
   - Security focuses on protecting systems, while privacy focuses on user rights.
   - Excessive security (like constant monitoring) can violate privacy, while too much privacy may weaken security.
   - AI-based models can help find the best balance for each application.
3. Zero-trust security [71]
   - Do not assume any user is permanently trusted.
   - Require repeated authentication at different stages instead of a one-time login.
4. Decentralized architectures [72]
   - Use blockchain for transparency and strong protection, but explore hybrid models to reduce privacy risks.
5. AI-driven security and privacy [73]
   - Employ AI to detect anomalies, malicious devices, or unusual traffic.
   - AI is also critical to defending against deepfakes and generative AI–based attacks.
6. Homomorphic encryption [74]
   - Promising technique allows computations on encrypted data without decryption.
   - Still limited to simple operations, but it could significantly enhance both security and privacy.
7. Quantum-resistant cryptography [75]
   - While quantum computing threatens traditional encryption, it also offers opportunities to create ultra-secure cryptographic methods.

8. Ethical innovation [76]

   o Establish policies for AI-generated content and ownership.

   o Develop verification and certification tools (e.g., NFT-based or government-backed platforms) to ensure accountability.

9. Digital fairness [77]

   o Ensure equitable access to modern technologies and reduce the digital divide.

   o Promote inclusive participation while safeguarding privacy and security.

10. Privacy-by-design is important for IIoT/CPS.

    o Proposed a new Software Development Life Cycle that focuses on security and privacy in the analysis and design phases, not only in the implementation phase [78].

Briefly, current techniques provide some level of protection, but evolving threats demand adaptive, AI-enhanced, and ethical solutions. Future approaches must balance security, privacy, trust, and fairness to ensure technology remains a tool for human progress rather than a source of new risks.

9. Conclusion

In conclusion, the rapid transformations brought by the Fourth and Fifth Industrial Revolutions demonstrate both the promise and the peril of emerging technologies. Industry 4.0 revolutionized efficiency and productivity through automation, connectivity, and data-driven intelligence, but it also introduced unprecedented challenges to security, privacy, and trust. The rise of interconnected devices, big data, AI, cloud computing, and digital twins has expanded the attack surface and exposed individuals, organizations, and societies to new forms of cyber and privacy risks.

Industry 5.0 seeks to go beyond these limitations by re-centering technology around the human being. It emphasizes sustainability, ethical innovation, and the integration of human creativity with intelligent machines. Yet, this new paradigm also magnifies risks by relying on massive amounts of sensitive personal, behavioral, and even biological data, requiring stronger safeguards to protect individuals and maintain trust.

Existing security and privacy techniques (such as encryption, authentication, intrusion detection, anonymization, and blockchain) offer critical foundations, but they are increasingly insufficient in the face of evolving threats powered by AI, advanced malware, and quantum computing. This reality demands forward-looking strategies that balance security and privacy, leverage adaptive and AI-driven defenses, employ zero-trust models, and promote ethical governance frameworks.

Ultimately, ensuring a secure and human-centric digital future will depend on harmonizing innovation with protection, efficiency with fairness, and automation

DOI: http://dx.doi.org/10.5772/intechopen.1013221

with ethics. The success of Industry 5.0 will not be measured solely by technological advancement but by its ability to build sustainable, trustworthy, and inclusive systems that safeguard both individuals and society at large.

Acknowledgment

I acknowledge the use of ChatGPT for English language polishing and paraphrasing of this chapter. I remain solely responsible for the content and its interpretation.

Author details

Ahmed Abi Sen Adnan
Hussein ElSayyed Research Center, Deanship of Graduate Studies and Scientific Research, University of Prince Mugrin, Madinah, Saudi Arabia

*Address all correspondence to: A.ahmed@upm.edu.sa Adnanmnm@hotmail.com

References

[1] Chou SY. The fourth industrial revolution. Journal of International Affairs. 2018;**72**(1):107–120

[2] Martinelli A, Mina A, Moggi M. The enabling technologies of industry 4.0: Examining the seeds of the fourth industrial revolution. Industrial and Corporate Change. 2021;**30**(1):161–188

[3] Monteiro ACB, Bonacin R, França RP, Arthur R. An overview of IIoT related to the modern panorama of the industrial field. Smart Cities: Innovations, Challenges and Future Perspectives. 2024;**1.0**:115–134. DOI: 10.1007/978-3-031-59846-3_5

[4] Subashini P, and Krishnaveni M. Artificial intelligence-based assistive technology. In: Kaliraj P, Devi T. Artificial Intelligence Theory, Models, and Applications. New York: Auerbach Publications. 2021. p. 217–240. DOI: 10. 1201/9781003175865

[5] Do Livramento Gonçalves G, Leal Filho W, da Silva Neiva S, Borchardt Deggau A, de Oliveira Veras M, Ceci F, . . . Salgueirinho Osório de Andrade Guerra JB. The impacts of the fourth industrial revolution on smart and sustainable cities. Sustainability. 2021;**13**(13):7165

[6] Kauf S. Smart City in the era of the fourth industrial revolution. Zeszyty Naukowe. Organizacja I Zarządzanie/Politechnika Śląska. 2020;**145**:211–220

[7] Xu M, David JM, Kim SH. The fourth industrial revolution: Opportunities and challenges. International Journal of Financial Research. 2018;**9**(2):90–95

[8] Onik MMH, Chul-Soo KIM, Jinhong YANG (2019, February). Personal data privacy challenges of the fourth industrial revolution. *2019 21st International Conference on Advanced Communication Technology (ICACT)*. (pp. 635–638). IEEE

[9] Zheng T, Ardolino M, Bacchetti A, Perona M. The applications of Industry 4.0 technologies in manufacturing context: A systematic literature review. International Journal of Production Research. 2021;**59**(6):1922–1954

[10] Sima V, Gheorghe IG, Subić J, Nancu D. Influences of the industry 4.0 revolution on the human capital development and consumer behavior: A systematic review. Sustainability. 2020;**12**(10):4035

[11] Ziatdinov R, Atteraya MS, Nabiyev R. The fifth industrial revolution as a transformative step towards society 5.0. Societies. 2024;**14** (2):19

[12] Gauri PRATIK, Van Eerden J. A 5th industrial revolution?(What it is, and why it matters). World Economic Forum. 2019;**3**:1-6

[13] Akubo AA, Odiji OL, Muhammed MB. Fifth industrial revolution and an overview of its impact on human resources. Human Capital Analytics: Exploring the HR Spectrum in Industry 5.0. 2025;**1.0**:65–80. DOI: 10.1002/9781394238354.ch3

[14] Briken K, Moore J, Scholarios D, Rose E, Sherlock A. Industry 5 and the human in human-centric manufacturing. Sensors. 2023;**23** (14):6416

[15] Palle RR, Kathala KCR. Balance between security and privacy. In: Privacy in the Age of Innovation: AI Solutions for Information Security. Berkeley, CA: Apress. 2024. p. 129–135 DOI: 10.1007/979-8-8688-0461-8_11

[16] Ghobakhloo M, Iranmanesh M, Tseng ML, Grybauskas A, Stefanini A, Amran A. Behind the definition of Industry 5.0: A systematic review of technologies, principles, components, and values. Journal of Industrial and Production Engineering. 2023;**40** (6):432–447

[17] Ayub MF, Li X, Mahmood K, Shamshad S, Saleem MA, Omar M. Secure consumer-centric demand response management in resilient smart grid as industry 5.0 application with blockchain-based authentication. IEEE Transactions on Consumer Electronics. 2023;**70**(1):1370–1379

[18] Golovianko M, Terziyan V, Branytskyi V, Malyk D. Industry 4.0 vs. Industry 5.0: Co-existence, transition, or a hybrid. Procedia Computer Science. 2023;**217**:102–113

[19] Aveni A. Industry 5.0 and industry 4.0 comparative definitions. Revista Processus de Estudos de Gestão, Jurídicos E Financeiros. 2023;**14** (46):01–15

[20] Alabdulatif A, Thilakarathne NN, Lawal ZK. A review on security and privacy issues pertaining to cyber-physical systems in the industry 5.0 Era. Computers, Materials & Continua. 2024;**80**(3):3917. DOI: 10.32604/cmc.2024.054150

[21] Saeed MM, Saeed RA, Ali ES, Ahmed ZE, Mokhtar RA. Privacy and security challenges of AI integration in industry 4.0 and 5.0. In: Joseph Bamidele, Awotunde, Kamalakanta, Muduli, Biswajit, Brahma. Computational Intelligence for Analysis of Trends in Industry 4.0 and 5.0. New York: Auerbach Publications; 2025. p. 353–372

[22] Adeniyi AE, Jimoh RG, Awotunde JB, Ninan DO, Aworinde HO, Oyebade A, Babatunde AOS. Privacy, trust, and other issues in industries 4.0 and 5.0. In: Joseph Bamidele, Awotunde, Kamalakanta, Muduli, Biswajit, Brahma 1. Computational Intelligence in Industry 4.0 and 5.0 Applications. New York: Auerbach Publications; 2025. p. 132–160. DOI: 10.1201/9781003581963

[23] Shaikh ZA, Hajjej F, Uslu YD, Yüksel S, Dınçer H, Alroobaea R, … Chinta U. A new trend in cryptographic information security for industry 5.0: A systematic review. IEEE Access. 2024;**12**:7156–7169

[24] Dmitrieva E, Balmiki V, Bhardwaj N, Kumar K, Sharma A, Shruthi CM (2024). Security and privacy in AI-driven Industry 5.0: Experimental insights and threat analysis. *BIO Web of Conferences* (Vol. **86**, p. 01097). EDP Sciences

[25] Abi Sen AA, Eassa FA, Jambi K, Yamin M. Preserving privacy in internet of things: A survey. International Journal of Information Technology. 2018;**10**(2):189–200.

[26] Diller ER, Williamson L. Supporting one health for pandemic prevention: The need for ethical innovation. Journal of Bioethical Inquiry. 2023;**20**(3):345–352

[27] Julakanti SR, KiranmayeeSattiraju NS, Julakanti R.

Data protection through governance frameworks. arXiv. 2025; **arXiv**: 2502.10404. *preprint*

[28] Yee CK, Zolkipli MF. Review on confidentiality, integrity and availability in information security. Journal of ICT in Education. 2021;**8**(2):34–42

[29] Bahbouh N, Basahel A, Sendra S, Abi Sen AA. Tokens shuffling approach for privacy, security, and reliability in IoHT under a pandemic. Applied Sciences. 2022;**13**(1):114

[30] Wilshire HG. The significance of the trust. In: Peter J, Cain, Mark Harrison. Imperialism. London: Routledge; 2023. p. 208–222 DOI: 10.4324/9781003101512

[31] Irmak E, Erkek İ. (2018, March). An overview of cyber-attack vectors on SCADA systems. In: *2018 6th international symposium on digital forensic and security (ISDFS)*. (pp. 1–5). IEEE

[32] Kumar R, Kandpal B, Ahmad V (2023, March). Industrial IoT (IIOT): Security threats and countermeasures. In: *2023 International Conference on Innovative Data Communication Technologies and Application (ICIDCA)*. (pp. 829–833). IEEE

[33] Sollins KR. IoT big data security and privacy versus innovation. IEEE Internet of Things Journal. 2019;**6**(2):1628–1635

[34] Hossain MT, Afrin R, Biswas MAA. A review on attacks against artificial intelligence (AI) and their defence image recognition and generation machine learning, artificial intelligence. Control Systems and Optimization Letters. 2024;**2**(1):52–59

[35] Averin A, Averina O (2019, October). Review of blockchain technology vulnerabilities and blockchain-system attacks. *2019 International Multi-Conference on Industrial Engineering and Modern Technologies (FarEastCon)*. (pp. 1–6). IEEE

[36] Alouffi B, Hasnain M, Alharbi A, Alosaimi W, Alyami H, Ayaz M. A systematic literature review on cloud computing security: Threats and mitigation strategies. IEEE Access. 2021;**9**:57792–57807

[37] Alcaraz C, Lopez J. Digital twin: A comprehensive survey of security threats. IEEE Communications Surveys & Tutorials. 2022;**24**(3):1475–1503

[38] El-Sayegh S, Romdhane L, Manjikian S. A critical review of 3D printing in construction: Benefits, challenges, and risks. Archives of Civil and Mechanical Engineering. 2020;**20**(2):34

[39] Yi KJ, Jeong YS. Smart factory: Security issues, challenges, and solutions. Journal of Ambient Intelligence and Humanized Computing. 2022;**13**(10):4625–4638

[40] Abishek BA, Kavyashree T, Jayalakshmi R, Tharunkumar S, Raffik R (2023, June). Collaborative robots and cyber security in industry 5.0. *2023 2nd International Conference on Advancements in Electrical, Electronics, Communication, Computing and Automation (ICAECA)*. (pp. 1–6). IEEE

[41] Xu L, Li Y, Yao Q. Information security investment and purchase decision for personalized products. Managerial and Decision Economics. 2022;**43**(6):2619–2635

[42] Wei J, Lin Y, Yao X, Zhang J, Liu X. Differential privacy-based genetic matching in personalized medicine. IEEE Transactions on Emerging Topics in Computing. 2020;**9**(3):1109–1125

[43] Edu JS, Such JM, Suarez-Tangil G. Smart home personal assistants: A security and privacy review. ACM Computing Surveys. 2020;**53**(6):1–36

[44] Walker KL, Bodendorf K, Kiesler T, de Mattos G, Rostom M, Elkordy A. Compulsory technology adoption and adaptation in education: A looming student privacy problem. Journal of Consumer Affairs. 2023;**57** (1):445–478

[45] Kumar A (2023). When AI fails: The risks of over-reliance on machine learning for healthcare security

[46] Huang Y, Li YJ, Cai Z. Security and privacy in metaverse: A comprehensive survey. Big Data Mining and Analytics. 2023;**6**(2):234–247

[47] Balasubramanian V, Aloqaily M, Guizani M, Ouni B. Security challenges and solutions for autonomous vehicles and drones in the AI age. IEEE Internet of Things Magazine. 2025;**8**(2):129–136

[48] Cilliers L. Wearable devices in healthcare: Privacy and information security issues. Health Information Management Journal. 2020;**49** (2-3):150–156

[49] Dziundziuk VB, Kotukov OA, Hryn DV, Kotukh EV (2021). Spread of virtual communities as a potential threat to state security and sustainable well-being. *Rivista Di Studi Sulla Sostenibilita'*. (**2020**/2 suppl.)

[50] Singh S, Sharma PK, Moon SY, Park JH. Advanced lightweight encryption algorithms for IoT devices: Survey, challenges and solutions. Journal of Ambient Intelligence and Humanized Computing. 2024;**15** (2):1625–1642

[51] Alicea M, Alsmadi I. Misconfiguration in firewalls and network access controls: Literature review. Future Internet. 2021;**13**(11):283

[52] Samociuk D. Antivirus evasion methods in modern operating systems. Applied Sciences. 2023;**13**(8):5083

[53] Sodhro AH, Awad AI, van de Beek J, Nikolakopoulos G. Intelligent authentication of 5G healthcare devices: A survey. Internet of Things. 2022;**20**:100610

[54] Abdulganiyu OH, Ait Tchakoucht T, Saheed YK. A systematic literature review for network intrusion detection system (IDS). International Journal of Information Security. 2023;**22** (5):1125–1162

[55] Ghaffari F, Gilani K, Bertin E, Crespi N. Identity and access management using distributed ledger technology: A survey. International Journal of Network Management. 2022;**32**(2):e2180

[56] Maher R, Nasr OA. DropStore: A secure backup system using multi-cloud and fog computing. IEEE Access. 2021;**9**:71318–71327

[57] Bisheh-Niasar M, Azarderakhsh R, Mozaffari-Kermani M. Cryptographic accelerators for digital signature based on Ed25519. IEEE Transactions on Very Large Scale Integration (VLSI) Systems. 2021;**29**(7):1297–1305

[58] Alkodre AB, Bahbouh NM, Sendra S, Sen AAA, Alsaawy Y,

Alqahtany SS, . . . Almoamari H. A shuffling-steganography algorithm to protect data of drone applications. Computers, Materials & Continua. 2024;**81**(2):2727–2751

[59] Sen AAA, Bahbouh NM, Sendra S, Tayan O. PoPR: A novel blockchain consensus mechanism for reliable crowdsourced data. International Journal of Information Technology. 2025;**1**:1–13. DOI: 10.1007/s41870-025-02547-4

[60] Alsaawy Y, Alkhodre A, Abi Sen A. NiPA: Nicknames pool approach for protecting privacy of user data in intelligent transportation systems. WSEAS Transactions on Computers. 2022;**21**:211–220

[61] Abi Sen AA, Alnsour A, Aljwair SA, Aljwair SS, Alnafisah HI, Altamimi BA (2021, March). Fog mix-zone approach for preserving privacy in IOT. *2021 8th International Conference on Computing for Sustainable Global Development (INDIACom)*. (pp. 405–408). IEEE

[62] Yamin M, Abi Sen AA. A new method with swapping of peers and fogs to protect user privacy in IoT applications. IEEE Access. 2020;**8**:210206–210224

[63] Albouq SS, Abi Sen AA, Namoun A, Bahbouh NM, Alkhodre AB, Alshanqiti A. A double obfuscation approach for protecting the privacy of IoT location based applications. IEEE Access. 2020;**8**:129415–129431

[64] Almutairi MM, Apostolopoulou D, Halikias G, Abi Sen AA, Yamin M (2022, March). Enhancing privacy and security in crowds using fog computing. *2022 9th International Conference on Computing for Sustainable Global Development (INDIACom)*. (pp. 57–62). IEEE

[65] Yamin M, Alsaawy Y, B. Alkhodre A, Abi Sen AA. An innovative method for preserving privacy in Internet of Things. Sensors. 2019;**19**(15):3355

[66] Bahbouh NM, Alkhodre AB, Sendra S, Abi Sen AA, Alsaawy Y, Benaida M, Almoamari H. Double cloak area approach for preserving privacy and reliability of crowdsourcing data. IEEE Access. 2024;**12**:100529–100543

[67] Abi Sen AA, Nazar SKA, Osman NA, Bahbouh NM, Aloufi HF, Alawfi IMM (2021, March). A new technique for managing reputation of peers in the cooperation approach for privacy protection. *2021 8th International Conference on Computing for Sustainable Global Development (INDIACom)*. (pp. 409–412). IEEE

[68] Almutairi MM, Abi Sen AA, Yamin M (2021, March). Survey of PIR Approach and its Techniques for Preserving Privacy in IoT. *2021 8th International Conference on Computing for Sustainable Global Development (INDIACom)*. (pp. 417–421). IEEE

[69] Abi Sen AA. A comprehensive privacy and security framework for dynamic protection (CPSF). International Journal of Information Technology. 2022;**14**(5):2477–2485

[70] Abi Sen AA, Mnaouer AB, Tayan O, Al-Hammouri A, Askool S (2025, February). A model for balancing data security and privacy in IoT applications. *2025 Global Information Infrastructure and Networking Symposium (GIIS)*. (pp. 1–6). IEEE

[71] He Y, Huang D, Chen L, Ni Y, Ma X. A survey on zero trust architecture: Challenges and future trends. Wireless Communications and Mobile Computing. 2022;**2022** (1):6476274

[72] Rathore S, Kwon BW, Park JH. BlockSecIoTNet: Blockchain-based decentralized security architecture for IoT network. Journal of Network and Computer Applications. 2019;**143**: 167–177

[73] Almasabi AM, Alkhodre AB, Khemakhem M, Eassa F, Abi Sen AA, Harbaoui A. Internet of things-based anomaly detection hybrid framework simulation integration of deep learning and blockchain. Information. 2025;**16** (5):406

[74] Marcolla C, Sucasas V, Manzano M, Bassoli R, Fitzek FH, Aaraj N. Survey on fully homomorphic encryption, theory, and applications. Proceedings of the IEEE. 2022;**110** (10):1572–1609

[75] Tambe-Jagtap SN. A survey of cryptographic algorithms in cybersecurity: From classical methods to quantum-resistant solutions. SHIFRA. 2023;**2023**:43–52

[76] Calvo P. Cryptoart: Ethical challenges of the NFT revolution. Humanities and Social Sciences Communications. 2024;**11**(1):1–10

[77] Helberger N, Kas B, Micklitz HW, Namysłowska M, Naudts L, Rott P, … Veale M. Towards digital fairness. Journal of European Consumer and Market Law. 2024;**13**(1):24–30

[78] Sen AAA, et al. "A new SDLC model for enhancing privacy and security in a design-based approach," 2025 12th International Conference on Computing for Sustainable Global Development (INDIACom). Delhi, India, 2025, pp. 1–6, DOI: 10.23919/INDIACom66777.2025.11115282

Chapter 17

Perspective Chapter: AI-Enhanced Digital Product Passports – Enabling ESG Transparency in Industry 5.0

Hans-Dieter Wehle

Abstract

This chapter explores how digital product passports (DPPs), enhanced by artificial intelligence (AI), can support greater environmental, social, and governance (ESG) transparency in the context of Industry 5.0. It introduces a modular software architecture that combines machine learning techniques for sustainability management – such as forecasting carbon emissions, detecting anomalies, and predicting environmental impacts – with advanced language models for interpreting sustainability documentation. The chapter also discusses current European regulatory drivers, including the Corporate Sustainability Reporting Directive (CSRD) and the European Green Deal, and highlights the practical challenges companies face in integrating sustainability data across complex supply chains. A special focus is placed on responsible data agents, which are AI-powered interfaces that allow human users to interact with sustainability data through natural language queries and receive transparent, traceable explanations. Drawing from applied research and pilot implementations, the proposed framework demonstrates how manufacturing companies can use digital tools and AI-based decision support to manage compliance more efficiently and advance sustainable industrial practices.

Keywords: digital product passport, ESG reporting, Industry 5.0, artificial intelligence in manufacturing, CO_2 emissions forecasting, RDAs, machine learning for sustainability

1. Introduction

In recent years, the discourse around industrial transformation has expanded beyond the boundaries of automation and digitalization. With the introduction of Industry 5.0, the European Commission proposed a complementary paradigm that places human-centricity, sustainability, and resilience at the core of industrial innovation [1]. While Industry 4.0 emphasized connectivity, cyber-physical systems, and data-driven efficiency, Industry 5.0 highlights the value of technology for society, bridging digital progress with environmental and social responsibility.

This shift comes at a time when regulatory frameworks, such as the European Green Deal, the Corporate Sustainability Reporting Directive (CSRD), and the proposed digital product passport (DPP), are reshaping how companies must document, evaluate, and communicate the environmental and social impact of their products and operations. Particularly for manufacturing companies and their supply chains, this creates new challenges in data integration, semantic interoperability, and compliance monitoring.

This chapter introduces a modular platform architecture that leverages artificial intelligence (AI) to support environmental, social, and governance (ESG) transparency in line with Industry 5.0 principles. The system integrates machine learning (ML) for sustainability (such as CO_2 forecasting and risk assessment) and large language models (LLMs) to interpret unstructured ESG documentation. Furthermore, it introduces responsible data agents (RDAs) – AI-based conversational interfaces that provide real-time, voice-enabled access to structured and unstructured ESG insights.

The approach is grounded in real-world use cases from small and medium-sized enterprises (SMEs) and demonstrates how digital tools can bridge the gap between regulatory complexity and practical implementation. By connecting data, compliance, and decision-making, this contribution illustrates a scalable path toward human-centered and sustainable industrial innovation.

2. EU policy and regulatory context

Industry 5.0 is more than a technological concept; it represents a broader political and societal ambition. Defined by the European Commission, it aims to align industrial innovation with sustainability, human-centric design, and resilience. While Industry 4.0 primarily focused on automation, data integration, and operational efficiency, Industry 5.0 emphasizes using technology responsibly and inclusively to benefit both people and the planet.

To achieve these goals, several European policy frameworks have been introduced. The European Green Deal [2] and the Circular Economy Action Plan promote transparent and data-driven methods to track material flows, energy consumption, and environmental impacts throughout production and supply chains. These efforts are supported by regulatory proposals such as the Eco-design for Sustainable Products Regulation (ESPR), which sets harmonized requirements for product design and lifecycle transparency [3].

A central instrument in this effort is the DPP [4]. This initiative will make product-specific sustainability data, covering ESG factors, available across the entire value chain.

Complementing this, the CSRD [5] requires companies to provide standardized and verifiable sustainability reports. For many businesses, particularly in manufacturing, these regulations create an urgent need for digital systems that can collect, analyze, and communicate ESG-related information accurately and efficiently, ensuring compliance with evolving sustainability requirements.

Figure 1 illustrates how these European policy initiatives are connected and how they collectively define ESG requirements for companies. These requirements – such as transparency in data handling, energy efficiency, comprehensive ESG reporting, and responsible use of AI – directly support the transition toward Industry 5.0.

The presented approach is based on applied research and real-world implementation experience within a SME context. It aims to demonstrate scalable

Figure 1.
Relationship between the European Green Deal, CSRD, and the digital product passport (DPP). These policy instruments jointly define ESG requirements for companies – such as data transparency, energy efficiency, and the use of AI – which support the transition toward Industry 5.0.

pathways for integrating ESG requirements into industrial systems through digital innovation.

This chapter responds to these regulatory and technological developments with a practical and scalable AI-based solution designed to operationalize ESG transparency across industrial contexts.

3. System architecture and AI components

To translate the regulatory and strategic objectives of Industry 5.0 into actionable digital systems, this section presents a modular architecture that integrates AI into the design and operation of DPPs. It outlines the system components required to ensure scalable ESG data handling, semantic interoperability, and real-time responsiveness – factors that are essential for both academic research and practical implementation.

3.1 System overview

To address the regulatory, technological, and ethical requirements outlined in the context of Industry 5.0, the proposed system architecture follows a modular and layered design.

It enables the implementation of AI-enhanced DPPs that combine structured ESG data, real-time process monitoring, and human-centric interaction capabilities.

The architecture consists of four core layers:

Data acquisition layer: responsible for collecting information from diverse sources such as production systems, supply chain interfaces, and external regulatory databases. This includes both structured data (e.g., material flows, CO_2 values) and unstructured documents (e.g., sustainability declarations, audit reports).

AI processing layer: integrates domain-specific ML models for sustainability forecasting (e.g., emissions trends), as well as LLMs to semantically interpret ESG documentation. It also includes anomaly detection mechanisms for identifying inconsistencies or signs of greenwashing.

Application and interaction layer: features **RDAs** that enable voice-based interaction, multilingual access, and contextual explanations of ESG metrics. These agents are designed to support human users in interpreting DPP information and making informed decisions, thereby fostering transparency and trust.

Governance and compliance layer: supports rule-based validation, audit trails, and adaptive compliance mechanisms. This layer enhances organizational resilience by enabling proactive responses to changing ESG regulations or reputational risks across the value chain.

By decoupling data ingestion, intelligent processing, human-centered interaction, and compliance logic, the system is both scalable and adaptable to different industrial contexts. It facilitates the operationalization of ESG principles in real-time production environments while allowing for interpretability, traceability, and user empowerment – key aspects aligned with the vision of Industry 5.0.

Figure 2 illustrates the multilayered architecture of the AI-enhanced DPP system, showing how data ingestion, intelligent processing, human interaction, and compliance logic are structurally decoupled and integrated.

3.2 Data model and sources

The effectiveness of DPPs in industrial environments depends fundamentally on the quality, granularity, and interoperability of the underlying data. In the context of Industry 5.0, where human-centricity, sustainability, and resilience are key pillars, data models must be capable of capturing not only technical and economic parameters but also social and environmental indicators across the entire product life cycle.

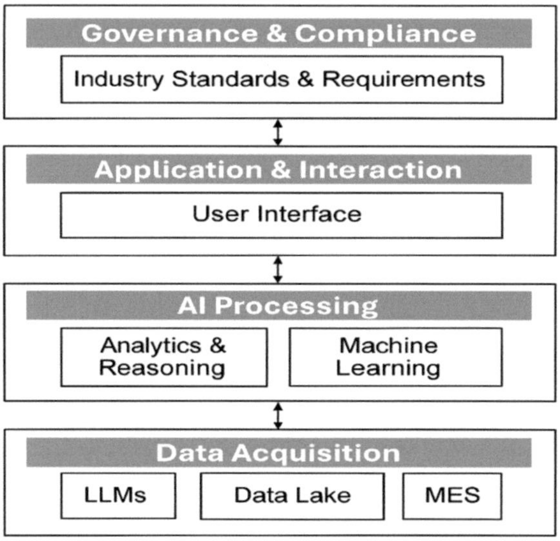

Figure 2.
Layered architecture of the FLX:DPPplus system. The architecture comprises four layers – data acquisition, AI processing, application and interaction, and governance and compliance – each designed to address specific functional, technical, and human-centric requirements in the implementation of ESG-compliant digital product passports.

The proposed system differentiates between two main data categories: static master data and dynamic operational data.

Master data includes product-specific information such as bills of materials (BOMs), technical drawings, and supply chain structures, which are typically stored in ERP, PLM, or SCM systems. These provide the structural foundation for ESG attribution, for example, by linking material content to environmental impact or supplier compliance ratings.

Operational data, in contrast, refers to real-world consumption figures and process-specific measurements. These may include energy usage during manufacturing, transport emissions, or assembly-related resource consumption. Such data is vital for calculating actual carbon footprints and for validating predictive models trained on historical patterns.

To ensure continuous and scalable data integration, the architecture supports a wide range of digital interfaces, including:

- **OPC UA** and **MQTT** for industrial machine and sensor connectivity
- **REST APIs** for platform integration (e.g., MES, cloud services)
- **CSV/XML/JSON** imports for structured data exchange

At the same time, recognizing the diversity of digital maturity levels across suppliers and manufacturing environments, the system includes options for manual data entry via structured web forms. This ensures that even organizations without fully digitized operations can contribute to and benefit from ESG-based product documentation.

All data inputs are normalized and semantically tagged upon ingestion to ensure interoperability, traceability, and compliance readiness. This enables downstream AI components to perform reliable analysis, prediction, and certification generation.

3.3 Machine learning components

A core innovation of the FLX:DPPplus system lies in its integrated use of ML to derive actionable insights from sustainability-related data. The architecture combines supervised learning techniques for environmental forecasting with large-scale language models for semantic document analysis, creating a hybrid AI framework tailored to the complexities of ESG compliance in industrial settings.

3.3.1 CO_2 emission forecasting with gradient boosting

To enable predictive sustainability management, FLX:DPPplus applies **gradient boosting rRegressor (GBR)** models for the forecasting of CO_2 emissions [6]. These models iteratively improve their accuracy by combining multiple weak learners, optimized via gradient descent, and are trained on historical energy consumption, production volumes, and process emissions.

The model's performance is continuously monitored through **mean absolute error (MAE)** and **R^2 scores**, ensuring robustness and reliability in production-oriented environments. This allows manufacturers to simulate sustainability scenarios, anticipate compliance deviations, and implement corrective measures proactively – a key requirement under Industry 5.0, where resilience and foresight are as important as efficiency.

3.3.2 Natural language processing for ESG documentation

In parallel, the system integrates **LLMs** – such as transformer-based architectures – to interpret unstructured documents like audit reports, supplier declarations, or regulatory texts. These models are fine-tuned on ESG-specific corpora and support tasks like:

- Semantic classification of document contents,
- Extraction of relevant sustainability indicators
- Risk flagging for noncompliant statements

This approach significantly reduces the manual effort of ESG reporting and helps mitigate greenwashing risks by linking linguistic content to structured ESG data fields. To ensure fairness and reliability, all ML models undergo regular bias detection and validation procedures. Prediction outputs are monitored for systematic deviations that might disadvantage certain suppliers, product types, or lifecycle phases. These checks are incorporated into periodic model retraining and hyperparameter tuning to maintain ethical, trustworthy decision support in sustainability assessments.

3.3.3 Data handling via Lambda architecture

Both ML components are embedded within a Lambda architecture, allowing for a combination of batch processing (e.g., historical CO_2 trends) and stream processing (e.g., real-time sensor or logistics data). As shown in **Figure 3**, this ensures that ML pipelines can react to both planned reporting cycles and unexpected deviations in production or supply chain behavior.

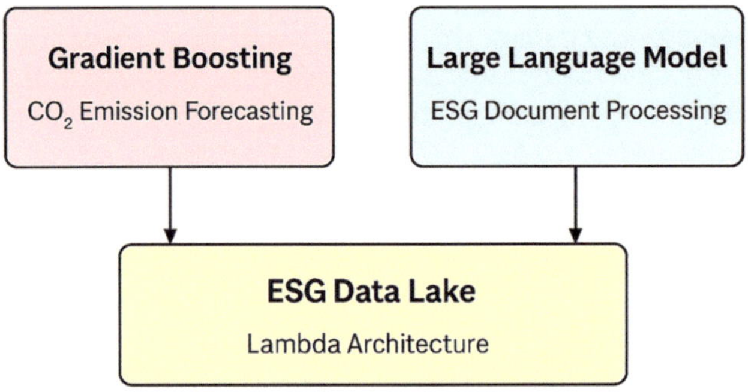

Figure 3.
Machine learning components of the FLW:DPPplus platform. The system integrates predictive models (gradient boosting) for CO_2 emission forecasting and NLP-driven ESG document processing using fine-tuned large language models. Both streams feed into a unified ESG data lake structured according to a Lambda architecture.

3.4 Responsible data agents (RDA)

To ensure the accessibility, explainability, and trustworthiness of ESG-related data within industrial systems, FLX:DPPplus incorporates a modular subsystem of **RDAs**. These agents serve as intelligent intermediaries between technical ESG infrastructures and human users, enabling both voice-based interaction and retrieval-augmented reasoning across structured and unstructured data spaces.

Technically, RDAs are based on a multilayered architecture that combines:

- **Automatic speech recognition (ASR)** and transcription pipelines
- **Semantic classification models** for ESG-relevant topics
- **Retrieval-augmented generation (RAG)** techniques using fine-tuned transformer-based LLMs [7]
- **Certificate generation engine** for formalized ESG and product documentation

3.4.1 Core capabilities and modules

The RDA framework integrates four tightly coupled functional domains:

1. **ESG expertise encoding**
   Unstructured domain knowledge (e.g., expert interviews, field notes, supplier calls) is converted into semantically annotated, machine-readable knowledge units using LLM-powered classification pipelines. This process builds a corpus of ESG-specific insights, enabling the system to answer queries and generate context-aware outputs.

2. **Interactive retrieval and explanation**
   Using RAG principles, RDAs retrieve relevant knowledge from the indexed ESG corpus and generate explanatory responses tailored to the user's intent. This supports interpretability, a key factor in establishing user trust in AI-supported systems under the principles of Industry 5.0.

3. **Voice-enabled input and assistance**
   Users can interact via voice or text interfaces to query ESG performance indicators, request product-related sustainability certificates, or validate supplier compliance data. The integration of speech-to-text and LLM annotation modules allows for hands-free, natural interaction with the system, making it usable even on the shop floor or during audits.

4. **Actionable output generation**
   Validated insights are structured into formal outputs, such as DPPs or supplier audit PDFs, and optionally enriched with CO_2 profile annotations and source documentation. The system's output is always traceable and auditable, ensuring compliance with the CSRD and DPP frameworks.

3.4.2 Human-centered design and explainability

Aligned with the human-centricity principle of Industry 5.0, RDAs are explicitly designed not to replace human decision-making but to augment expert roles with context-sensitive recommendations, explanations, and dialogue-based guidance. Explainability is achieved through:

- Source citations
- Visualization of ESG score logic
- Optional user feedback loops for response correction
- This ensures that ESG insights are not only technically correct but also comprehensible and trusted by diverse stakeholders, from sustainability officers to production engineers. Privacy and data protection are fundamental design principles of the RDAs. All supplier and operational data processed within the system adhere to the General Data Protection Regulation (GDPR) of the European Union. The agents do not permanently store any personal identifiers and provide auditable logs for every interaction, ensuring transparency and accountability in compliance-sensitive industrial environments [8].

3.5 Integration and interfaces

The integration layer of the **FLX:DPPplus** platform is designed for high adaptability across heterogeneous industrial IT and OT environments. It enables both automated and manual data flows, ensuring that the platform can be deployed in digitally advanced as well as partially digitized manufacturing contexts, while supporting transparency and accountability in compliance-sensitive environments.

To support machine connectivity and real-time data exchange, the system provides support for commonly used industrial protocols and interfaces, including:

- **OPC UA** and **MQTT** for communication with shop floor assets, such as machines, sensors, and edge gateways
- **RESTful APIs** for integration with external systems like MES, ERP, PLM, or ESG databases.
- And structured import/export formats like **CSV**, **JSON**, or **XML** for bulk data transfer.

In addition, the platform provides manual input options via structured web forms for cases where digital interfaces are unavailable, particularly relevant for supplier declarations or legacy system data.

All data entering the system is normalized and semantically enriched upon ingestion to ensure consistency, interoperability, and compliance-readiness. This enables downstream modules, such as ML components and RDAs, to operate on unified and interpretable data structures.

4. Use case/application scenario

AI-based certification of sustainable products via digital product passports

This use case illustrates how the FLX:DPPplus platform can be applied to generate trustworthy, machine-verifiable sustainability certificates for industrial products – anchored in the concept of the DPP and enriched through AI-driven analysis and human-AI collaboration.

4.1 Objective

The goal is to create a digitally certifiable sustainability profile for a product, based on verifiable ESG indicators, lifecycle data, and real-time production inputs. The certification process must be transparent, repeatable, and explainable in order to meet the expectations of regulators, supply chain partners, and consumers alike.

4.2 Data collection and processing

The product's sustainability profile is built from a combination of:

- **Master data**: product name, manufacturer, category, and production context
- **Operational data**: energy consumption, emissions, supply chain routes, and material origin
- **ESG declarations**: provided by suppliers or generated through audits and NLP-assisted analysis
- **External criteria sets**: aligned with frameworks such as the EU Taxonomy or CSRD [5]

ML models, for example, gradient boosting and linear regression, forecast expected emissions and ESG performance. Simultaneously, LLMs interpret unstructured sources, such as audit reports or supplier declarations, and extract key sustainability factors.

4.3 Certificate composition

The platform generates a structured, digitally signed ESG certificate embedded within the DPP, containing:

1. **Product identity** and manufacturer information
2. **Sustainability criteria** met (e.g., renewable energy use, recycled content)
3. **Quantitative sustainability score** (e.g., 82/100, based on multicriteria evaluation)

4. **Certification authority** (internal or third-party instance)
5. **Validity period and revision history**
6. **Unique certificate ID** and QR-based verification
7. **Impact comparison** with conventional product alternatives
8. **Data transparency section**, documenting sources, models and assumptions

The entire process is automated and explainable, supported by RDAs that guide users through data submission, review results, and answer follow-up questions via voice or text interfaces.

The following illustration (**Figure 4**) exemplifies how a DPP certificate can be structured for an industrial product in the mechanical engineering domain. It summarizes the core ESG attributes, verification elements, and data sources that are automatically generated and validated within the FLX:DPPplus platform.

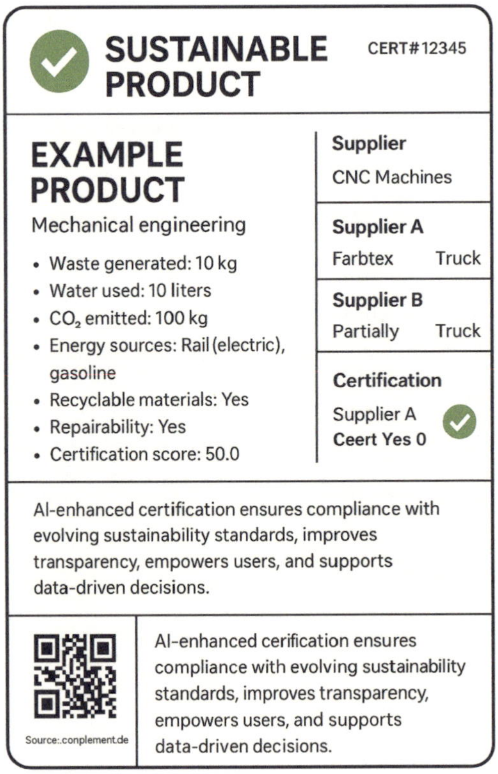

Figure 4.
Exemplary digital product passport (DPP) certificate for a mechanical engineering product. The certificate integrates key ESG data, including product-specific sustainability indicators, data provenance, audit status, verification authority, and lifecycle impact scores. It illustrates how AI-assisted components contribute to automated generation, validation, and traceability of machine-verifiable sustainability documentation within the FLX:DPPplus platform.

4.4 Practical impact

Pilot implementations of the FLX:DPPplus platform in SMEs have demonstrated significant efficiency gains. Initial certificate creation efforts can be reduced by over 30%, while recertification or follow-up audits achieve workload reductions of more than 60%. Furthermore, automated checks and guided data entry have led to a marked decrease in human errors, improving the accuracy and reliability of ESG documentation across the supply chain.

4.5 Industry 5.0 relevance

This scenario demonstrates how AI-enhanced certification mechanisms contribute to the realization of Industry 5.0 objectives. By combining machine intelligence with structured product and ESG data, the system supports resilient compliance with evolving regulatory frameworks, such as the CSRD and EU DPP requirements. At the same time, it increases transparency and trust in product-related sustainability communications, both within industrial supply chains and toward end-users.

Furthermore, the platform empowers human decision-makers by offering explainable AI components that support the interpretation, validation, and adjustment of ESG metrics. Rather than replacing expert judgment, the system augments it with contextualized insights, verifiable sources, and interactive guidance. This hybrid interaction fosters data-driven decision-making across the product lifecycle while preserving accountability and interpretability.

In the context of mechanical engineering, where products are often complex, long-living, and heavily regulated, this approach enables manufacturers to manage ESG performance not only at the component level but across global supply networks. The integration of predictive models, audit trails, and RDAs facilitates traceability and adaptability even under uncertain regulatory or market conditions, thereby contributing directly to organizational resilience.

Ultimately, this use case reflects the core vision of Industry 5.0 [1], drawing from prior research on AI-based ESG certification [9]: to balance technological automation with human-centric oversight, environmental responsibility, and socio-economic resilience. The DPP thus becomes more than a compliance tool; it evolves into a standardized interface connecting technology, regulation, and society.

5. Outlook

The FLX:DPPplus architecture presented in this chapter demonstrates how AI can be systematically applied to enhance ESG transparency in industrial contexts. By integrating predictive analytics, semantic document processing, and interactive RDAs, the system responds directly to the operational and regulatory challenges posed by the European Green Deal, CSRD, and the DPP [4] initiative.

From a research perspective, the modular combination of ML and RAG-based LLM technologies offers a promising foundation for scalable, domain-specific ESG intelligence. At the same time, the human-centered interaction design ensures that AI outputs remain interpretable and actionable, thereby reinforcing trust, a key pillar of Industry 5.0.

Overall, AI-enhanced DPPs offer a practical and forward-looking approach to operationalizing sustainability – bridging the gap between technical innovation, regulatory compliance, and human values.

Acknowledgments

The author would like to thank the Strategy, Research, and Development team at FLX-Business for their invaluable contributions throughout the conceptualization and technical implementation of the FLX:DPPplus platform. Special recognition goes to Laura Wehle and Luisa Wehle for their dedication and insight during the design, validation, and evaluation phases.

Sincere thanks are also extended to the **Steinbeis Transfer Center for Industrial Digitalization** for its scientific guidance and methodological input, which significantly enriched the research quality and practical relevance of this work.

Glossary of terms

Term	Definition
AI	*Artificial Intelligence* – A broad field of computer science focused on building systems capable of performing tasks that typically require human intelligence, such as decision-making, pattern recognition, and natural language understanding.
DPP	*Digital Product Passport* – A structured and standardized digital data record that contains product-specific information related to sustainability, environmental impact, material composition, and lifecycle performance, mandated under the EU's Ecodesign framework.
ESG	*Environmental, Social, and Governance* – A framework for evaluating a company's operations and products based on environmental responsibility, social impact, and ethical governance practices.
CSRD	*Corporate Sustainability Reporting Directive* – A European Union directive that requires companies to disclose standardized ESG performance data, applicable to large and mid-sized businesses.
LLM	*Large Language Model* – A type of deep learning model trained on massive text datasets to perform tasks such as text generation, summarization, and classification. Examples include GPT-based models.
RAG	*Retrieval-Augmented Generation* – A machine learning architecture that enhances language model outputs by retrieving relevant contextual documents and integrating them into the generation process.
RDA	*Responsible Data Agent* – A voice- and interface-enabled AI component designed to interact with users in a human-centered way, explain ESG metrics, and assist in data interpretation and certification processes.
Lambda architecture	A data processing architecture that combines batch and stream processing to enable real-time responsiveness and historical analysis within one unified system.
CO_2 forecasting	The use of predictive analytics to estimate future carbon emissions is based on historical process data, energy consumption, and production volumes.
Greenwashing	A deceptive practice in which companies overstate or falsify their environmental performance or sustainability claims in order to appear more responsible than they actually are.

Author details

Hans-Dieter Wehle
CEO, FLX-Business, Research Lead, FLX:DPPplus Platform, Böblingen, Germany

*Address all correspondence to: hdw@flx-business.de

References

[1] European Commission. Industry 5.0 – Towards a Sustainable, Human-centric and Resilient European Industry. Brussels; 2021. Available from: https://op.europa.eu/en/publication-detail/-/publication/468a892a-5097-11eb-b59f-01aa75ed71a1/

[2] European Commission. The European Green Deal. COM(2019) 640 Final; 2019. Available from: https://eur-lex.europa.eu/legal-content/EN/TXT/?uri=CELEX:52019DC0640

[3] European Commission. Proposal for a Regulation on Ecodesign for Sustainable Products. COM(2022) 142 Final; 2022. Available from: https://environment.ec.europa.eu

[4] European Commission. Corporate Sustainability Reporting Directive (CSRD). Directive (EU) 2022/2464; 2022. Available from: https://eur-lex.europa.eu/legal-content/EN/TXT/?uri=CELEX:32022L2464

[5] Chen T, Guestrin C. XGBoost: A Scalable Tree Boosting System. In: Proceedings of the 22nd ACM SIGKDD International Conference on Knowledge Discovery and Data Mining; 2016. p. 785–794

[6] Lewis M, et al. Retrieval-Augmented Generation for Knowledge-Intensive NLP Tasks. Advances in Neural Information Processing Systems (Neurips) 2020

[7] Wehle H-D. FLX:DPPplus – AI-based ESG Certification in Industrial Applications. ResearchGate; 2025. Available from: https://www.researchgate.net/publication/389513716_FLX-DPPplus_AI-supported_Sustainability_Monitoring_for_CO_Emission_Forecasting_and_ESG_Evaluation

[8] European Commission. Digital Product Passport: Enabling Circular Economy and Sustainable Value Chains. Brussels; 2023. Available from: https://single-market-economy.ec.europa.eu

[9] Doshi-Velez F, Kim B. Towards A Rigorous Science of Interpretable Machine Learning. arXiv preprint arXiv:1702.08608; 2017

Chapter 18

Perspective Chapter: Social Innovation, Ethical Digital Services, and Personal Data in an AI Era – A Case Study from Switzerland

Athanasios Priftis and Catherine Equey

Abstract

The Swiss platform e-rentes.ch showcases how digital tools can drive social innovation by giving individuals secure, transparent access to their pension information and personal data. Created by HES-SO Geneva (University of Applied Sciences Western Switzerland), this prototype allows users to estimate retirement pensions from Switzerland's three-pillar pension system – state, occupational, and private – in one place, without storing any personal data. Instead, it uses real-time connections to trusted sources, ensuring privacy and data control remain with the user. e-rentes.ch aims to simplify the complexity of old-age pension planning, especially for those who lack financial expertise or professional advice. e-rentes.ch has no commercial objectives (i.e., no aim to sell financial products). It empowers users to better understand their financial retirement outlook and simulate different life scenarios, fostering financial inclusion and informed decision-making. By promoting data sovereignty and usability, the platform highlights how digital services can serve the public good – respecting privacy while addressing real social needs. This case offers a model for transparent, user-centered innovation in public services. The Swiss platform e-rentes.ch is a promising example of social innovation, offering users real-time access to their old-age pension estimation while respecting privacy. However, its cautious stance on artificial intelligence (AI) raises important questions. Unlike many digital services that rush to implement AI, e-rentes.ch deliberately avoids it – for now. This restraint reflects a growing concern: AI in personal finance often lacks transparency, introduces bias, and can erode user trust. AI use would go against the e-rentes.ch objectives, which are explained later in this article. By choosing a non-AI, user-centric design, e-rentes.ch prioritizes data sovereignty, citizen empowerment, and ethical responsibility. It resists the trend of opaque algorithms making life-altering predictions based on incomplete or misunderstood data. While AI promises personalization and predictive power, it also risks turning sensitive financial planning into a black box – one where users no longer understand or control the outcomes. In this sense, e-rentes.ch serves not only as a tool for retirement pension insight but also as a critique of premature AI adoption in public digital services.

IntechOpen

Keywords: social innovation, digital public services, data sovereignty, artificial intelligence, e-rentes.ch, Industry 4.0

1. Introduction

Industry 4.0 is often understood through its technological foundations – automation, data analytics, artificial intelligence (AI), and cyber-physical systems – largely within manufacturing. Yet, scholars increasingly recognize that its transformative potential reaches far beyond production, reshaping the delivery of public services and the relationship between citizens and institutions [1, 2]. The social dimension of Industry 4.0 remains underexplored: While digitalization can enhance efficiency and sustainability, it also raises pressing questions of equity, data sovereignty, and ethical responsibility [3, 4]. For the public sector, the challenge is therefore not only technological adoption but also how digital infrastructures can be designed to advance trust, transparency, and social innovation.

Social innovation provides a useful lens for analyzing these developments. Defined as new solutions that address societal needs more effectively than existing practices, social innovation in digital services can foster inclusion, empower individuals, and strengthen democratic values [5]. Recent research highlights that countries with higher levels of digital integration also exhibit greater capacity for social innovation [6]. At the same time, experiences from European pension-tracking systems illustrate the governance complexities of cross-institutional collaboration and the need for citizen trust in sensitive domains such as retirement planning [7]. This interplay between digital infrastructures and social innovation is especially salient in contexts where citizens depend on accurate, secure, and transparent information for long-term financial decisions.

A growing body of work further warns against the risks of premature or uncritical adoption of AI in public digital services. Comparative reviews of AI ethics frameworks converge on principles such as transparency, fairness, and accountability [8], yet their translation into practice remains uneven [9]. Moreover, algorithmic opacity and the potential for bias can erode trust in public systems, particularly in high-stakes domains such as personal finance or taxation, undermining the concept of digital self-determination, which calls for preserving individual agency and sovereignty over personal data [10].

This chapter examines the Swiss prototype e-rentes.ch, a platform that consolidates information from the old-age provision system through secure real-time connections to trusted sources. It should be noted that the Swiss pension system is complex, mainly due to the multitude of players involved. Retirees receive two or three pensions from at least three different institutions (which are not linked to each other). This system is known as the three-pillar system. The pension received upon retirement consists of a state pension, a pension linked to their professional employment, and their personal savings. This complexity gives e-rentes.ch an informational and educational role. Importantly, the platform does not store user data, thereby reinforcing individual control and data sovereignty. By simplifying the complexity of pension planning and allowing citizens to simulate life scenarios, e-rentes.ch exemplifies how digital services can promote inclusion and informed decision-making. Unlike many contemporary digital services, however, the platform deliberately avoids AI for the moment. This cautious stance reflects a critical position within Industry 4.0: resisting "solutionism" by privileging transparency, usability, education, and ethical responsibility over algorithmic prediction. The paradigm is to give insured individuals back control and understanding of their data. e-rentes.ch was built using prior research into European pension-tracking systems [7] and frameworks for citizen-centered digital service design [7].

This chapter is organized as follows: Section 2 discusses social innovation in the specific case of public administration. The third part examines the government's use of AI. Section 4 focuses on the governance of personal data and its challenges, while Section 5 presents our case study. Finally, the last part summarizes our choices not to use AI for e-rentes.ch before concluding. This article aims to advance the discussion by situating e-rentes.ch within the broader Industry 4.0 debate, using it both as a case of social innovation and as a critique of unreflective AI adoption in the public sector. In doing so, it contributes to ongoing debates on how digital infrastructures can foster trust, inclusion, and ethical responsibility – demonstrating that the future of Industry 4.0 lies not only in technological sophistication but also in its capacity to serve the public good.

2. Social innovation in the public sector

Social innovation is broadly defined as the development and implementation of new ideas, services, or models that address social needs more effectively than existing solutions, while simultaneously fostering new social relationships and collaborations [5, 11]. In the public sector, social innovation plays a particularly critical role because it links technological change to societal outcomes, ensuring that innovation is not merely efficiency-driven but also oriented toward inclusivity, equity, and citizen empowerment [12]. Digital public services exemplify this potential: By leveraging emerging technologies, they can improve access to information, foster transparency, and strengthen citizens' trust in institutions.

However, realizing this potential is not without challenges. Trust remains a central issue in the digital transformation of public services. Citizens are often asked to share highly sensitive data, yet experiences of data breaches, opaque algorithms, and limited accountability can undermine confidence [13, 14]. Inclusion requires not only technical accessibility but also cultural and linguistic adaptability, particularly in countries with diverse populations. These challenges highlight that digital transformation in the public sector must be accompanied by governance models that foreground ethics, accountability, and citizen-centered design.

Switzerland's pension system illustrates why such an approach is both complex and necessary. Based on a three-pillar model – state, occupational, and private savings – it represents one of the most intricate pension systems in Europe [15]. Navigating this system requires financial literacy, sustained record-keeping across multiple institutions, and the ability to interpret long-term projections. For many citizens, particularly those without financial expertise or access to professional advisors, these demands are prohibitive. This makes the Swiss retirement pension domain an ideal test bed for digital social innovation: Solutions that can integrate fragmented information, present it transparently, and empower users to simulate life scenarios directly respond to a pressing societal need. At the same time, the sensitivity of pension data underscores the importance of designing digital services that respect privacy, guarantee data sovereignty, and foster trust.

In this context, the case of e-rentes.ch – a prototype pension-tracking platform developed in Switzerland by a consortium of private, public, and academic partners – exemplifies how social innovation can be realized in digital public services. By consolidating information from across the three pillars without storing personal data, it directly addresses (and decreases) the complexity of the system while simultaneously embedding ethical safeguards. This dual orientation – toward both usability and ethical responsibility – illustrates the distinctive contribution of social innovation to the digital transformation of public services.

3. AI in the public sector

Artificial intelligence is increasingly promoted as a transformative technology for public digital services. Governments and public agencies experiment with AI for tasks ranging from fraud detection and taxation to predictive policing, welfare allocation, and personalized citizen services [16]. The appeal of AI lies in its ability to process large volumes of data, generate insights, and deliver tailored outputs in ways that promise greater efficiency and responsiveness [17]. In citizen-facing services, AI could enhance personalization by adapting information or recommendations to an individual's circumstances. Predictive analytics could offer the possibility of anticipating needs or risks, enabling proactive interventions that may improve service delivery and reduce costs.

Yet, alongside these opportunities, AI adoption in the public sector introduces significant risks. A growing body of research highlights concerns about the opacity of algorithmic systems, often described as "black boxes" that limit citizens' and policymakers' ability to understand or contest outcomes [18]. This lack of transparency can erode trust, especially in high-stakes domains such as taxation, social welfare, or pensions. Furthermore, AI systems are prone to reproducing or amplifying biases present in the data on which they are trained [19, 20]. Such biases can disproportionately affect vulnerable populations, undermining equity and inclusivity – the very goals digital public services are meant to advance. For now, AI still provides quite a few incorrect responses and requires users to have the ability to analyze the accuracy of the answers. Ethical dilemmas also arise concerning accountability: When decisions are mediated by AI, it is often unclear whether responsibility lies with the technology provider, the public institution, or the policy framework under which the system operates [21].

These tensions have prompted scholars and practitioners to reflect on two contrasting approaches. On the one hand, an AI-driven model prioritizes predictive capacity and efficiency, using algorithms to guide or automate decision-making. While such systems may appear technologically sophisticated, they risk alienating users by removing transparency and agency from the process. On the other hand, a user-centric, non-AI model emphasizes transparency, usability, and citizen sovereignty over data. Instead of relying on opaque predictive systems, these approaches foreground clarity, human interpretability, and individual control. In sensitive areas such as personal finance or retirement pensions, this latter model may better align with principles of trust, accountability, and ethical responsibility.

The case of e-rentes.ch illustrates the deliberate choice of a user-centric, non-AI approach. e-rentes.ch challenges users to better understand the pension provision system. By resisting the premature adoption of AI, the platform prioritizes data sovereignty, clarity, and user empowerment, thereby challenging the assumption that technological sophistication must equate to algorithmic automation. This contrast highlights an important dimension of social innovation in the public sector: In certain contexts, innovation may consist not in deploying the most advanced technology but in exercising restraint to safeguard trust and inclusivity.

4. Personal data: Public/private and ownership

In the context of Industry 4.0, data are often referred to as the "new oil" of the digital economy, underpinning innovations across manufacturing, health care, and public services. Yet, unlike oil, data are not a finite resource; its value is relational, dependent on how it is collected, processed, and shared [22]. This raises fundamental questions of data sovereignty, privacy, and user control, particularly in public services where the data in question often concern sensitive personal information. Data sovereignty refers to the right of individuals, organizations, or states to control data in accordance with ethical norms and legal frameworks [23]. The governance of personal data thus emerges as a critical issue that links technological design choices to broader societal values such as autonomy, responsibility, dignity, and trust.

Different models of data governance highlight these tensions. Centralized storage models typically consolidate large amounts of personal information in institutional or third-party repositories, where they can be analyzed for predictive insights or operational efficiency. While such systems may enable powerful applications of AI, they also introduce risks: The concentration of data makes systems more vulnerable to breaches, misuse, and surveillance [24]. Moreover, centralized storage can erode user agency by shifting control from the individual to the institution. By contrast, real-time secure access models – such as that adopted by e-rentes.ch – eschew data storage in favor of live connections to trusted sources. This design enables users to access accurate and up-to-date information without ceding ownership or control to an intermediary. Such models operationalize the principle of privacy by design [25], aligning digital services with the ethical imperative of protecting user sovereignty.

Debates about data as a public good versus a private asset further complicate the landscape. On the one hand, treating data as a public good emphasizes its potential to generate collective benefits, from better policy design to improved social services [26]. On the other hand, conceptualizing personal data as a private asset highlights individual rights of ownership, consent, and compensation [27]. Both perspectives have limitations: a purely public good approach risks overlooking individual rights and privacy, while a purely private property model risks commodifying sensitive information in ways that exacerbate inequality. The challenge, therefore, is to navigate a middle ground in which personal data are governed in ways that maximize social value while safeguarding individual autonomy.

The e-rentes.ch prototype contributes to this debate by demonstrating how public services can provide societal value without compromising individual data sovereignty. By consolidating fragmented pension information (held by three different types of institutions) while avoiding central storage, the platform positions data as both *useful for the public good* and *fully under individual control*. In this sense, it exemplifies a form of digital social innovation that reconfigures the relationship between personal data, public institutions, and individual users – advancing a model where ethical responsibility is embedded into the very architecture of digital infrastructures.

Most importantly, the e-rentes.ch prototype is deployed respecting specific privacy-by-design features that include live authenticated connections to trusted systems only, no database creation or storage after the session ends, and the introduction of various consent checkpoints before accessing data and before exporting/sharing. This goes hand in hand with a declared data minimization and purpose limitation policy: only the data needed, only when needed, aiming to strengthen user control and default privacy.

5. Our case study – e-rentes.ch

The Swiss pension system, structured around its three-pillar model (state, occupational, and private savings), provides an exemplary test bed for digital social innovation due to its complexity and fragmentation [15]. Citizens are required to navigate multiple institutions and reporting mechanisms, often without financial expertise or professional guidance. This makes old-age pensions an area where digital solutions must not only integrate diverse sources of information but also build trust, usability, and ethical safeguards into their core design.

The e-rentes.ch prototype, developed at HES-SO Geneva, was conceived to address precisely these challenges. Building on a citizen-centered service design methodology [7], the project triangulated three perspectives: (1) citizens' needs, particularly clarity, accessibility, and empowerment; (2) institutional requirements, including interoperability with existing pension funds and regulatory frameworks; and (3) technological feasibility, emphasizing privacy-by-design and secure data flows. This participatory and iterative approach ensured that the service architecture was not dictated by technology alone but coshaped through stakeholder dialog and alignment [28].

The design of e-rentes.ch followed an eight-phase framework derived from the DIBS research project, ranging from ecosystem mapping to prototype testing [7]. Each phase was associated with distinct methodologies: surveys and interviews to capture user expectations, focus groups and codesign sprints to iteratively refine functionalities, and specialized documentation reviews to address legal and technical requirements. While quantitative tools were used at specific junctures – such as defining service scope and measuring user needs – the process relied predominantly on qualitative inputs. This methodological choice proved effective, as it allowed the project team to surface nuanced user concerns, institutional constraints, and governance sensitivities that would likely have been overlooked in purely statistical studies [29, 30].

Stakeholder engagement was structured around a steering committee of five core partners (public authorities, private firms, and academia) and a larger coordination circle of roughly 25 entities representing government agencies, insurers, pension funds, banks, professional associations, and civil society organizations. This arrangement ensured that both decision-making and consultation were institutionalized from the outset. Importantly, users themselves were involved at critical stages – needs assessment, persona building, and prototype testing – where their perspectives shaped the usability and trustworthiness of the service.

The result was a process in which almost all categories of actors, from government to citizens, contributed at different phases, reflecting the project's grounding in the quadruple-helix model of social innovation [31, 32]. All steps to create a concrete data protection action plan with registered activities are initiated by the project leader (HES-SO, academia) and then transformed into separate data protection agreements with all data stakeholders. Through these agreements, all stakeholders appoint a person responsible for the execution of the action plan.

Technically, e-rentes.ch differs from many pension-tracking systems by avoiding centralized storage of personal data. Instead, it operates through real-time secure access to institutional databases, allowing users to consolidate their pension information without relinquishing data control to the platform itself. This architectural choice responds to the ethical concerns identified in the literature: Centralized data systems amplify risks of breaches, surveillance, and institutional overreach [24], whereas real-time architectures operationalize the principle of data sovereignty [25]. The prototype thus embodies a hybrid model of digital service design that combines transparency, usability, and trust with respect to privacy.

Early outcomes of the pilot suggest several contributions. Firstly, e-rentes.ch has been successful in simplifying complexity: Users in focus groups and prototype tests report that they better understand their pension situation when data from all three pillars are presented in one interface. e-rentes.ch fulfilled its role of simplification and training. Secondly, the platform will support scenario simulation, enabling individuals to model life events (e.g., career breaks, changes in income, or retirement age) and assess their long-term implications. This feature fosters financial inclusion by empowering those who may lack professional advisory services. e-rentes.ch provides the tools needed to plan financially for retirement in a context where poverty affects elderly people.

e-rentes.ch demonstrates a deliberate non-AI design choice: While predictive analytics could, in theory, offer personalized recommendations, the project team intentionally refrained from premature AI adoption. As Equey et al. [7] argue, citizen-centered design requires participation, explainability, and control over algorithmic opacity. Avoiding uncritical AI adoption allows individuals to understand and maintain control of a complex field through e-rentes.ch. To ensure that the pilot's impact and public value are measurable, we have integrated selected indicators to evaluate the impact of our pilot:

- User trust and understanding (satisfaction rate, perceived clarity).
- Consent rates for data sharing via APIs.
- Time-to-information (average time compared to traditional channels).
- Zero central storage confirmed (compliance with the decentralized architecture).
- Complaints or other "pain points" that may occur during the pilot.
- Successful scenario simulations (number and relevance of projections shown to users).

In this way, e-rentes.ch illustrates how Industry 4.0 technologies can be harnessed for public good without defaulting to opaque or extractive models. Rather than treating data as a commodity, the platform positions data sovereignty and citizen empowerment as central design principles. The case thus reinforces the broader argument that, in public digital services, innovation lies as much in governance, simplicity, accessibility, and restraint as in technological sophistication.

6. Discussion on AI

As already demonstrated, one of the most distinctive features of e-rentes.ch is its prioritization of transparency, interpretability, and user data control. This choice reflects not a technological limitation but a conscious design decision, rooted in the principles of citizen-centered service design [7]. This stance constitutes a clear countertrend to what has been described as "AI solutionism" – the belief that complex societal problems can and should be solved through algorithmic optimization [33, 34]. In the context of public digital services, AI solutionism risks reframing human needs in ways that fit technological capabilities rather than designing technologies around citizen priorities. Automated predictions of retirement readiness, for example, might appear efficient but could reproduce existing biases in labor market data or obscure the assumptions embedded in various models.

The inclusive approach also highlighted the main obstacles to implementing such a service. The project revealed the high level of complexity generated by numerous partners working together, as well as the asymmetry of involvement – Government, academia, and software providers were consistently active, while other private actors were more peripheral. Most critically, governance emerged as the decisive bottleneck: Certain stakeholders were reluctant to share or relinquish control over pension data, given the financial and political stakes or interests. This resistance confirmed that technological feasibility alone is insufficient; legal frameworks and enforceable governance mechanisms are indispensable to sustain a citizen-centered digital pension service [35, 36].

The broader implications of this approach are twofold. Firstly, it challenges the prevailing narrative that innovation in Industry 4.0 must always equate to increased automation or algorithmic sophistication. By showing that the absence of AI can itself be innovative, e-rentes.ch expands the meaning of digital transformation beyond technical novelty to include governance, ethics, and social legitimacy. Secondly, it underscores that social innovation often lies in aligning technology with values such as inclusion, autonomy, and accountability.

In this light, not using AI is not a regression but a deliberate form of progress: It maintains citizens' ability to interpret and act upon their data without surrendering agency to opaque systems. This raises the key question of whether AI can be used by everyone. The case of e-rentes.ch shows that, even without considering potential errors and biases, future retirees need to understand the system to plan their financial revenues at retirement. It includes potential savings but also life choices (i.e., part-time job), which are crucial decisions to maintain one's standard of living in retirement.

Framed within the wider discourse on responsible Industry 4.0 transformation, e-rentes.ch highlights the need for a pluralistic approach to digital innovation.

Industry 4.0 technologies – automation, AI, big data – should be critically assessed for their societal consequences. In domains where trust, privacy, and long-term security are paramount, restraint may be more innovative than acceleration.

7. Conclusions

The case of e-rentes.ch demonstrates how Industry 4.0 can be mobilized in the public sector not merely to deliver technological sophistication but to foster transparency, trust, inclusion, and sovereignty. Digital transformation in the public sector is not just technological but societal. By consolidating Switzerland's fragmented pension system into a simplified, user-friendly, secure, and citizen-centered platform, the prototype illustrates the value of social innovation in addressing complex societal needs. Importantly, it shows that design choices grounded in ethical responsibility – such as real-time secure access in lieu of centralized data storage – can produce both social legitimacy and user empowerment.

A key lesson emerging from this analysis is that innovation in public digital services must be critically balanced with ethics. Industry 4.0 technologies, such as AI and big data, offer powerful capabilities for personalization and predictive analytics, yet their uncritical deployment risks opacity, bias, and the erosion of citizen trust. e-rentes.ch exemplifies a counter-narrative: Innovation is not defined solely by the sophistication of technology but by its alignment with democratic values, citizen agency, and social inclusion. In this sense, exercising technological restraint can itself be an innovative act.

Looking ahead, the question arises whether responsible forms of AI could eventually be integrated into e-rentes.ch. The platform's current non-AI design reflects a deliberate prioritization of trust and interpretability. However, if future developments in responsible AI – such as explainable algorithms, bias-mitigation techniques, and participatory governance frameworks – prove robust, they may complement the platform's existing architecture without undermining its ethical foundations. In such a scenario, AI could serve as an optional layer to support simulation or tailored advice, provided it remains transparent, auditable, and fully under user control.

The e-rentes.ch experience can be replicated in other domains by following a simple formula built on open standards, a minimal stack, and a phased pilot plan. At its core, services should adopt open APIs and standardized data models, such as OpenFinance schemas, combined with OAuth2.1/OpenID Connect for identity and consent management. Privacy is ensured through privacy-by-design defaults: ephemeral in-memory computation, no data persistence, and minimal cryptographic logging of consent events. This creates trust with users, regulators, and data providers while enabling interoperability across systems.

From a technical and operational standpoint, the service can run on a lightweight stack: a frontend guiding users through login, consent, live data fetch, computation, and results; a stateless backend orchestrating API calls; and an integration layer with live connectors to registries, payroll, and tax systems. The rollout follows a pilot-by-phases plan: start narrow with one trusted source (e.g., pension registry), test the flows with a small pilot group, then incrementally expand to additional sources such as payroll or insurance. With feedback loops for usability and compliance, the model scales into a trusted, transparent entitlement service that can be reused in contexts such as health care, insurance, or education benefits.

Ultimately, the lessons of e-rentes.ch extend beyond old-age pensions. They underscore that Industry 4.0 in public services must not only transform infrastructures but also safeguard the values that underpin democratic societies. Responsible digital transformation requires recognizing that technology is never neutral: It either strengthens or weakens the relationship between institutions and citizens. By foregrounding transparency, trust, inclusion, and sovereignty, e-rentes.ch offers a replicable model of how public digital services can navigate the promises and perils of Industry 4.0 – innovating responsibly for the common good.

Author details

Athanasios Priftis* and Catherine Equey
HEG-Genève, HES-SO, University of Applied Sciences and Arts Western Switzerland, Geneva, Switzerland

*Address all correspondence to: athanasios.priftis@hesge.ch

References

[1] Kovács O. Industry 4.0 in public sector innovation: A critical review. Society and Economy. 2020;**42**(2):183–203. DOI: 10.1556/204.2019.014

[2] Stentoft J, Rajkumar C, Mikkelsen OS. Industry 4.0 in the public sector: Implications for digital transformation and service innovation. Government Information Quarterly. 2021;**38**(4):101592. DOI: 10.1016/j.giq.2021.101592

[3] Kamble SS, Gunasekaran A, Gawankar SA. Sustainable Industry 4.0 framework: A systematic literature review. Sustainable Production and Consumption. 2018;**17**:278–292. DOI: 10.1016/j.spc.2018.11.001

[4] Birkel H, Veile JW, Müller JM, Hartmann E, Voigt K-I. Industry 4.0 and sustainability: A systematic literature review of past achievements and future challenges. Journal of Cleaner Production. 2023;**382**:135375. DOI: 10.1016/j.jclepro.2022.135375

[5] Howaldt J, Schwarz M. Social Innovation: Concepts, Research Fields, and International Trends. Cham: Springer; 2017

[6] Nagy J, Somosi MV. Digital transformation and social innovation capacity: An empirical analysis of European countries. Technological Forecasting and Social Change. 2022;**182**:121785. DOI: 10.1016/j.techfore.2022.121785

[7] Equey C, Priftis A, Trabichet J-P, Hutzli V. Designing a digital citizen-centered service. Technological Forecasting and Social Change. 2024;**202**:123280. DOI: 10.1016/j.techfore.2024.123456

[8] Jobin A, Ienca M, Vayena E. 2019. The global landscape of AI ethics guidelines. Nature Machine Intelligence. **1**(9):389–399. DOI: 10.1038/s42256-019-0088-2

[9] Veale M, Van Kleek M, Binns R (2018). Fairness and accountability design needs for algorithmic support in high-stakes public sector decision-making. Proceedings of the 2018 CHI Conference on Human Factors in Computing Systems. 1–14. 10.1145/3173574.3174014

[10] Cohen J, Dolin R, Dubljević V. 2020. Digital self-determination. Ethics and Information Technology. **22** (3):179–190. DOI: 10.1007/s10676-020-09542-2

[11] Murray R, Caulier-Grice J, Mulgan G. The Open Book of Social Innovation. London: The Young Foundation/NESTA; 2010

[12] Cajaiba-Santana G. Social innovation: Moving the field forward. A conceptual framework. Technological Forecasting and Social Change. 2014;**82**:42–51. DOI: 10.1016/j.techfore.2013.05.008

[13] Meijer A. 2015. E-governance innovation: Barriers and strategies. Government Information Quarterly. **32** (2):198–206. DOI: 10.1016/j.giq.2015.01.001

[14] Criado JI, Sandoval-Almazan R, Gil-Garcia JR. Government innovation through social media. Government

Information Quarterly. 2013;**30**(4): 319–326. DOI: 10.1016/j.giq.2013.10.003

[15] OECD. OECD Pensions at a Glance 2021: OECD and G20 Indicators. Paris: OECD Publishing; 2021. DOI: 10.1787/ca401ebd-en

[16] Wirtz BW, Weyerer JC, Geyer C. Artificial intelligence and the public sector—Applications and challenges. International Journal of Public Administration. 2019;**42**(7):596–615. DOI: 10.1080/01900692.2018.1498103

[17] Sun TQ, Medaglia R. Mapping the challenges of artificial intelligence in the public sector: Evidence from public healthcare. Government Information Quarterly. 2019;**36**(2):368–383. DOI: 10.1016/j.giq.2018.09.008

[18] Burrell J. How the machine “thinks”: Understanding opacity in machine learning algorithms. Big Data & Society. 2016;**3**(1):1–12. DOI: 10.1177/2053951715622512

[19] O’Neil C. Weapons of Math Destruction: How Big Data Increases Inequality and Threatens Democracy. New York: Crown; 2016

[20] Noble SU. Algorithms of Oppression: How Search Engines Reinforce Racism. New York: New York University Press; 2018

[21] Zouridis S, van Eck M, Bovens M. Algorithmic regulation in the public sector: The black box challenge. Regulation & Governance. 2020;**14**(4):578–593. DOI: 10.1111/rego.12228

[22] Mayer-Schönberger V, Cukier K. Big Data: A Revolution that Will Transform How We Live, Work, and Think. Boston: Houghton Mifflin Harcourt; 2013

[23] van der Sloot B, Zuiderveen Borgesius F. The right to protection of personal data: Numerical criteria as an instrument of data protection. Computer Law & Security Review. 2020;**37**:105429. DOI: 10.1016/j.clsr.2020.105429

[24] Zuboff S. The Age of Surveillance Capitalism: The Fight for a Human Future at the New Frontier of Power. New York: PublicAffairs; 2019

[25] Cavoukian A. Privacy by Design: The 7 Foundational Principles. Information and Privacy Commissioner of Ontario; 2011. https://www.ipc.on.ca/wp-content/uploads/resources/7foundationalprinciples.pdf

[26] OECD. Enhancing Access to and Sharing of Data: Reconciling Risks and Benefits for Data Reuse across Societies. Paris: OECD Publishing; 2019. DOI: 10.1787/276aaca8-en

[27] Posner EA, and Weyl EG. Radical Markets: Uprooting Capitalism and Democracy for a Just Society. Princeton: Princeton University Press; 2018

[28] Equey C, Hutzli V, Bayat-Ricard M, Priftis A, and Schaffter S (2024, December 4). Digital social innovation: Lessons learned from European pension tracking systems. ISPIM Conference. Osaka

[29] Ippoliti N, Sekamana M, Baringer L, Hope R. 2021. Using human-centered design to develop, launch, and evaluate a national digital health platform to improve reproductive health for Rwandan youth. Global Health: Science and Practice. **9** (Suppl 2):S244–S260. DOI: 10.9745/GHSP-D-21-00220

[30] Abras C, Maloney-Krichmar D, Preece J. User-centered design. In Bainbridge WS, editor. Berkshire Encyclopedia of Human-computer Interaction. Great Barrington: Berkshire Publishing Group; 2004. p. 445–456

[31] Carayannis EG, Campbell DFJ. 2021. Democracy of climate and climate for democracy: The evolution of quadruple and quintuple helix innovation systems. Journal of the Knowledge Economy. **12**(4):2050–2082. DOI: 10.1007/s13132-021-00778-x

[32] Aksoy L, Alkire L, Choi S, Kim PB, Zhang L. 2019. Social innovation in service: A conceptual framework and research agenda. Journal of Service Management. **30**(3):429–448. DOI: 10.1108/JOSM-11-2018-0376

[33] Morozov E. To Save Everything, Click Here: The Folly of Technological Solutionism. New York: PublicAffairs; 2013

[34] Pasquale F. New Laws of Robotics: Defending Human Expertise in the Age of AI. Cambridge: Harvard University Press; 2020

[35] Malodia S, Dhir A, Mishra M, Bhatti ZA. Future of e-government: An integrated conceptual framework. Technological Forecasting and Social Change. 2021;**173**:121102. DOI: 10.1016/j.techfore.2021.121102

[36] Teece DJ, Pundziene A, Heaton S, Vadi M. 2022. Managing multi-sided platforms: Platform origins and go-to-market strategy. California Management Review. **64**(4):5–19. DOI: 10.1177/00081256221109961